Educational Institute
American Hotel & Motel Association

Keep this book. You will
need it and use it throughout
your career.

HOUSEKEEPING MANAGEMENT

Educational Institute Courses

Introductory

INTRODUCTION TO THE HOSPITALITY INDUSTRY
Third Edition
Gerald W. Lattin

AN INTRODUCTION TO HOSPITALITY TODAY
Second Edition
Rocco M. Angelo, Andrew N. Vladimir

TOURISM AND THE HOSPITALITY INDUSTRY
Joseph D. Fridgen

Rooms Division

FRONT OFFICE PROCEDURES
Fourth Edition
Michael L. Kasavana, Richard M. Brooks

HOUSEKEEPING MANAGEMENT
Margaret M. Kappa, Aleta Nitschke, Patricia B. Schappert

Human Resources

HOSPITALITY SUPERVISION
Second Edition
Raphael R. Kavanaugh, Jack D. Ninemeier

HOSPITALITY INDUSTRY TRAINING
Second Edition
Lewis C. Forrest, Jr.

HUMAN RESOURCES MANAGEMENT
Robert H. Woods

Marketing and Sales

MARKETING OF HOSPITALITY SERVICES
Revised Edition
Christopher W. L. Hart, David A. Troy

HOSPITALITY SALES AND MARKETING
Second Edition
James R. Abbey

CONVENTION MANAGEMENT AND SERVICE
Leonard H. Hoyle, David C. Dorf, Thomas J. A. Jones

MARKETING IN THE HOSPITALITY INDUSTRY
Third Edition
Ronald A. Nykiel

Accounting

UNDERSTANDING HOSPITALITY ACCOUNTING I
Third Edition
Raymond Cote

BASIC FINANCIAL ACCOUNTING FOR THE HOSPITALITY INDUSTRY
Raymond S. Schmidgall, James W. Damitio

FINANCIAL ACCOUNTING FOR THE HOSPITALITY INDUSTRY II
Second Edition
Raymond Cote

MANAGERIAL ACCOUNTING FOR THE HOSPITALITY INDUSTRY
Third Edition
Raymond S. Schmidgall

Food and Beverage

FOOD AND BEVERAGE MANAGEMENT
Second Edition
Jack D. Ninemeier

QUALITY SANITATION MANAGEMENT
Ronald F. Cichy

FOOD PRODUCTION PRINCIPLES
Jerald W. Chesser

FOOD AND BEVERAGE SERVICE
Anthony M. Rey, Ferdinand Wieland

HOSPITALITY PURCHASING MANAGEMENT
William P. Virts

BAR AND BEVERAGE MANAGEMENT
Lendal H. Kotschevar, Mary L. Tanke

FOOD AND BEVERAGE CONTROLS
Third Edition
Jack D. Ninemeier

General Hospitality Management

HOTEL/MOTEL SECURITY MANAGEMENT
Raymond C. Ellis, Jr., Security Committee of AH&MA

HOSPITALITY LAW
Third Edition
Jack P. Jefferies

RESORT MANAGEMENT
Second Edition
Chuck Y. Gee

INTERNATIONAL HOTEL MANAGEMENT
Chuck Y. Gee

HOSPITALITY INDUSTRY COMPUTER SYSTEMS
Second Edition
Michael L. Kasavana, John J. Cahill

MANAGING FOR QUALITY IN THE HOSPITALITY INDUSTRY
Robert H. Woods, Judy Z. King

Engineering and Facilities Management

FACILITIES MANAGEMENT
David M. Stipanuk, Harold Roffman

HOSPITALITY INDUSTRY ENGINEERING SYSTEMS
Michael H. Redlin, David M. Stipanuk

HOSPITALITY ENERGY AND WATER MANAGEMENT
Robert E. Aulbach

HOUSEKEEPING MANAGEMENT

Margaret M. Kappa, CHHE
Aleta Nitschke, CHA
Patricia B. Schappert, CHHE

EDUCATIONAL INSTITUTE
of the American Hotel & Motel Association

Disclaimer

This publication is designed to provide accurate and authoritative information in regard to the subject matter covered. It is sold with the understanding that the publisher is not engaged in rendering legal, accounting, or other professional service. If legal advice or other expert assistance is required, the services of a competent professional person should be sought.

—*From the Declaration of Principles jointly adopted by the American Bar Association and a Committee of Publishers and Associations*

The authors, Margaret M. Kappa, Aleta Nitschke, and Patricia B. Schappert, are solely responsible for the contents of this publication. All views expressed herein are solely those of the authors and do not necessarily reflect the views of the Educational Institute of the American Hotel & Motel Association (the Institute) or the American Hotel & Motel Association (AH&MA).

Nothing contained in this publication shall constitute a standard, an endorsement, or a recommendation of the Institute or AH&MA. The Institute and AH&MA disclaim any liability with respect to the use of any information, procedure, or product, or reliance thereon by any member of the hospitality industry.

ISBN 0-86612-047-5
ISBN 0-86612-126-9 (pbk.)

Project Editor: Ann M. Halm
Editors: Ann M. Halm
        Jean A. Raber
        Priscilla J. Wood

Contents

'Chap. 1-7
First Quiz
Friday Oct 9/98

Chap 8-15
Second Quiz
Friday Oct 16/98

TEST OCT 19/98
MONDAY

Congratulations. . .

You have a running start on a fast-track career!

Developed through the input of industry and academic experts, this course gives you the know-how hospitality employers demand. Upon course completion, you will earn the respected American Hotel & Motel Association certificate that ensures instant recognition worldwide. It is your link with the global hospitality industry.

You can use your AH&MA certificate to show that your learning experiences have bridged the gap between industry and academia. You will have proof that you have met industry-driven learning objectives and that you know how to apply your knowledge to actual hospitality work situations.

By earning your course certificate, you also take a step toward completing the highly respected learning programs—Certificates of Specialization, the Hospitality Operations Certificate, and the Hospitality Management Diploma—that raise your professional development to a higher level. Certificates from these programs greatly enhance your credentials, and a permanent record of your course and program completion is maintained by the Educational Institute.

We commend you for taking this important step. Turn to the Educational Institute for additional resources that will help you stay ahead of your competition.

Preface

NOTHING SENDS A STRONGER MESSAGE than cleanliness in a hospitality operation. No level of service, friendliness, or glamour can equal the sensation a guest has upon entering a spotless, tidy, and conveniently arranged room.

To send this message of quality, housekeeping must be endowed with the same professionalism as other hospitality functions. *Housekeeping Management* gives managers the tools to systematically achieve the standards expected by today's guests in today's lodging and food service establishments.

While primarily written for the executive housekeeper, this text can be a resource for any professional who makes housekeeping decisions on a daily basis. The book is also designed to provide important technical information for persons seeking careers in this pivotal area. Every attempt has been made to thoroughly cover the day-to-day complexities of the housekeeping profession—from planning and organizing to budgeting, to supervising and performing the work itself.

To do so, the book is organized into four parts. Part I introduces the role of housekeeping in hospitality operations and focuses on the planning and organization of various housekeeping tasks. Part II emphasizes the importance of quality housekeeping staff by examining human resource management in the housekeeping department. Part III illustrates the various challenges and management responsibilities facing the executive housekeeper. Chapters focus on managing inventories, controlling expenses, and monitoring safety and security functions. For properties with on-premises laundries, a chapter is included which discusses how to effectively oversee the various laundry operations.

Finally, the details of housekeeping tasks are showcased in Part IV. This section consists of five technical/reference chapters which cover the how-to's of cleaning. Chapters focus on the basics of cleaning guestrooms, public areas, ceilings, walls, floors, carpets—as well as considerations in selecting and cleaning furniture, fixtures, linens, and other special features or conveniences. Step-by-step procedures follow most of the technical/reference chapters. While designed as guidelines only, these procedures show the actual application of many of the concepts narrated in the text. The "bookends" of this technical/reference section consist of a chapter on the management and use of cleaning chemicals and a concluding chapter on interior design.

To promote understanding, discussion questions and key terms appear at the end of most chapters. An extensive glossary places industry terms—particularly those related to housekeeping—at the reader's fingertips.

We hope this text meets its intended purpose as a practical resource for the executive housekeeper—and as a vehicle for promoting the professionalism of this important segment of hospitality.

Textbooks of this scope could not be written without the continual support of a great many people. Several industry and academic professionals have contributed time and expertise by writing and reviewing chapters; others by advising and shaping the book's content based on their years of professional experience. Particular thanks are extended to **Michael T. Floyd**, National Sales Manager—On-Premise Laundry, Speed Queen Company, Ripon, Wisconsin; **Melissa Frechen**, Executive Housekeeper, Holiday Inn—University Place, East Lansing, Michigan; **Sheryl Fried**, Assistant Professor, Widener University, School of Hotel and Restaurant Management, Chester, Pennsylvania; **Al Norwood**, Deephaven, Minnesota; **Jon M. Owens, CHHE**, Director of Housekeeping, Clarion Hotel & Conference Center, Lansing, Michigan; **Carolyn Rockefellow, CHA**, Executive Housekeeper, Radisson Hotel, Lansing, Michigan; and **James H. Simpson**, Vice President of Business Development, Flagship Cleaning Services, Newtown Square, Pennsylvania. We would also like to extend a special thanks to **Robert Di Leonardo, Ph.D.**, and his staff at Di Leonardo International, Inc., Hospitality Design, Warwick, Rhode Island, for contributing the concluding chapter on interior design.

Margaret M. Kappa, CHHE
Consultant
Hospitality Housekeeping
Wabasha, Minnesota

Aleta Nitschke, CHA
Corporate Director of Rooms
Radisson Hotel Corporation
Minneapolis, Minnesota

Patricia B. Schappert, CHHE
Director of Housekeeping
Opryland Hotel
Nashville, Tennessee

Study Tips for Users of
Educational Institute Courses

Learning is a skill, like many other activities. Although you may be familiar with many of the following study tips, we want to reinforce their usefulness.

Your Attitude Makes a Difference

If you want to learn, you will: it's as simple as that. Your attitude will go a long way in determining whether or not you do well in this course. We want to help you succeed.

Plan and Organize to Learn

- Set up a regular time and place for study. Make sure you won't be disturbed or distracted.

- Decide ahead of time how much you want to accomplish during each study session. Remember to keep your study sessions brief; don't try to do too much at one time.

Read the Course Text to Learn

- *Before* you read each chapter, read the chapter outline and the learning objectives. If there is a summary at the end of the chapter, you should read it to get a feel for what the chapter is about.

- Then, go back to the beginning of the chapter and *carefully* read, focusing on the material included in the learning objectives and asking yourself such questions as:

  —Do I understand the material?

  —How can I use this information now or in the future?

- Make notes in margins and highlight or underline important sections to help you as you study. Read a section first, then go back over it to mark important points.

- Keep a dictionary handy. If you come across an unfamiliar word that is not included in the textbook glossary, look it up in the dictionary.

- Read as much as you can. The more you read, the better you read.

Testing Your Knowledge

- Test questions developed by the Educational Institute for this course are designed to measure your knowledge of the material.

- End-of-the-chapter Review Quizzes help you find out how well you have studied the material. They indicate where additional study may be needed. Review Quizzes are also helpful in studying for other tests.

- Prepare for tests by reviewing:

  —learning objectives

  —notes

  —outlines

  —questions at the end of each assignment

- As you begin to take any test, read the test instructions *carefully* and look over the questions.

We hope your experiences in this course will prompt you to undertake other training and educational activities in a planned, career-long program of professional growth and development.

Part I

Introduction to Housekeeping

Chapter Outline

Types of Hotels
 Economy/Limited-Service Hotels
 Mid-Range-Service Hotels
 World-Class-Service Hotels
Hotel Management
Hotel Divisions and Departments
 The Rooms Division
 The Engineering and Maintenance
  Division
 The Human Resources Division
 The Accounting Division
 The Security Division
 The Food and Beverage Division
 The Sales and Marketing Division
Housekeeping and the Front Office
Housekeeping and Engineering/
 Maintenance
Teamwork

Learning Objectives

1. Describe the role of the housekeeping department in hotel operations.

2. Classify hotels according to the level of service provided.

3. Explain the function of a hotel organization chart.

4. Define the major divisions and departments of a hotel.

5. Explain the importance of effective communication between housekeeping and the front office.

6. Explain how the executive housekeeper uses a daily occupancy report.

7. Explain the function of a housekeeping rooms status report.

8. Describe two methods of tracking room status.

9. Explain the importance of effective communication between housekeeping and the engineering and maintenance division.

10. Describe how a work order system functions.

The Role of Housekeeping in Hospitality Operations

Efficiently managed housekeeping departments ensure the cleanliness, maintenance, and aesthetic appeal of lodging properties. The housekeeping department not only prepares, on a timely basis, clean guestrooms for arriving guests, it also cleans and maintains everything in the hotel so that the property is as fresh and attractive as the day it opened for business.[1] These are no small tasks, especially in light of the following statistics.[2]

There are an estimated 44,300 lodging properties in the United States with a total of 2.82 million guestrooms available for sale each day of the year. Assuming that, on the average, 63% of the rooms available are actually occupied by guests, hotel housekeeping departments would be responsible for cleaning 1,776,600 guestrooms each day. If, on the average, a room attendant cleans 15 rooms a day, then there are at least 118,440 room attendants employed each day in housekeeping departments across the United States. Add to this figure the management staff of housekeeping departments; the housekeeping employees assigned to clean public spaces, back-of-the-house areas, meeting rooms, banquet rooms; and the other housekeeping employees working in the hotel's linen and laundry rooms—and it's easy to see why there are usually more employees working in the housekeeping department than in any other hotel department.

Estimates are that, in the lodging industry, more than 1.5 million employees work approximately 2 billion hours each year to serve more than 270 million guests. The amount of cleaning necessary to provide for the needs of 270 million guests is staggering. Imagine the stacks of linens needed to make up 270 million beds. The truckloads of bath soap, tissue, and amenities such as shampoos and colognes that must be distributed to guestrooms. The thousands of miles of carpeting, floors, walls, and ceilings that need to be cleaned and maintained. The countless pieces of furniture that must be dusted and polished. And, the millions of barrels of cleaning compounds along with the thousands of special tools and equipment that housekeeping departments use in order to clean, clean, clean.

The tasks performed by a housekeeping department are critical to the smooth daily operation of any hotel. This chapter begins by briefly describing the roles that housekeeping performs within different types of hotels. Next, the structure of hotel management is described and housekeeping's place within the overall organization of hotel operations is identified. This chapter goes on to describe the basic functions of various hotel divisions and departments and briefly examines housekeeping's relationship to them. The chapter ends by stressing the kind of teamwork

Exhibit 1   Size Classifications of Hotels

Number of Guestrooms (A88)	Property*	Rooms**
Under 75	72.0%	20.9%
75–149	17.1%	29.8%
150–299	8.1%	26.3%
300 and over	2.8%	23.0%

* Based on a total of 44,300 properties.
** Based on a total of 2.82 million rooms.

Source: "Lodging Industry Profile," American Hotel & Motel Association, June 1989.

that is crucial to successful hotel operations. Detailed examples are provided of the teamwork that must exist between housekeeping and front office personnel and between housekeeping, engineering, and maintenance personnel.

Types of Hotels

Classifying hotels into types is not easy. The lodging industry is so diverse that many hotels do not fit into any single well-defined category. Some of the categories used to classify hotels are location, the types of guests (or markets) attracted, the kind of ownership structure or chain affiliation, size, and service level. From the point of view of housekeeping, the size and service level of a property are its most important characteristics. However, size and service level are not dependent on each other. The size of a property often has little to do with the level of service it offers.

The size of a property gives only a general idea of the amount of work performed by the housekeeping staff. Size characteristics may include the number of guestrooms, meeting rooms, and banquet rooms within the property; the square footage of public areas; and the number of divisions or departments within the hotel requiring housekeeping services. Exhibit 1 focuses on the number of guestrooms and provides statistics on four hotel size categories.

A more precise measure of the work performed by a hotel's housekeeping staff is the property's level of service. Indicators of service level include the kinds of furnishings and fixtures in the different types of guestrooms; the decor of public areas; and special features of other facilities. While the levels of service offered by hotels vary tremendously across the lodging industry, properties can, for the sake of simplicity, be classified in terms of three basic service level categories: economy/limited, mid-range, and world-class service.

Economy/Limited-Service Hotels

Economy/limited-service hotels are a growing segment of the lodging industry. These properties focus on meeting the most basic needs of guests by providing clean, comfortable, and inexpensive rooms. Economy hotels appeal primarily to budget-minded travelers who want rooms with all the amenities required for a comfortable stay, but without the extras they don't really need or want to pay for. The types of guests attracted to economy/limited-service hotels include families with children, bus tour groups, business travelers, vacationers, retirees, and groups of conventioneers.

The size of the economy/limited-service property has increased from the 40- to 50-room hotel of the 1960s. Some economy hotels now have as many as 600 guestrooms; however, managerial considerations keep most properties between 50 and 150 guestrooms. The staff of small economy hotels generally consists of a live-in couple as managers, several room attendants, front desk agents, and sometimes a maintenance person.

Low design and construction costs and low operating expenses are part of the reason why economy hotels can be profitable. They incorporate simple designs that can be built economically and maintained efficiently. Economy hotels are usually two or three floors of cinder block construction with double-loaded corridors (corridors with guestrooms on both sides). These structures are cheaper to build than the single-loaded corridors found in large hotels where guestrooms may overlook elaborate atriums.

In comparison to the early 1970s when the only amenity offered may have been a black-and-white TV, most economy properties now offer color TV (many with cable or satellite reception), swimming pools, limited food and beverage service, playgrounds, small meeting rooms, and other special features. However, many economy properties do not provide full food and beverage service, which means guests may need to eat at nearby restaurants. Also, economy properties do not usually offer room service, uniformed service, banquet rooms, health clubs, or any of the more elaborate services and facilities found at mid-range and world-class properties.

Mid-Range-Service Hotels

Hotels offering **mid-range service** probably appeal to the largest segment of the traveling public. Mid-range service is modest but sufficient and the staffing level is adequate without trying to provide overly elaborate service. Guests likely to stay at a mid-range hotel are business travelers on expense accounts, tourists, or families taking advantage of special children's rates. Special rates may be offered for military personnel, educators, travel agents, senior citizens, and corporate groups. Meeting facilities of the mid-range-service hotel are usually adequate for conferences, training meetings, and small conventions.

The typical hotel offering mid-range service is medium-sized (between 150 and 300 rooms). These hotels generally offer uniformed service, airport limousine service, and full food and beverage facilities. The property may have a specialty restaurant, coffee shop, and lounge that cater to local residents as well as to hotel guests. The management staff of a mid-range property usually consists of a general manager and several department managers. The executive housekeeper manages the housekeeping department whose staff generally outnumbers that of any other department in the hotel.

Suite hotels appeal to vacationing or relocating families. (Courtesy of Radisson Suite Resort, Marco Island, Florida)

A fast-growing segment of the mid-range-service category is the suite hotel. Typical hotel accommodations feature one room, an adjacent bathroom, a king-size bed or two double beds, a desk/dresser modular unit, and one or two chairs. A suite unit, on the other hand, offers a small living room or parlor area with a grouping of appropriate furniture (often including a sofa bed) and a small bedroom with a king-size bed. Suite hotels provide temporary living quarters for people who are relocating, serve as "homes-away-from-home" for frequent travelers, or appeal to families interested in non-standard hotel accommodations. Professionals such as accountants, lawyers, and executives find suite hotels particularly attractive since they can work or entertain in an area which is separate from the bedroom.

Some guest suites include a compact kitchenette complete with cooking utensils, refrigerator, microwave unit, and wet bar. These additional features mean that room attendants will need more time to clean a suite of rooms than to clean a standard guestroom. Therefore, housekeeping labor expenses may be higher for suite hotels than for other properties in the mid-range-service category. Due to these and other costs, suite hotels generally offer less public space and fewer guest services than other hotels.

World-Class-Service Hotels

World-class-service hotels provide upscale restaurants and lounges, exquisite decor, concierge service, and opulent meeting and private dining facilities. Primary markets for hotels offering these services are top business executives,

The lobby of a world-class hotel may be elaborately furnished and decorated. (Courtesy of Opryland Hotel, Nashville, Tennessee)

entertainment celebrities, high ranking political figures, and other wealthy people. To cater to these types of guests, the housekeeping staff is generally responsible for dispensing oversize bath towels, bars of scented soap, special shampoos and conditioners, shower caps, and other guestroom and bath amenities. Bath linens are typically replaced twice daily and a nightly turndown service is usually provided. In addition, these guestrooms contain more expensive furnishings, decor, and art work than guestrooms in the mid-range-service category.

Some mid-range-service hotels may dedicate certain floors (usually the top floors) of the property to world-class service. Entry to these floors may be restricted by the use of special elevator keys that allow access only by authorized guests. The rooms provided on the "executive floor" are normally very large and deluxe. Hotels will typically upgrade the furnishings and decor of these guestrooms and provide additional guest services and amenities. The room or suite may be stocked daily with fresh cut flowers and fresh fruit. Bath amenities are generally similar to those provided by world-class hotels.

Hotel Management

Management guides the operation of the hotel and regularly reports the property's overall operating results and other pertinent information to the owner. The management team achieves specific objectives and goals by planning, organizing, staffing, directing, controlling, and evaluating functional areas within the hotel. Top management executives coordinate the activities of the various division and department managers.

The use of the terms division and department is not standard throughout the lodging industry. Large properties may call their main functional areas divisions and smaller functional areas departments. Other properties may call their main areas departments and refer to smaller areas as sub-departments. Neither method is better than the other. For consistency, this chapter will refer to the main functional areas as **divisions** and to the areas within divisions as **departments**.

The highest ranking executive of a property is usually called the general manager, managing director, or director of operations. The general manager of a hotel reports directly to the owner or to an assigned person in the owner's company. Within hotel chain organizations, the general manager of a property may report to a district, area, or regional executive supervising the properties in a particular group.

While the general manager is responsible for supervising all the divisions of a hotel, he/she may assign specific divisions or departments to the resident manager to oversee. Typically, resident managers are assigned to supervise departments in the rooms division of large hotels. When the general manager is absent from the property, the resident manager becomes the acting general manager. A manager-on-duty is often appointed to take responsibility when the general manager and the resident manager are both absent from the property.

All organizations require a formal structure to carry out their mission and objectives. A common method of representing that structure is the **organization chart**. An organization chart diagrams the divisions of responsibility and lines of authority. Some organizations list each employee's name on the chart along with his/her position title. Since no two hotels are exactly alike, organizational structures must be tailored to fit the needs of each individual property. Exhibit 2 shows a sample organization chart for a midsize rooms-only hotel. Within this structure, all department managers report directly to the assistant manager.

Exhibit 3 shows a sample organization chart diagramming the management positions in a large hotel. Note that within this organizational structure, the executive housekeeper and the front office manager report directly to the rooms division manager. The rooms division manager ensures that the housekeeping and front office departments work as a team so that guestrooms are cleaned and made ready for arriving guests. The importance of effective communication between housekeeping and the front office is examined later in this chapter. The housekeeping department also works closely with the engineering and maintenance division. Since these functional areas do not usually report to the same manager, it is important that the executive housekeeper and the chief engineer establish a close working relationship. Communication between housekeeping and engineering and maintenance is also addressed later in this chapter.

Hotel Divisions and Departments

Departments within a hotel may be classified according to a variety of methods. According to one method, each department is classified as either a revenue center

Exhibit 2   Sample Organization Chart for a Midsize Rooms-Only Hotel

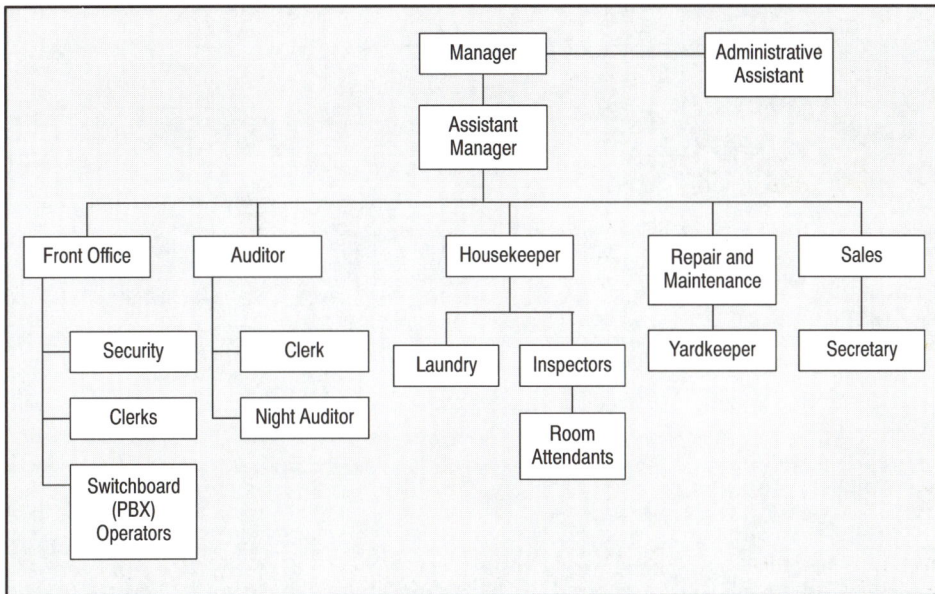

```
                    ┌──────────┐              ┌──────────────┐
                    │ Manager  │──────────────│Administrative│
                    └──────────┘              │  Assistant   │
                         │                    └──────────────┘
                    ┌──────────┐
                    │Assistant │
                    │ Manager  │
                    └──────────┘
                         │
   ┌──────────┬──────────┼──────────────┬──────────────┬──────────┐
┌────────┐ ┌────────┐          ┌────────────┐ ┌────────────┐ ┌────────┐
│ Front  │ │Auditor │          │Housekeeper │ │ Repair and │ │ Sales  │
│ Office │ │        │          │            │ │Maintenance │ │        │
└────────┘ └────────┘          └────────────┘ └────────────┘ └────────┘
```

Front Office	Auditor	Housekeeper	Repair and Maintenance	Sales
Security	Clerk	Laundry Inspectors	Yardkeeper	Secretary
Clerks	Night Auditor	Room Attendants		
Switchboard (PBX) Operators				

or a support center. This method is especially useful for accounting purposes and in relation to the property's recordkeeping and information system. A **revenue center** sells goods or services to guests and thereby generates revenue for the hotel. The front office and food and beverage outlets are examples of typical hotel revenue centers. **Support centers** do not generate revenue directly, but play a supporting role to the hotel's revenue centers. The housekeeping department is a major support center within the rooms division. Other hotel support centers include the areas of accounting, engineering and maintenance, and human resources.

The terms **front of the house** and **back of the house** may also be used to classify hotel departments and the personnel within them. Front-of-the-house functional areas are those in which employees have a great deal of guest contact, such as the front office and food and beverage facilities. Back-of-the-house functional areas are those in which employees have less direct guest contact, such as accounting, engineering and maintenance, and human resources. Although members of the housekeeping department have some contact with hotel guests, the department is generally considered a back-of-the-house functional area.

The following sections briefly describe the major divisions and departments typically found in a large hotel.

The Rooms Division

The rooms division is composed of departments and functions which play essential roles in providing the services that guests expect during their stay. In most hotels, the rooms division generates more revenue than any other area in the hotel.

Exhibit 3   Sample Organization Chart for a Large Hotel

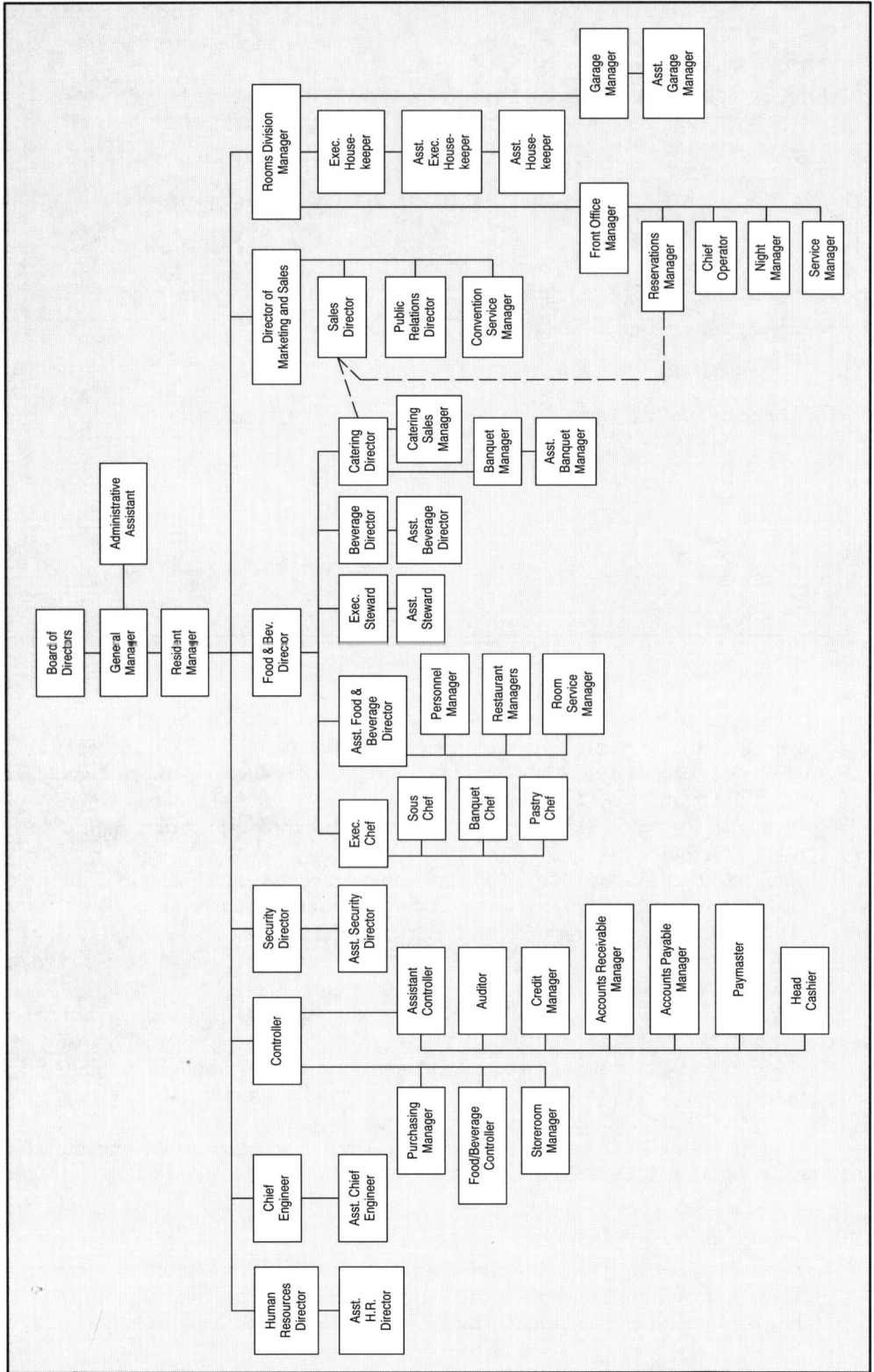

Board of Directors

General Manager
— Administrative Assistant

Resident Manager

Human Resources Director
— Asst. H.R. Director

Chief Engineer
— Asst. Chief Engineer

Controller
— Purchasing Manager
— Food/Beverage Controller
— Storeroom Manager
— Assistant Controller
  — Auditor
  — Credit Manager
  — Accounts Receivable Manager
  — Accounts Payable Manager
  — Paymaster
  — Head Cashier

Security Director
— Asst. Security Director

Food & Bev. Director
— Exec. Chef
  — Sous Chef
  — Banquet Chef
  — Pastry Chef
— Asst. Food & Beverage Director
  — Personnel Manager
  — Restaurant Managers
  — Room Service Manager
— Exec. Steward
  — Asst. Steward
— Beverage Director
  — Asst. Beverage Director
— Catering Director
  — Catering Sales Manager
  — Banquet Manager
    — Asst. Banquet Manager

Director of Marketing and Sales
— Sales Director
— Public Relations Director
— Convention Service Manager

Rooms Division Manager
— Exec. Housekeeper
  — Asst. Exec. Housekeeper
  — Asst. Housekeeper
— Front Office Manager
  — Reservations Manager
  — Chief Operator
  — Night Manager
  — Service Manager
— Garage Manager
  — Asst. Garage Manager

A VIP suite may be located on an executive floor. (Courtesy of Radisson Plaza Hotel at Austin Centre, Austin, Texas)

The revenue center of the rooms division is the front office department. This department is usually the most important revenue center in a hotel. Other departments within the rooms division serve as support centers for the front office. These may include the housekeeping, reservations, telephone, and uniformed service departments.

The front office is the most visible department in a hotel and has the greatest amount of direct guest contact. The front desk, cashier, and mail and information sections of the front office are located in the busiest area of the hotel's lobby. The front desk itself is the focal point of activity within the front office department. Guests are registered, assigned rooms, and checked out at the front desk.[3]

In some properties, the reservations and switchboard functions may be separate departments within the rooms division. The reservations area is responsible for receiving and processing reservations for future accommodations. Reservations agents must maintain accurate records and closely track the availability of rooms to ensure that no date is overbooked. Many departments within the hotel—especially housekeeping—use reservations data and other rooms forecast information to properly schedule personnel.

Hotel switchboard operators, sometimes referred to as PBX (private branch exchange) operators, answer calls and connect them to the appropriate extension. These operators relay telephone charges to the front office cashier for posting to the proper guest account, and, in some properties, place wake-up calls, monitor auto-mated systems, and coordinate emergency communication systems.

The hotel's uniformed service staff may include parking attendants, door at-tendants, porters, limousine drivers, and bellpersons. Uniformed service staff meet and greet guests and help them upon their arrival and departure.

The Engineering and Maintenance Division

A hotel's engineering and maintenance division is responsible for maintaining the appearance of the interior and exterior of the property and keeping its equipment operational.[4] This division is also typically responsible for swimming pool sanita-tion and the landscaping and upkeep of the property's grounds. Some hotels, how-ever, staff a grounds division or an outdoor and recreation division to perform these and other tasks. Not all engineering and maintenance work can be handled by the hotel's staff. Often, problems or projects require outside contracting.

The housekeeping department works closely with the engineering and main-tenance division to ensure that proper preventive maintenance procedures are carried out effectively. Since daily cleaning duties require that housekeeping per-sonnel enter almost every guestroom every day, the housekeeping department is in a position to identify maintenance needs and initiate work orders for engineering.

The Human Resources Division

In recent years, hotels have increased investment in and dependence on human re-sources management. The size and budgets of human resources divisions have grown steadily, along with their responsibility and influence. This expanded role is mirrored by the growing preference for the broader term *human resources* over *per-sonnel*. Recently, the scope of human resources management has changed in re-sponse to new legislation, the shrinking labor market, and the growing pressures of competition. Human resources functions may include employment (including external recruiting and internal reassignment), orientation and training, employee relations, compensation, benefits, labor relations, and safety.[5]

Many properties are not large enough to justify the creation of a human re-sources division or department. In these properties, the general manager and de-partment managers share many of the duties and responsibilities connected with the human resources function.

The Accounting Division

A hotel's accounting division is responsible for monitoring the financial activities of the property. Some hotels employ off-premises accounting services to comple-ment the work of their internal accounting division. In this case, the hotel's staff collects and transmits data to a service bureau or to chain headquarters. A hotel that performs all of its accounting work in-house will employ a larger accounting staff with a high level of responsibility.

The hotel's controller manages the accounting division. Accounting activities include paying invoices owed, distributing statements and collecting payments, processing payroll information, accumulating operating data, and compiling financial statements. In addition, the accounting staff may be responsible for making bank deposits, securing cash, and performing other control and processing functions required by the hotel's management.

In some properties, the purchasing manager and the storeroom manager may report to the hotel's controller. The executive housekeeper must often work closely with these managers because the housekeeping department maintains inventories of cleaning supplies, equipment, linens, uniforms, and other items.

The controller and the general manager are responsible for finalizing the budgets prepared by division and department managers.[6]

The Security Division

Security staff might include in-house personnel, contract security officers, or retired or off-duty police officers. Security responsibilities may include patrolling the property, monitoring surveillance equipment, and, in general, ensuring that guests, visitors, and employees are safe and secure at the hotel.[7] The cooperation and assistance of local law enforcement officials is critical to the security division's effectiveness.

A hotel's security program is most effective when employees who have primary responsibilities other than security also participate in security efforts. For example, housekeeping room attendants should follow the key control procedures of their properties. Also, when cleaning guestrooms, room attendants are usually responsible for locking and securing sliding glass doors, connecting doors, and windows. All employees should be wary of suspicious activities anywhere in the property and report such activities to an appropriate security authority. Since housekeeping personnel work in every area of the hotel, they are in a position to significantly contribute to the hotel's security efforts.

The Food and Beverage Division

A major revenue center in most hotels is the food and beverage division. There are almost as many varieties of food and beverage operations as there are hotels.[8] Many hotels offer guests more than a single food and beverage outlet. Possible outlet types include quick service, table service, specialty restaurants, coffee shops, bars, lounges, and clubs. The food and beverage division typically supports other hotel functions such as room service, catering, and banquet planning.

The executive steward supervises most of the kitchen sanitation and cleaning duties. However, the housekeeping department may be responsible for cleaning specific areas of the hotel's dining rooms, banquet rooms, and some back-of-the-house food and beverage areas.

The Sales and Marketing Division

The sales and marketing staff in a hotel can vary from one part-time person to a dozen or more full-time employees. These personnel typically have four functions: sales, convention services, advertising, and public relations.[9] The marketing arm of the division researches the marketplace, competing products, guest needs and expectations, and then develops sales action plans by which to attract guests to the

Exhibit 4   Reasons Guests Give for Returning to a Hotel

Question: What is the most important factor in your decision to return to a hotel/motel?

Finding:

Reason for Returning	By Total Travelers	By Frequent Travelers
Cleanliness/Appearance	63%	63%
Good Service	42%	45%
Facilities	35%	41%
Convenience/Location	32%	38%
Price/Reasonable Rates	39%	35%
Quiet and Private	9%	8%

Source: "Bringing in the Business and Keeping It." Study done for Procter & Gamble by Market Facts.

property. The primary goal of the division is to sell the products and services offered by the hotel.

Housekeeping's important contribution to the primary goal of the sales and marketing division often goes unrecognized. Successful sales departments maintain high percentages of repeat business. Exhibit 4 provides some data on the reasons guests have for returning to a hotel. Note that the single most important reason a traveler returns to a hotel is the cleanliness and appearance of the property. The data also indicate that good service is second in importance. Housekeeping staff are among the most visible hotel representatives in this regard. Therefore, an important contribution of the housekeeping staff to hotel sales is the repeat business obtained by providing the level of cleanliness and service that meets or exceeds guest expectations.

Housekeeping and the Front Office

Within the rooms division, housekeeping's primary communications are with the front office department, specifically with the front desk area. At most properties, the front desk agent is not allowed to assign guestrooms until the rooms have been cleaned, inspected, and released by the housekeeping department. Typically, rooms are recycled for sale according to the following process.

Each night, a front desk agent produces an **occupancy report**. The occupancy report lists rooms occupied that night and indicates those guests expected to check out the following day. The executive housekeeper picks up this list early the next morning and schedules the occupied rooms for cleaning. As guests check out of the hotel, the front desk notifies housekeeping. Housekeeping ensures that these rooms are given top priority so that clean rooms are available for arriving guests.

At the end of the shift, the housekeeping department prepares a **housekeeping status report** (see Exhibit 5) based on a physical check of each room in the property. This report indicates the current housekeeping status of each room. It is

Exhibit 5   Sample Housekeeping Rooms Status Report

| Housekeeper's Report | | | | | | A.M. | |
| Date _____ , 19 _____ | | | | | | P.M. | |

ROOM NUMBER	STATUS	ROOM NUMBER	STATUS	ROOM NUMBER	STATUS	ROOM NUMBER	STATUS
101		126		151		176	
102		127		152		177	
103		128		153		178	
104		129		154		179	
105		130		155		180	
106		131		156		181	
107		132		157		182	
108		133		158		183	
120		145		170		195	
121		146		171		196	
122		147		172		197	
123		148		173		198	
124		149		174		199	
125		150		175		200	

Remarks:

_____
_____
_____

Legend:

✓	–	Occupied
000	–	Out-of-Order
———	–	Vacant
B	–	Slept Out (Baggage Still in Room)
X	–	Occupied, No Baggage
C.O.	–	Slept In but Checked Out Early A.M.
E.A.	–	Early Arrival

_____
               Housekeeper's Signature

compared to the front desk occupancy report, and any discrepancies are brought to the attention of the front office manager. A **room status discrepancy** is a situation in which the housekeeping department's description of a room's status differs from the room status information being used by the front desk to assign guestrooms. Room status discrepancies can seriously affect a property's ability to satisfy guests and maximize rooms revenue.

To ensure efficient rooming of guests, housekeeping and the front office must inform each other of changes in a room's status. Knowing whether a room is occupied, vacant, on-change, out-of-order, or in some other condition is important to rooms management. For example, if a guest checks out before the stated departure date, the front desk must notify the housekeeping department that the room is no longer a stayover, but is now a check-out. Exhibit 6 defines typical room status terms used in the lodging industry. While the guest is in the hotel, the housekeeping status of the guestroom changes several times. However, not every room status will occur for each guestroom during every stay.

Exhibit 6   Room Status Definitions

Occupied: A guest is currently registered to the room.

Complimentary: The room is occupied, but the guest is assessed no charge for its use.

Stayover: The guest is not checking out today and will remain at least one more night.

On-change: The guest has departed, but the room has not yet been cleaned and readied for resale.

Do not disturb: The guest has requested not to be disturbed.

Sleep-out: A guest is registered to the room, but the bed has not been used.

Skipper: The guest has left the hotel without making arrangements to settle his/her account.

Sleeper: The guest has settled his/her account and left the hotel, but the front office staff has failed to properly update the room's status.

Vacant and ready: The room has been cleaned and inspected, and is ready for an arriving guest.

Out-of-order: The room cannot be assigned to a guest. A room may be out-of-order for a variety of reasons, including the need for maintenance, refurbishing, and extensive cleaning.

Lock-out: The room has been locked so that the guest cannot re-enter until he/she is cleared by a hotel official.*

DNCO (did not check out): The guest made arrangements to settle his/her account (and thus is not a skipper), but has left without informing the front office.

Due out: The room is expected to become vacant after the following day's check-out time.

Check-out: The guest has settled his/her account, returned the room keys, and left the hotel.

Late check-out: The guest has requested and is being allowed to check out later than the hotel's standard check-out time.

*A hotel should not employ a lock-out without consulting legal counsel; relevant laws vary from state to state. See also Jack P. Jefferies, *Understanding Hospitality Law,* 3rd ed. (East Lansing, Mich.: Educational Institute of the American Hotel & Motel Association, 1995).

Promptly notifying the front desk of the housekeeping status of rooms is a tremendous aid in getting early-arriving guests registered, especially during high-occupancy or sold-out periods. Keeping room status information up-to-date requires close coordination and cooperation between the front desk and the housekeeping department. The two most common systems for tracking current room status are mechanical room rack systems and computerized status systems.

The front desk may use a **room rack** to track the status of all rooms. A room rack slip containing the guest's name and other relevant information is normally completed during the registration process and placed in the room rack slot corresponding to the assigned room number. The presence of a room rack slip indicates that the room is occupied. When the guest checks out, the rack slip is removed and the room's status is changed to on-change. An on-change status indicates that the room is in need of housekeeping services before it can be made available to arriving guests. As unoccupied rooms are cleaned and inspected, the housekeeping department notifies the front desk, which updates the room's status to available for sale.

The cumbersome nature of tracking and comparing housekeeping and front desk room status information often leads to mistakes. For example, if a room rack slip is mistakenly left in the rack even though the guest has checked out, front desk agents may falsely assume that a vacant room is still occupied. This is an example of a room status discrepancy called a sleeper: the rack slip is "asleep" in the rack, and the potential revenue from the sale of the room is lost.

Problems may also arise from communication delays between the housekeeping department and the front desk. Communication between these areas may be spoken, written, or conveyed by means of a telewriter. Spoken communication—over the telephone—relays information quickly, but without supporting documentation. A written report has the advantage of documenting the information, but is time-consuming since it must be hand-delivered. A telewriter, on the other hand, communicates and documents information quickly, without requiring anyone to be on its receiving end. Telewriters are especially helpful when front desk agents or housekeepers are busy with other responsibilities and do not have time to place a call or answer the telephone.

In a computerized room status system, housekeeping and the front desk often have instantaneous access to room status information. When a guest checks out, a front desk agent enters the departure into a computer terminal. Housekeeping is then alerted that the room needs cleaning through a remote terminal located in the housekeeping department. Next, housekeeping attendants clean the room and notify the housekeeping department when it is ready for inspection. Once the room is inspected, housekeeping enters this information into its departmental terminal. This informs the front office computer that the room is available for sale.

While room occupancy status within a computerized system is almost always current, reporting of housekeeping status may lag behind. For example, the housekeeping supervisor may inspect several rooms at once, but may not update the computer's room status files until the end of a long inspection round. In a large operation, calling the housekeeping department after each room is inspected is generally inefficient, since answering the phone can be a constant interruption. A delay may also occur when a list of clean, inspected rooms is furnished to the housekeeping office but not immediately entered into the computer system.

The problems in promptly reporting housekeeping status to the front office can be eliminated when the computer system is directly connected to the guestroom telephone system. With such a network, supervisors can inspect rooms, determine their readiness for sale, and then enter a designated code on the room telephone to change the room's status in the hotel's computer system. No one needs to answer the phone, since the computer automatically receives the relay, and there is little chance for error. Within seconds, the room's updated status can be displayed on the screen of a front desk computer terminal. This procedure can significantly reduce not only the number of guests forced to wait for room assignment, but also the length of their wait.

Teamwork between housekeeping and the front office is essential to daily hotel operations. The more familiar housekeeping and front office personnel are with each other's procedures, the smoother the relationship between the two departments is likely to be.

Housekeeping and Engineering/Maintenance

In most non-lodging commercial buildings, housekeeping, engineering, and maintenance personnel generally report to the same department manager. This makes a great deal of sense because these functional areas have similar goals and methods and must have a close working relationship. In most midsize and large lodging operations, however, housekeeping reports to the rooms division manager, while engineering and maintenance constitute a separate division. Different reporting responsibilities can become barriers to effective communication between these important support centers of a hotel.

In fact, it is unfortunate that support centers often seem to have an almost adversarial relationship. For example, housekeeping personnel sometimes resent having to clean up after various types of maintenance, while engineering personnel may be upset if the misuse of chemicals and equipment by housekeeping results in additional work for them. In order to ensure the smooth operation of both departments, housekeeping and engineering managers need to devote attention to improving the relationship between their departments.

The housekeeping department often takes the first steps in relation to maintenance functions for which engineering is ultimately responsible. There are three kinds of maintenance activities: routine maintenance, preventive maintenance, and scheduled maintenance.

Routine maintenance activities are those which relate to the general upkeep of the property, occur on a regular (daily or weekly) basis, and require relatively minimal training or skills. These are maintenance activities which occur outside of a formal work order system and for which no specific maintenance records (time or materials) are kept. Examples include sweeping carpets, washing floors, cleaning readily accessible windows, cutting grass, cleaning guestrooms, shoveling snow, and replacing burned-out light bulbs. Many of these routine maintenance activities are carried out by the housekeeping department. Proper care of many surfaces and materials by housekeeping personnel is the first step in the overall maintenance program for the property's furniture and fixtures.

Preventive maintenance consists of three parts: inspection, minor corrections, and work order initiation. For many areas within the hotel, inspections are performed by housekeeping personnel in the normal course of their duties. For example, room attendants and inspectors may regularly check guestrooms for leaking faucets, cracked caulking around bathroom fixtures, and other items that may call for action by engineering staff. Attending to leaking faucets and improper caulking around sinks and tubs can control maintenance costs by preventing greater problems, such as ceiling or wall damage in the bath below. Such maintenance protects the physical plant investment and contributes to guest satisfaction.

Communication between housekeeping and engineering should be efficient so that most minor repairs can be handled while the room attendant is cleaning the guestroom. In some properties, a full-time maintenance person may be assigned to inspect guestrooms and to perform the necessary repairs, adjustments, or replacements.

Preventive maintenance, by its nature, sometimes identifies problems and needs beyond the scope of a minor correction. These problems are brought to the attention of engineering through the work order system. The necessary work is then scheduled by the building engineer. This type of work is often referred to as **scheduled maintenance**.

Exhibit 7   Sample Maintenance Work Order

DELTA FORMS - MILWAUKEE U.S.A.

(414) 461-0086

HYATT HOTELS ⊕

MAINTENANCE REQUEST

TIME _____ **1345239**

BY _____ DATE _____

LOCATION _____

PROBLEM _____

_____

_____

_____

_____

_____

_____

ASSIGNED TO _____

DATE COMPL. _____ TIME SPENT _____

COMPLETED BY _____

REMARKS _____

_____

_____

_____

_____

_____

_____

RPHK-04

HYATT HOTELS MAINTENANCE CHECK LIST
Check (☒) Indicates Unsatisfactory Condition
Explain Check In Remarks Section

BEDROOM - FOYER - CLOSET

☐ WALLS   ☐ WOODWORK   ☐ DOORS
☐ CEILING   ☐ TELEVISION   ☐ LIGHTS
☐ FLOORS   ☐ A.C. UNIT   ☐ BLINDS
☐ WINDOWS   ☐ DRAPES

REMARKS :_____

_____

_____

BATHROOM

☐ TRIM   ☐ SHOWER
☐ DRAINS   ☐ LIGHTS
☐ WALL PAPER   ☐ PAINT
☐ TILE OR GLASS ☐ DOOR
☐ ACCESSORIES   ☐ WINDOW

REMARKS :_____

_____

_____

_____

Courtesy of Hyatt Corporation, Chicago, Illinois

Scheduled maintenance activities are initiated at the property based on a formal work order or similar document. Work orders are a key element in the communication between housekeeping and engineering. A sample work order is shown in Exhibit 7. In many properties work orders are numbered three-part forms. Each part of the form is color-coded for its recipient.

For example, when a member of the housekeeping department fills out a work order form, one copy is sent to the executive housekeeper, and two copies to engineering. The chief engineer gets one of these copies and gives the other to the tradesperson assigned to the repair. The individual completing the task indicates the number of hours required to complete the work, any parts or supplies required, and other relevant information. When the job is completed, a copy of the tradesperson's completed work order is sent to the executive housekeeper. If this copy is not returned within an appropriate amount of time, housekeeping issues another work order, which signals engineering to provide a status report on the requested repair.

Engineering generally keeps data cards and history records on all equipment operated by housekeeping personnel. Equipment data cards contain basic information about pieces of equipment. This information can include technical data,

Exhibit 8   Sample Equipment History Record Card

DATE	W.O. NO.	DESCRIPTION OF REPAIRS	DOWN TIME	MAN HOURS	MATERIAL COST

HISTORY OF REPAIRS

TAG NO.   DESCRIPTION   WEEKLY CONTROL   MONTHLY INSPECTION CONTROL

TYPIST PLEASE NOTE — START ALL TYPING AT SAME POINT ON SCALE. THEN REMOVE THIS STUB. BE SURE YOU HAVE A WELL INKED RIBBON. CARE USED IN TYPING WILL IMPROVE REFERENCE DURING THE ENTIRE LIFE OF THE INDEX. TRY A FEW IN THE POCKETS TO SEE HOW THEY LOOK BEFORE TYPING THE ENTIRE LIST.

RE-ORDER FORM NO. 60P01   ACME VISIBLE RECORDS   PRINTED IN U.S.A.
                               CROZET VIRGINIA

Courtesy of Acme Visible Records

manufacturers' information, cost, special instructions, warranty information, and references to other information as well (such as the storage location of manuals and drawings). Equipment history records (see Exhibit 8) are logs of the inspection and maintenance work performed on a given piece of equipment. History records may be separate cards or may be incorporated into the equipment data card. Their purpose is to provide documentation of all maintenance activity on a given piece of equipment. Many properties have computerized these recordkeeping functions, making it easier for the executive housekeeper to retrieve pertinent information when requesting replacement or new equipment items.

Teamwork

Teamwork is the key to successful hotel operations. Housekeeping must work closely not only with the front office and engineering but also with every other department in the hotel. Although the general manager is responsible for coordinating and implementing the teamwork philosophy, each department and every employee can help.

Endnotes

1. As it is used in this chapter, hotel is a broad generic term for all types of lodging operations including luxury hotels, motels, motor inns, and inns.

2. These statistics are published in "Lodging Industry Profile," a pamphlet prepared by the Communications Department of the American Hotel & Motel Association.

3. For further information regarding the front office, see Michael L. Kasavana and Richard M. Brooks, *Managing Front Office Operations*, 4th ed. (East Lansing, Mich.: Educational Institute of the American Hotel & Motel Association, 1995).

4. For further information regarding the engineering and maintenance division of hotels, see Michael H. Redlin and David M. Stipanuk, *Managing Hospitality Engineering Systems* (East Lansing, Mich.: Educational Institute of the American Hotel & Motel Association, 1987).

5. For further information regarding human resources functions, see Robert Woods, *Managing Hospitality Human Resources* (East Lansing, Mich.: Educational Institute of the American Hotel & Motel Association, 1992).

6. For further information about financial responsibilities of department managers, see Raymond S. Schmidgall, *Hospitality Industry Managerial Accounting,* 3rd ed. (East Lansing, Mich.: Educational Institute of the American Hotel & Motel Association, 1995).

7. For further information about hotel security, see Raymond C. Ellis, Jr., and the Security Committee of AH&MA, *Security and Loss Prevention Management* (East Lansing, Mich.: Educational Institute of the American Hotel & Motel Association, 1986).

8. For further information about food and beverage service, see Jack D. Ninemeier, *Management of Food and Beverage Operations*, 2d ed. (East Lansing, Mich.: Educational Institute of the American Hotel & Motel Association, 1995).

9. For further information about sales and marketing functions, see James R. Abbey, *Hospitality Sales and Advertising,* 2d ed. (East Lansing, Mich.: Educational Institute of the American Hotel & Motel Association, 1993); and Christopher W. L. Hart and David A. Troy, *Strategic Hotel/Motel Marketing*, rev. ed. (East Lansing, Mich.: Educational Institute of the American Hotel & Motel Association, 1986).

Key Terms

- back of the house
- department
- division
- economy/limited service
- front of the house
- housekeeping status report
- mid-range service
- occupancy report
- organization chart
- preventive maintenance
- revenue center
- room rack
- room status discrepancy
- routine maintenance
- scheduled maintenance
- support center
- world-class service

Discussion Questions

1. Name the three basic service level categories of hotels. What are typical characteristics of each?

2. What is the purpose of an organization chart?

3. What is the difference between a revenue center and a support center? What hotel departments typically fall under each category?

4. What is meant by the terms front of the house and back of the house? What typical functional areas are classified under each term?

5. What seven major divisions are typically found in a large hotel?

6. What important contribution does housekeeping make to a property's sales effort? How?

7. Why are two-way communications necessary between the front desk and housekeeping?

8. What are the two most common systems used by the front desk and house-keeping to track current room status?

9. Compare the three kinds of maintenance activities.

10. What is the ideal relationship between housekeeping and maintenance? What is the actual situation in some properties?

REVIEW QUIZ

When you feel you have covered all of the material in this chapter, answer these questions. Choose the *best* answer.

True (T) or False (F)

T (F) 1. Economy/limited-service hotels appeal to the largest segment of the traveling public.

(T) F 2. The management staff of a mid-range-service property usually has several department managers reporting to a general manager.

T (F) 3. The housekeeping department is generally considered to be a front-of-the-house functional area.

T (F) 4. In most hotels, the food and beverage division generates more revenue than any other revenue center.

T (F) 5. Guests are assigned rooms and checked out in the reservations area of the hotel.

T (F) 6. The controller and the director of marketing and sales are responsible for finalizing budgets prepared by department managers.

(T) F 7. The executive housekeeper is responsible for supervising most of the kitchen sanitation and cleaning duties in food and beverage outlets within the hotel.

T (F) 8. Producing a nightly occupancy report is the responsibility of the executive housekeeper.

T (F) 9. A room rack utilizes computer technology to track the status of guestrooms.

T (F) 10. The term check-out refers to a room that is expected to become vacant after the following day's check-out time.

Alternate/Multiple Choice

11. Housekeeping labor expenses are likely to be _____ in suite hotels than in other mid-range-service hotels.

    a. higher
    b. lower

12. Housekeeping is a _____ within the rooms division.

    a. revenue center
    b. support center

13. Concierge service, private dining facilities, and nightly turndown service are generally provided by:

   a. economy/limited-service hotels.
   b. mid-range-service hotels.
   c. world-class-service hotels.
   d. suite hotels.

14. The general manager of a hotel reports to the:

   a. division manager.
   b. manager-on-duty.
   c. resident manager.
   d. owner.

15. A room status discrepancy occurs when the housekeeping status report does not match the:

   a. balance sheet.
   b. records at the front desk.
   c. nightly revenue report.
   d. monthly budget.

14/15

Chapter Outline

Identifying Housekeeping's Responsibilities
Planning the Work of the Housekeeping
   Department
      Area Inventory Lists
      Frequency Schedules
      Performance Standards
      Productivity Standards
      Equipment and Supply Inventory
        Levels
Organizing the Housekeeping Department
      The Department Organization Chart
      Job Lists and Job Descriptions
Other Management Functions of the
   Executive Housekeeper
      Coordinating and Staffing
      Directing and Controlling
      Evaluating

Learning Objectives

1. Identify typical cleaning responsibilities of the housekeeping department.

2. Explain the function of area inventory lists.

3. Describe how frequency schedules are used.

4. Explain the function of performance standards.

5. Describe how productivity standards are established for employees.

6. Distinguish between recycled and non-recycled inventories.

7. Distinguish between job lists, job descriptions, job breakdowns.

8. Describe the function of the housekeeping department's operating budget.

9. Define the seven basic management functions of the executive housekeeper.

2

Planning and Organizing the Housekeeping Department

Lᴉᴋᴇ ᴀʟʟ ᴏᴛʜᴇʀ ᴍᴀɴᴀɢᴇʀs in a hotel, the executive housekeeper uses available resources to attain objectives set by top management executives. Resources include people, money, time, work methods, materials, energy, and equipment. These resources are in limited supply, and most executive housekeepers will readily admit that they rarely have all the resources they would like. Therefore, an important part of the executive housekeeper's job is planning how to use the limited resources available to attain the hotel's objectives.

The executive housekeeper uses objectives set by the general manager as a guide in planning more specific, measurable goals for the housekeeping department. For example, one of the first planning activities of the executive housekeeper is to clarify the department's cleaning responsibilities and to map strategies for carrying out these responsibilities effectively. Strategies will identify the types of cleaning tasks and how frequently the tasks must be performed.

This chapter begins by identifying some of the executive housekeeper's most important planning functions. Major cleaning responsibilities of the housekeeping department are identified and suggestions for planning work within the department are presented. In addition, the chapter examines the organizational structure of several housekeeping departments and presents sample job descriptions for executive housekeeper positions. Job descriptions are also presented for typical housekeeping positions in a mid-range-service hotel. The chapter closes by showing how other important management functions of the executive housekeeper fit into the overall process of management.

Identifying Housekeeping's Responsibilities

Regardless of the size and structure of a housekeeping department, it is typically the responsibility of the hotel's general manager to identify which areas of the property housekeeping will be responsible for cleaning. Most housekeeping departments are responsible for cleaning the following areas:

- Guestrooms
- Corridors
- Public areas, such as the lobby and public restrooms
- Pool and patio areas

Housekeeping Success Tip

Wai Kai Au-Yeung, CHHE
Group Executive Housekeeper
New World Hotels International
Hong Kong

❝Guest expectations have become higher and higher as more new and better-equipped hotels have opened or expanded. As a result, the hospitality industry has changed considerably over the years.

In the past, guests looked for a clean, safe, and pleasant environment. The main function of housekeeping was to provide cleaning service only. But nowadays, guests take a clean, safe, and pleasant environment for granted. Now they look for more sophisticated and personalized service. Therefore, the housekeeping department must provide many other services, such as turndown service, shoe shine service, room butler service, etc.

If we don't meet the rising expectations of our guests, we will be out of business. Moreover, their expectations must be fulfilled by people, not machines. So it is important that housekeeping management plan and organize the work so that quality services flow naturally out of the department's activities. Planning and organizing in the housekeeping department, however, is like an appetizer before a meal; there must be more to follow to make the meal a success. Training and motivating are what must follow planning and organizing to make the efforts of the housekeeping department a success.

A key element of motivating is to provide advancement opportunities for staff. I always tell my staff that I will not let them stay in the same position forever unless they prefer to do so. I believe that there is always room for advancement for those who deserve it. The point is whether the staff is prepared or not.

Preparation means keeping up-to-date with a changing world. Continuing education is therefore important. I encourage my staff to attend training courses such as those organized by the Educational Institute. As a department head, I try my best to set a good example for my staff. I attended a few courses at the Center for Professional Development, School of Hotel Administration, Cornell University.

Besides encouraging my staff to attend training courses, I also provide lots of on-the-job training for them. On-the-job training offers very good practical exercises for staff. This could include cross-training room attendants to be floor supervisors or role playing. These kinds of activities offer other opportunities such as:

- Helping managers observe and develop staff potential

- Helping the staff to understand and appreciate other people's jobs so better coordination can be achieved

- Improving the relationship and working atmosphere among employees in different positions

---

Housekeeping Success Tip *(continued)*

I think the key to successful continuing education is the word 'continuing.' One-shot training only affords short-term results. If we want to have long-lasting results, we must provide continuing training, especially for junior staff. We need to help them build good work habits. If all staff developed good habits, our guests would certainly show their appreciation. That, in turn, builds staff morale. And this creates a positive service-feedback cycle. As an executive housekeeper, I think this is the greatest success—to see happy and efficient staff serving guests to their greatest satisfaction. 99

- Management offices
- Storage areas
- Linen and sewing rooms
- Laundry room
- Back-of-the-house areas, such as employee locker rooms

Housekeeping departments of hotels offering mid-range and world-class service are generally responsible for additional areas, such as:

- Meeting rooms
- Dining rooms
- Banquet rooms
- Convention exhibit halls
- Hotel-operated shops
- Game rooms
- Exercise rooms

Housekeeping's cleaning responsibilities in the food and beverage areas vary from property to property. In most hotels, housekeeping has very limited responsibilities in relation to food preparation, production, and storage areas. The special cleaning and sanitation tasks required for maintaining these areas are usually carried out by kitchen staff under the supervision of the chief steward. In some properties, the dining room staff cleans service areas after breakfast and lunch periods; housekeeping's night cleaning crew does the in-depth cleaning after dinner service or early in the morning before the dining room opens for business. The executive housekeeper and the dining room managers must work closely together to ensure that quality standards are maintained in the guest service and server station areas.

The same cooperation is necessary between housekeeping and banquet or convention services. The banquet or convention staff generally sets up function and meeting rooms and is responsible for some cleaning after the rooms are used.

The final in-depth cleaning is left to the housekeeping crew. This means that the final responsibility for the cleanliness and overall appearance of these areas falls squarely on the shoulders of the housekeeping staff.

The general manager typically designates which areas housekeeping will be responsible for cleaning. However, if areas of responsibility cross department lines, the managers of those departments must get together and settle among themselves any disputes about cleaning responsibilities. The agreement among the managers is then reported to the general manager for his/her approval. A good housekeeping manager can effectively solve problems on his/her level with other managers, thereby relieving the general manager of day-to-day operational problems.

It is a good idea for the executive housekeeper to obtain a floor plan of the hotel and color in those areas for which housekeeping is responsible. Different colors can be used to designate those areas for which other department managers are responsible. To ensure that all areas of the property have been covered—and to avoid future misunderstandings about responsibilities—copies of this color-coded floor plan should be distributed to the general manager and to all department managers. This way, everyone can see at a glance who is responsible for cleaning each area in the hotel. The color-coded floor plan also presents a clear and impressive picture of the housekeeping department's role in cleaning and maintaining the hotel.

Once housekeeping's areas of responsibility have been identified, planning focuses on analyzing the work required for cleaning and maintaining each area.

Planning the Work of the Housekeeping Department

Planning is probably the executive housekeeper's most important management function. Without competent planning, every day may present one crisis after another. Constant crises lower morale, decrease productivity, and increase expenses within the department. Also, without the direction and focus that planning provides, the executive housekeeper can easily become sidetracked by tasks which are unimportant or unrelated to accomplishing the hotel's objectives.

Since the housekeeping department is responsible for cleaning and maintaining so many different areas of the hotel, planning the work of the department can seem like an enormous task. Without a systematic, step-by-step approach to planning, the executive housekeeper can easily become overwhelmed and frustrated by the hundreds of important details that must be addressed in order to ensure that the work is not only done—but done correctly, efficiently, on time, and with the least cost to the department.

Exhibit 1 shows how the executive housekeeper can plan the work of the department. The exhibit lists the initial questions that guide the general planning activities of the executive housekeeper, and identifies the end result of each step in the planning process. The resulting documents form the fundamental plans that must be in place for the housekeeping department to run smoothly. The following sections examine each step in the planning process.

Area Inventory Lists

Planning the work of the housekeeping department begins with creating an inventory list of all items within each area that will need housekeeping's attention.

Exhibit 1   Basic Planning Activities

INITIAL PLANNING QUESTIONS	RESULTING DOCUMENTS
1. What items within this area must be cleaned or maintained?	Area Inventory List
2. How often must the items within this area be cleaned or maintained?	Frequency Schedules
3. What must be done in order to clean or maintain the major items within this area?	Performance Standards
4. How long should it take an employee to perform an assigned task according to the department's performance standards?	Productivity Standards
5. What amounts of equipment and supplies will be needed in order for the housekeeping staff to meet performance and productivity standards?	Inventory Levels

Preparing area inventory lists is the first planning activity because the lists ensure that the rest of the planning activities address every item for which housekeeping will be held accountable. Inventory lists are bound to be long and extremely detailed. Since most properties offer several different types of guestrooms, separate inventory lists may be needed for each room type.

When preparing a guestroom area inventory list, it is a good idea to follow the same system that room attendants will use as the sequence of their cleaning tasks and that supervisors will use in the course of their inspections. This enables the executive housekeeper to use the inventory lists as the basis for developing cleaning procedures, training plans, and inspection checklists. For example, items within a guestroom may appear on an inventory list as they are found from right to left and from top to bottom around the room. Other systematic techniques may be used, but the point is that *some* system should be followed—and this system should be the same one used by room attendants and inspectors in the daily course of their duties.

Frequency Schedules

Frequency schedules indicate how often items on inventory lists are to be cleaned or maintained. Items that must be cleaned on a daily or weekly basis become part of a routine cleaning cycle and are incorporated into standard work procedures. Other items (which must be cleaned or maintained biweekly, monthly, bimonthly, or according to some other cycle) are inspected on a daily or weekly basis, but they

Exhibit 2   Sample Frequency Schedule

_	PUBLIC AREA #2—LIGHT FIXTURES		
LOCATION	**TYPE**	**NO.**	**FREQ.**
Entrance #1	Sconce	2	1/W
Lobby	Chandelier	3	1/M
Entrance #2	Crown Sconce	2	1/M
Behind Fountain	Sconce	3	1/W
Catwalk	Pole Light	32	1/M
Lower Level	Pole Light	16	1/M
Fountain Area	Pole Light	5	1/M
Restaurant Courtyard	Pole Light	10	1/M
Restaurant Courtyard	Wall Light	5	1/M
Restaurant Patio	Half Pole Light	16	1/W
Restaurant Entrance	White Globe Pole Light	6	1/W
Crystal Gazebo	White Globe Pole Light	8	1/W
2nd Stairs to Catwalk	White Globe Pole Light	2	1/W
Fountain	White Globe Pole Light	4	1/W
Lounge Patio	Wall Light	4	1/W
Restaurant Entrance	Chandelier	1	1/W

become part of a **general (or deep) cleaning** program and are scheduled as special cleaning projects. Exhibit 2 presents a sample frequency schedule for light fixtures found in a public area of a large convention hotel. Exhibit 3 presents a sample frequency list for special project duties carried out by housekeeping's night cleaning crew.

Items on an area's frequency schedule that are made part of housekeeping's general cleaning program should be transferred to a calendar plan and scheduled as a special cleaning project. The calendar plan guides the executive housekeeper in scheduling the appropriate staff to perform the necessary work. The executive housekeeper must take into account a number of factors when scheduling general cleaning of guestrooms or other special projects. For example, whenever possible, days marked for guestroom general cleaning should coincide with low occupancy periods. Also, the general cleaning program must be flexible in relation to the activities of other departments. For example, if the engineering department schedules extensive repair work for several guestrooms, the executive housekeeper should make every effort to coordinate a general cleaning of these rooms with engineering's timetable. Careful planning will produce good results for the hotel with the least possible inconvenience to guests or to other departments.

Exhibit 3   Sample Frequency List for Night Cleaning Projects

		Frequency	
Special Projects		**Wkly**	**Mthly**
1.	Wash down tile walls in restrooms	1	
2.	Strip and wax the following:		
	Restrooms (as necessary)		1
	Basement hallway	1	
	Lounge, lobby, and stairs		1
3.	Shampoo the following:		
	Registration area		1
	Stairs		1
	Restrooms		1
	All dining rooms		2
	All lounges		1
	Coffee shop		1
	Meeting rooms		1
	Guest elevators		1
	Employee cafeteria (as needed)		2
4.	Spot shampoo the following:		
	Front entrance		2
	Side entrance		2
	Front desk area		2
5.	Wash windows in pool area		1
6.	Dust louvers in pool area		1
7.	Clean guest and service elevator tracks	1	
8.	Polish kitchen equipment		1
9.	Polish drinking fountains	1	
10.	Clean outside of guest elevators	2	

Performance Standards

The executive housekeeper can begin to develop **performance standards** by answering the question: What must be done in order to clean or maintain the major items within this area? Standards are required levels of performance that establish the quality of the work that must be done. Performance standards state not only *what* must be done; they also describe in detail *how* the job must be done.

One of the primary objectives of planning the work of the housekeeping department is to ensure that all employees carry out their cleaning tasks in a consistent manner. The keys to consistency are the performance standards which the executive housekeeper develops, communicates, and manages. Although these standards will vary from one housekeeping department to another, executive housekeepers can ensure consistency of cleaning by demanding 100% conformity to the standards established by their departments. When performance standards are not properly developed, effectively communicated, and consistently managed, the productivity of the housekeeping department suffers because employees will not be performing their tasks in the most efficient and effective manner.

The most important aspect of developing standards is gaining consensus on how cleaning and other tasks are to be carried out. Consensus can be achieved by having individuals who actually perform the tasks contribute to the standards that are eventually adopted by the department.

Performance standards are communicated through ongoing training programs. Many properties have developed performance standards and have included them between the covers of impressive housekeeping procedure manuals. However, all too often, these manuals simply gather dust on shelves in the offices of executive housekeepers. Well-written standards are useless unless they are applied. The only way to get standards in the workplace is through effective training programs.

After communicating performance standards through ongoing training activities, the executive housekeeper must manage those standards. Managing standards means ensuring conformity to standards by constant inspection. Experienced housekeepers know the truth of the adage: "You can't expect what you don't inspect." Daily inspections and periodic performance evaluations should be followed up with specific on-the-job coaching and retraining. This ensures that all employees are consistently performing their tasks in the most efficient and effective manner. The executive housekeeper should review the department's performance standards at least once a year and make appropriate revisions as new work methods are implemented.

Productivity Standards

While performance standards establish the expected "quality" of the work to be done, **productivity standards** determine the acceptable "quantity" of work to be done by department employees. An executive housekeeper begins to establish productivity standards by answering the question: How long should it take for a housekeeping employee to perform an assigned task according to the department's performance standard? Productivity standards must be determined in order to properly staff the department within the limitations established by the hotel's operating budget plan.

Since performance standards vary in relation to the unique needs and requirements of each hotel, it is impossible to identify productivity standards that would apply across the board to every housekeeping department. Since the duties of room attendants vary widely among economy/limited-service, mid-range-service, and world-class-service hotels, the productivity standards for room attendants will also vary.

When determining realistic productivity standards, an executive housekeeper does not have to carry around a measuring tape, stopwatch, and clipboard and conduct time and motion studies on all the tasks necessary to clean and maintain each item on an area's inventory list. The labor of the executive housekeeper and other management staff is also a precious department resource. However, housekeeping managers must know how long it should take a housekeeping employee

Exhibit 4   Sample Productivity Standard Worksheet

Step 1

> **Determine how long it should take to clean one guestroom according to the department's performance standards.**
>
> Approximately 27 minutes*

Step 2

> **Determine the total shift time in minutes**
>
> 8 hours × 60 minutes = 480 minutes

Step 3

> **Determine the time available for guestroom cleaning.**
>
> Total Shift Time .... 480 minutes
> Less:
>   Beginning-of-Shift Duties .... 20 minutes
>   Morning Break .... 15 minutes
>   Afternoon Break .... 15 minutes
>   End-of-Shift Duties .... 20 minutes
> Time Available for Guestroom Cleaning ..... 410 minutes

Step 4

> **Determine the productivity standard by dividing the result of Step 3 by the result of Step 1.**
>
> $$\frac{410 \text{ minutes}}{27 \text{ minutes}} = 15.2 \text{ guestrooms per 8-hour shift}$$

* Since performance standards vary from property to property, this figure is used for illustrative purposes only. It is not a suggested time figure for cleaning guestrooms.

to perform the major tasks identified on the cleaning frequency schedules—such as guestroom cleaning. Once this information is known, productivity standards can be developed.

Let's assume that, at a hotel offering mid-range service, the executive housekeeper determines that a room attendant can meet performance standards and clean a typical guestroom in approximately 27 minutes. Exhibit 4 presents a sample productivity standards worksheet and shows how a productivity standard can then be established for room attendants working 8-hour shifts. Calculations within the exhibit assume that room attendants take a half-hour unpaid lunch period. The exhibit shows that the productivity standard for room attendants should be to clean 15 guestrooms per 8-hour shift.

Quality and quantity can be two sides of a double-edged sword. On one side, if the quality expectations (performance standards) are set too high, the quantity of work that can be done accordingly may be unacceptably low. This forces the executive housekeeper to add more and more staff to ensure that all the work gets done. However, sooner or later (and probably sooner than expected), the general manager will use the double-edged sword to cut the high labor expense of the housekeeping department. This action would force the executive housekeeper to reduce the size of the department staff and to realign quality and quantity by redefining performance standards in light of more realistic productivity standards.

On the other side, if performance standards are set too low, the quantity of work that can be done accordingly will be unexpectedly high. At first, the general manager may be delighted. However, as complaints from guests and staff increase and the property begins to reflect dingy neglect, the general manager may, once again, wield the double-edged sword. This time, the general manager may choose to replace the executive housekeeper with a person who will establish higher performance standards and monitor department expenses more closely.

The challenge is to effectively balance performance standards and productivity standards. Quality and quantity need not be a double-edged sword; instead, each can serve to check and balance the other. A concern for productivity may not necessarily lower performance standards—it can sharpen and refine current work methods and procedures. If room attendants are constantly returning to the housekeeping area for cleaning and guestroom supplies, there is something wrong with the way they set up and stock their carts. Wasted motion is wasted time, and wasted time depletes the most important and most expensive resource of the housekeeping department: labor. The executive housekeeper must be constantly on the alert for new and more efficient work methods.

Remember, an executive housekeeper will rarely have all the resources necessary to do everything he/she may want to accomplish. Therefore, labor must be carefully allocated to achieve acceptable performance standards and realistic productivity standards.

Equipment and Supply Inventory Levels

After planning what must be done and how the tasks are to be performed, the executive housekeeper must ensure that employees have the necessary equipment and supplies to get their jobs done. The executive housekeeper plans appropriate inventory levels by answering the following question: What amounts of equipment and supplies will be needed for the housekeeping staff to meet the performance and productivity standards of the department? The answer to this question ensures smooth daily housekeeping activities and forms the basis for planning an effective purchasing system. A purchasing system must consistently maintain the needed amounts of items stored within inventories controlled by the housekeeping department.

Essentially, the executive housekeeper is responsible for two types of inventories. One type stores items which are recycled during the course of hotel operations; the other type stores non-recyclable items. Non-recyclable items are consumed or used up during routine activities of the housekeeping department. Due to limited storage facilities and management's desire not to tie up cash in overstocked inventories, the executive housekeeper must establish reasonable inventory levels for both recyclable and non-recyclable items.

Recycled Inventories. Recycled inventories include linens, most equipment items, and some guest supplies. Recycled equipment includes room attendant carts, vacuum cleaners, carpet shampooers, floor buffers, and many other items. Recycled guest supplies include such items as irons, ironing boards, cribs, and refrigerators that guests may need during the course of their stay. Housekeeping is responsible for storing and maintaining these items as well as issuing them as they are requested by guests.

The number of recycled items that must be on hand to ensure smooth operations is expressed as a **par number**. Par refers to the standard number of items that must be on hand to support daily, routine housekeeping operations. For example, one par of linens is the total number of items needed to outfit all the hotel guestrooms once; two par of linens is the total number of items needed to outfit all the hotel guestrooms twice; and so on.

Non-Recycled Inventories. Non-recycled inventories include cleaning supplies, guestroom supplies (such as bath soap), and guest amenities (which may range from toothbrushes and shampoos and conditioners, to scented bath powders and colognes). Since non-recyclable items are used up in the course of operations, inventory levels are closely tied to the purchase ordering system used at the property. A purchase ordering system for non-recyclable inventory items establishes a par number that is based on two figures—a minimum quantity and a maximum quantity. The **minimum quantity** is the fewest number of purchase units that should be in stock at any time. Purchase units are counted in terms of normal size shipping containers, such as cases, drums, and so on. The inventory level should never fall below the minimum quantity. When the inventory level of a non-recyclable item reaches the minimum quantity, additional supplies must be ordered.

The actual number of additional supplies that must be ordered is determined by the **maximum quantity**. The maximum quantity is the greatest number of purchase units that should be in stock at any time. This maximum quantity must be consistent with available storage space and must not be so high that large amounts of the hotel's cash resources are tied up in an overstocked inventory. The shelf-life of an item also affects the maximum quantity of purchase units that can be stored.

Organizing the Housekeeping Department

Organizing refers to the executive housekeeper's responsibility to structure the department's staff and to divide the work so that everyone gets a fair assignment and all the work can be finished on time.

Structuring the department's staff means establishing the lines of authority and the flow of communication within the department. Two important principles that should guide the organization of a department are:

- Each employee should have only one supervisor.

- Supervisors should have the authority and information necessary to guide the efforts of employees under their direction.

The executive housekeeper delegates authority to supervisors and must ensure that each employee recognizes the authority structure of the department. While the

Exhibit 5   Organization Chart for a Small Economy/Limited-Service Hotel

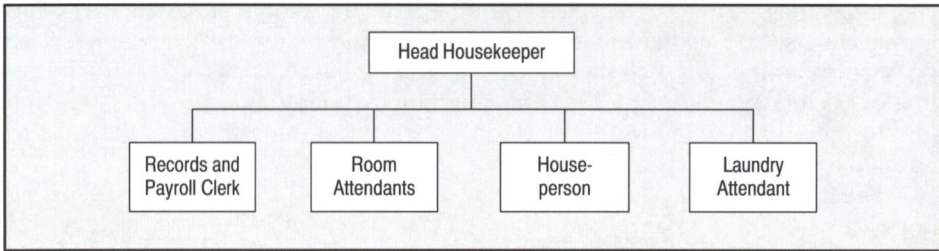

```
                        ┌──────────────────┐
                        │  Head Housekeeper │
                        └──────────────────┘
         ┌──────────────┬──────────┴──────────┬──────────────┐
 ┌──────────────┐ ┌──────────────┐ ┌──────────────┐ ┌──────────────┐
 │  Records and │ │     Room     │ │    House-    │ │   Laundry    │
 │ Payroll Clerk│ │  Attendants  │ │    person    │ │  Attendant   │
 └──────────────┘ └──────────────┘ └──────────────┘ └──────────────┘
```

executive housekeeper may delegate authority, he/she cannot delegate responsibility. The executive housekeeper is ultimately responsible for the actions of department supervisors. Therefore, it is important that supervisors be well-informed about hotel policies, procedures, and the limits of their authority.

The Department Organization Chart

An **organization chart** provides a clear picture of the lines of authority and the channels of communication within the department. Exhibits 5 through 7 present sample organization charts for housekeeping departments of different sizes and service levels. At small, economy/limited-service properties, the title of the housekeeping department manager depends on the specific duties and responsibilities of the position. The position title is often head housekeeper, or, simply, housekeeper. When compared to the housekeeping departments of other types of hotels, the housekeeping staff of the economy/limited-service property seems small. However, within the economy/limited-service property, the housekeeping staff may account for nearly half of the total number of employees at the hotel. Exhibit 6 suggests that properties offering mid-range service generally have a large housekeeping staff supervised by an executive housekeeper. Exhibit 7 shows that very large properties offering world-class service may have an entire housekeeping division with several managers led by a director of housekeeping.

The organization chart of the department not only provides for a systematic direction of orders, but also protects employees from being over-directed. The chart shows that each employee takes orders only from the person who is directly above him/her in the department's organization. An organization chart also shows how grievances or other communications are channeled through the department.

A copy of the chart should be posted in an area so that all housekeeping employees can see where they fit into the overall organization of the department. Some housekeeping departments post organization charts that show employees at the top and the executive housekeeper at the bottom. Posting this type of chart emphasizes the importance of the work performed by the majority of employees; it conveys that employees are "at the top of the chart." Such a chart also illustrates how the entire department balances on the managerial talents of the executive housekeeper and other department managers.

Exhibit 6   Organization Chart for a Large Mid-Range-Service Hotel

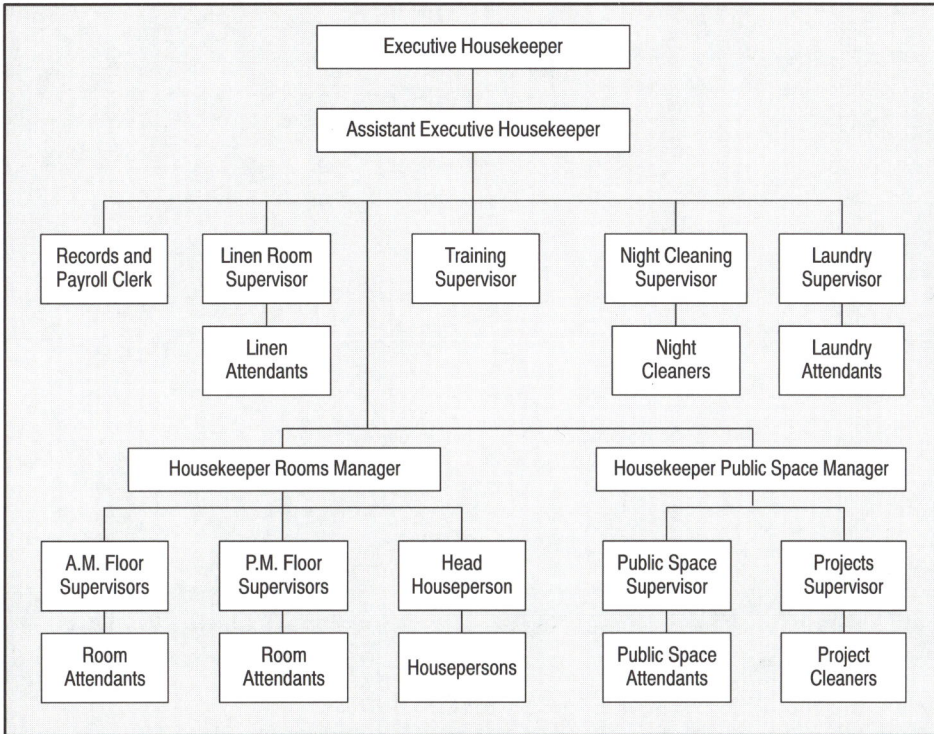

Job Lists and Job Descriptions

If the executive housekeeper has planned the work of the housekeeping department properly, organizing the department staff becomes a relatively straightforward matter. Executive housekeepers use information gathered from earlier planning activities to identify the number and types of positions that are needed and to develop job lists and job descriptions for each of these positions.

A **job list** identifies the tasks that must be performed by an individual occupying a specific position within the department. The tasks on the job list should reflect the total job responsibilities of the employee. However, the list should not be a detailed breakdown of the procedures that the employee will follow in carrying out each task. The job list should simply state what the employee must be able to do in order to perform the job. Exhibit 8 presents a sample job list for the position of housekeeping manager at a small economy/limited-service hotel.

Some types of **job descriptions** simply add information to the appropriate job lists. This information may include reporting relationships, additional responsibilities, and working conditions, as well as equipment and materials to be used in the course of the job. Exhibit 9 presents sample job descriptions for typical housekeeping positions found at medium-size mid-range-service properties.

Exhibit 7   Organization Chart for a Large World-Class-Service Hotel

						Uniforms-Banquet Linens Asst. Housekeeper	
						Uniforms Banquet Linens Supervisor	Third Shift Asst. Supervisor — Restaurant Linen Clerks
							Second Shift Asst. Supervisor — Uniform Clerks
						First Shift Asst. Supervisor — Uniform Clerks / Seamstresses / Valet Inventory Cleaner	Valet Inventory Clerks / Seamstresses / Uniform Clerks

Director of Housekeeping — Secretary / Payroll Clerk

Executive Housekeeper Public Space — Administrative Asst. / Asst. Executive Housekeeper Public Space

Executive Housekeeper Rooms — Asst. Executive Housekeeper Rooms

Administrative Asst.

Administrative Asst. Housekeeper — Administrative Supervisor — Inventory and Supply Asst. Supervisor / Lost and Found Asst. Supervisor — Third Shift Lost and Found Clerks / Rooms Control / Rooms Rags/Linen Rejects / First Shift Inventory Supply / Second Shift Inventory Supply / Third Shift Inventory Supply

Projects Asst. Housekeeper — Projects Supervisor — Projects Asst. Supervisor — Project Persons

First Shift Asst. Housekeeper — Building Supervisors / Floor Supervisors / Trainers / Houseman Supervisor — Head Houseman / Asst. Head Houseman Trainer — Housekeepers / Linen Distributor / Housemen / Project Housemen / Housekeepers

Second Shift Asst. Housekeeper — Second Shift Rooms Supervisor / Second Shift Public Space Supervisor — Floor Supervisor / Public Space Asst. Supervisor — Housekeepers / Housemen / Lobby Attendants / Office Cleaners

Third Shift Asst. Housekeeper — Third Shift Area Supervisor / Third Shift Area Supervisor / First Shift Area Supervisors — First Shift Asst. Supervisors / First and Third Shift Trainers / Third Shift Asst. Supervisor / Third Shift Asst. Supervisor PAC Team — Lobby Attendants / A.M. PAC Team / Night Cleaners / PAC Team

Positions shown in the bottom two rows are all on the same organizational level, as are those on the third and fourth rows from the bottom. They have been separated to meet space constraints.

Courtesy of Opryland Hotel, Nashville, Tennessee

Exhibit 8   Sample Job List: Housekeeping Manager at a Small Economy/Limited-Service Hotel

Date Prepared: xx/xx/xx

JOB LIST

Position: Housekeeping Manager
Tasks: Manager must be able to:

1. Schedule employees
2. Adjust employee schedules daily, as needed
3. Schedule extra cleaning tasks
4. Prepare room attendant work assignment sheets
5. Prepare housekeeping room status report
6. Inspect clean rooms
7. Report clean, inspected rooms to the front desk
8. Clean rooms, if necessary
9. Train employees
10. Compute labor hours on housekeeping staff time cards
11. Compute room attendant time per rented room
12. Safeguard pass keys
13. Follow personnel management procedures
14. Prepare cleaning products for use by room attendants
15. Check unrented rooms
16. Check security of storage areas
17. Report maintenance problems
18. Conduct monthly inventory
19. List supplies for reorder
20. Replace inventoried items missing from guestrooms
21. Sort unusable linens
22. Re-cover pillows
23. Spot-clean carpets
24. Paint touch-ups on corridor and guestroom walls
25. Perform minor maintenance tasks
26. Perform duties as assigned by the general manager

To be most effective, job descriptions must be tailored to the specific operational needs of individual properties. Therefore, the form and content of job descriptions will vary among housekeeping departments. Exhibit 10 presents a sample job description for the position of executive housekeeper at a medium-size mid-range-service hotel. Exhibit 11 presents an executive housekeeper's job description at a large world-class-service hotel.

The range of duties and responsibilities of executive housekeepers at various sizes and types of properties varies enormously. This is because many of the housekeeping management functions at small, independent economy/limited-service hotels may be carried out by the general manager. In the case of chain-affiliated properties, many housekeeping management functions are performed by staff at corporate headquarters. This leaves the task of implementing standardized procedures to the general managers and head housekeepers at individual properties.

Since job descriptions may become inappropriate as work assignments change, they should be reviewed at least once a year for possible revision. Properly

Exhibit 9 Sample Job Description for Typical Housekeeping Positions

ROOM ATTENDANT

Basic Function

Performs routine duties in the cleaning and servicing of guestrooms and baths under close supervision of an inspector.

Duties and Responsibilities

1. Enters and prepares room for cleaning.
2. Makes the bed.
3. Dusts the room and furniture.
4. Replenishes guestroom and bath supplies.
5. Cleans the bathroom.
6. Cleans the closet.
7. Vacuums and rakes the carpet.
8. Checks and secures the room.

Relationships

Reports directly to inspector, assistant housekeeper, and executive housekeeper.

HOUSEPERSON

Basic Function

Performs any combination of the following tasks to maintain guestrooms, working areas, and the hotel premises in general, in a clean and orderly manner.

Duties and Responsibilities

1. Cleans rugs, carpets, and upholstered furniture using vacuum cleaner, broom, and shampoo machine.
2. Cleans rooms, hallways, and restrooms.
3. Washes walls and ceilings, moves and arranges furniture, and turns mattresses.
4. Sweeps, mops, scrubs, waxes, and polishes floors.
5. Dusts and polishes metal work.
6. Collects soiled linen for laundering.
7. Receives linen supplies.
8. Stores linen supplies in floor linen closets.
9. Maintains housekeeping carts.
10. Removes trash collected by room attendants.

Relationships

Reports to head houseperson and assistant housekeeper.

LOBBY ATTENDANT

Basic Function

Keeps all lobbies and public facilities (such as lobby restrooms, telephone areas, the front desk, and offices) in a neat and clean condition.

Duties and Responsibilities

1. Cleans and maintains all lobbies and public restrooms.
2. Sweeps carpets.
3. Empties ashtrays and urns.
4. Polishes furniture and fixtures.
5. Vacuums and polishes elevators.
6. Keeps the front of the hotel free from trash.

Relationships

Reports to housekeeping management and hotel assistant managers.

LINEN AND UNIFORM ATTENDANT

Basic Function

Stores and issues uniforms, bed linen, and table linen; also takes inventory and maintains linen room supplies.

Duties and Responsibilities

1. Sorts, counts, and records number of items soiled.
2. Places linen and uniforms in containers for transport to laundry.
3. Examines laundered items to ensure cleanliness and serviceability.
4. Sends torn articles to seamstress for repair.
5. Stores laundered linen and uniforms on shelves, after verifying numbers and types of articles.
6. Issues linen and uniforms, to be exchanged on a clean-for-soiled basis only.
7. Counts and records linen to fill requisitions.

Relationships

Reports to the linen room supervisor and assistant housekeeper.

written job descriptions can ease employee anxiety by specifying responsibilities, requirements, and peculiarities of their jobs. Employees should be involved in writing and revising job descriptions for their positions.

Each employee of the housekeeping department should be given a copy of the job description for his/her position. A job description may also be given to all final

Exhibit 10 Sample Job Description for the Executive Housekeeper at a Medium-Size Mid-Range-Service Hotel

JOB DESCRIPTION

JOB TITLE: Executive Housekeeper

IMMEDIATE SUPERVISOR: Resident Manager/Assistant Resident Manager

JOB SUMMARY	Supervises all housekeeping employees, has the authority to hire or discharge, plans and assigns work assignments, informs new employees of property regulations, inspects housekeeping personnel work assignments, and requisitions supplies.
DUTIES	Supervises all housekeeping employees, hires new employees as needed, discharges employees when necessary, and writes warning notices when policy has been violated. Evaluates employees in order to upgrade when openings arise.
	Plans the work for the Housekeeping Department and distributes assignments accordingly. Assigns Housemen, Inspectresses, and Linen Room Attendants to their regular duties, or any special assignments that need to be accomplished. Schedules employees and assigns extra days off according to the occupancy forecast. Maintains a time log record book of all employees within the department.
	Informs new employees of regulations. Trains and assigns new employees to work with experienced help. Checks on the work of these employees occasionally and observes the reports made by the Inspectress or Section Housekeeper.
	Inspects the housekeeping staff periodically to determine if they are on duty and checks the quantity and quality of their work, checking places likely to be overlooked.
	Approves all supply requisitions, such as spreads and bathroom rugs. Maintains a lost and found department and is responsible for all lost and found items. Determines the rightful owner and mails to appropriate address.
PREREQUISITES	Education

	Education	High school required.
	Experience	Minimum three (3) years as an Assistant Housekeeper or Inspectress.
	Skills	Ability to plan and implement housekeeping programs and policies and to work and communicate with management, associates, and subordinates.

Approved _____ Date _____

Courtesy of Best Western International, Phoenix, Arizona

job candidates before an employment offer is made. This is preferable to having someone accept the job and then decide the job is unsuitable because he/she was unaware of its requirements.

Job lists and job descriptions form the basis for developing **job breakdowns** (specific, step-by-step procedures for accomplishing a task), training plans, and effective performance evaluation forms. The final sections of this chapter examine other important functions of the executive housekeeper within the overall management process.

Exhibit 11   Sample Job Description for the Director of Housekeeping at a World-Class-Service Hotel

---

POSITION DESCRIPTION

Position: Director of Housekeeping	Job No:
Incumbent:	Date:
Reports to: Director of Rooms Division	Prepared by:
Org. Unit:	Approved By: (Incumbent)
Wage & Hour Class:	Approved By: (Supvr.)
Company:	Classification Code:

POSITION PURPOSE

To direct the administration of all housekeeping services for guestrooms, public and staff areas; to ensure the highest standards of sanitation, safety, comfort, and aesthetics, and to direct all of housekeeping's projects and programs.

NATURE AND SCOPE

This position reports to the Director of Rooms. The incumbent has the ultimate responsibility to ensure that the property provides salable, well furnished and maintained guestrooms, public space, and staff areas. The incumbent is responsible for developing departmental policies and procedures by which the highest possible degree of cleanliness, maintenance, and aesthetic value is achieved.

The incumbent will review all personnel actions of subordinates in the areas of discipline, termination, and promotion. The incumbent will be responsible for the development of all training programs within the Housekeeping Department and ensure proper implementation through the Executive Housekeepers in Rooms and Public Space.

The incumbent is responsible for smooth, timely communication between departments. The incumbent will participate in community activities, professional organizations, and supplier relations and programs.

The incumbent will provide to the Director of Rooms, expense and capital expenditure recommendations and reports. This input is included in the annual Rooms Division profit plan and submission. Monthly reports are submitted on preventive housekeeping maintenance and analysis of loss/usage per occupied room on every major expense category including but not limited to labor, chemicals, guestroom supplies, linens, laundry, uniforms, and contract cleaning.

The incumbent makes recommendations to the Rooms Division Director for capital expenditure and special repair and maintenance programs in all areas of the department.

The annual profit plan preparation involves complete analysis of Housekeeping labor productivity and projection of improvement for the future.

The incumbent is expected, through excellence in recordkeeping, professional purchasing, and inventory controls to forecast future expenditures including any potential enhancements in F.F.&E. in the rooms, staff, or public space areas. He/She will also be expected to be involved in any expansion plans by overall long-term planning as it relates to additional rooms as well as public space.

The incumbent should have a minimum of eight years of direct experience in Housekeeping management of which four years was in a major (800 rooms) or above hotel property. He/She must have demonstrated knowledge in personnel administration, laundry sanitation, and preventive maintenance. Complete knowledge should be possessed in uniform issue and control, fabrics and linens, cleaning chemicals, design, engineering, maintenance equipment, guestroom and public space cleaning procedures, scheduling, budget and office management. The incumbent must be familiar with the duties of the Front Office Manager, Assistant Rooms Division Manager, Reservations Manager, Room Service, Restaurant and Banquet Managers. He/She should have demonstrated knowledge in Security, Accounting, and Purchasing.

PRINCIPAL ACTIVITIES

1. To ensure well furnished and maintained guestrooms and public space areas.

2. To ensure excellence in Housekeeping sanitation, safety, comfort, and aesthetics for hotel guests.

3. To oversee coordination of and administer all housekeeping programs and projects.

4. To act as a source of contact in interdepartmental communications, vendors, professional agencies, etc.

5. To provide budget, budget control and forecasting to upper management.

THE USE OF PERSONAL PRONOUNS SHOULD NOT BE CONSIDERED SEX BASED

Courtesy of Opryland Hotel, Nashville, Tennessee

Exhibit 12   Overview of the Management Process

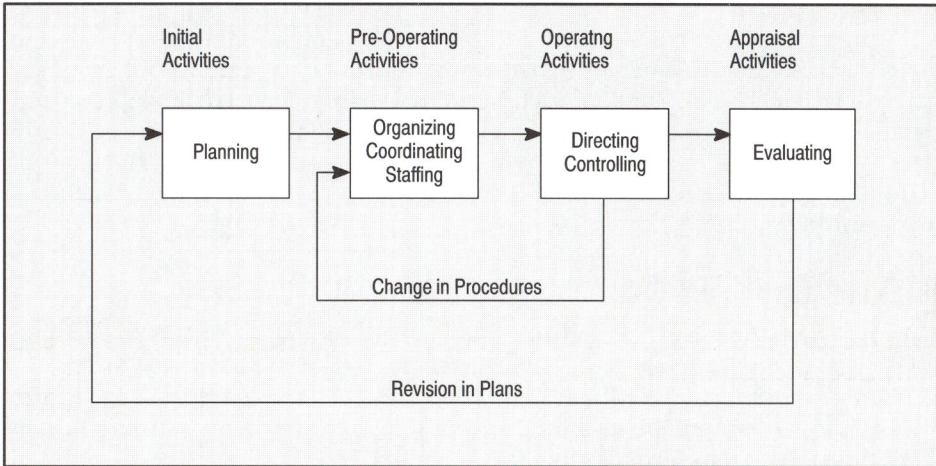

Source: Jack D. Ninemeier, *Planning and Control for Food and Beverage Operations,* 3d ed. (East Lansing, Mich.: Educational Institute of the American Hotel & Motel Association, 1991), p. 17.

Other Management Functions of the Executive Housekeeper

Exhibit 12 presents an overview of the management process and shows how each management function contributes to the overall success of a hotel or, for that matter, any sort of business. Top executives must plan what the hotel is to accomplish by defining its objectives. The desire to attain these objectives leads to organizing, coordinating, and staffing activities. Once members of the hotel staff are selected, management can direct the course of their work and implement control systems to protect the hotel's assets and to ensure smooth, efficient operating activities. Finally, management must evaluate the extent to which the objectives of the organization have been attained. An analysis of actual operating results may lead to changes in organizing, coordinating, or staffing procedures. Also, as a result of evaluating all planning and operating activities, management may find that revisions to the organization's plans or objectives are needed.

An important planning activity of the executive housekeeper is drafting the housekeeping department's operating budget. The housekeeping operating budget estimates expenses of the department for the upcoming year. Expenses include labor, linens, laundry operation, cleaning compounds, some types of equipment, and other supplies. Initial expense estimates are based on information supplied by the accounting department. This information includes expense reports for months of the past year and for the current year as well as monthly occupancy forecasts for the upcoming year.

The executive housekeeper's initial expense estimates are revised by top management executives in relation to the overall financial objectives of the hotel for the upcoming year. The hotel's owner, general manager, and controller coordinate and finalize the annual operating budget for the entire hotel. The resulting budget presents the executive housekeeper (and every other department manager) with a month-by-month plan by which to organize, coordinate, staff, direct, control, and evaluate operations.

Although specific management tasks vary from one management position to another, the same fundamental management functions are carried out by every manager within a hotel. Previous sections of this chapter focused on planning and organizing activities of the executive housekeeper. The following sections briefly examine the executive housekeeper's management responsibilities in the areas of coordinating, staffing, controlling, and evaluating the operation of the housekeeping department.

Coordinating and Staffing

Coordinating is the management function of implementing the results of planning and organizing at the level of daily housekeeping activities. Each day, the executive housekeeper must coordinate schedules and work assignments and ensure that the proper equipment, cleaning supplies, linens, and other supplies are on hand for employees to carry out their assignments.

Staffing involves recruiting applicants, selecting those best qualified to fill open positions, and scheduling employees to work. Since labor is housekeeping's largest expense item, properly scheduling employees is one of the most important management responsibilities of the executive housekeeper.

Most housekeeping departments use some type of staffing guidelines. These guidelines are usually based on formulas which are used to calculate the number of employees required to meet operational needs at specific occupancy levels. However, the management function of staffing goes beyond simply applying a formula. Staffing must be adequate to meet the general (or deep) cleaning schedules for various areas of the hotel and to meet the needs of other special cleaning projects. Therefore, the executive housekeeper must be flexible and creative when establishing staffing patterns that permit the department to reach its goals within the limits of the budget plan.

Directing and Controlling

Many people confuse the two very different management functions of directing and controlling. The easiest way to distinguish them is to remember that managers direct people and control things.

Directing is a complicated management skill which is exercised in a wide variety of situations. For an executive housekeeper, directing involves supervising, motivating, training, and disciplining individuals who work in the department. Motivating the housekeeping staff is a particularly important skill and is closely connected with the executive housekeeper's ability to lead the department. Motivation (or the lack of it) is contagious. Attitudes and work habits filter down to employees from their supervisors. The attitudes and work habits of supervisors are usually a reflection of the leadership provided by the executive housekeeper. A strong executive housekeeper personally expresses a genuine interest in everyone's performance and, thereby, creates an atmosphere in which motivation can thrive. On the other hand, an executive housekeeper who plays favorites with

Emiliana Barela, CHHE
Executive Housekeeper
Antlers Condominium Association
Vail, Colorado

❝The housekeeping staff I have put together in the 17 years I have been with the Antlers deserves all the credit for our department's success. Our exceptionally clean rooms and our nearly zero turnover rate show the pride our room attendants take in their work and their fellow workers.

In 1982, I received the Colorado-Wyoming Housekeeper of the Year award. Although this was an individual award, the entire housekeeping department felt that it was partly theirs—and rightly so. The Antlers has also honored five employees on the housekeeping staff for 10 years of outstanding employment. The sixth will be honored later this year.

As a supervisor, I think it is very important to let your staff know that you trust them with the work they do and let them know how important it is. My function is to be their leader and teacher, not to work against them.

I was fortunate enough to work under Mr. Bud Benedict, a past general manager here at the Antlers. He set the example that I try to follow with my staff, some of whom have worked with me for 16 years. When asked why they stay on here at the Antlers, their response is, 'Millie is very honest, friendly, and fair. But most of all, she treats us all equally. We're a family here, and we look after each other.'

We start every morning in the housekeeping department with a cup of coffee and a short staff meeting. We discuss the daily routine and anything special that might have to be done that day. We also talk about problems that have come up. It's a time for general recognition and reinforcement.

Then we are off on our own daily routines. The Antlers has 70 condominium units with fully equipped kitchens. I divide room attendants into pairs, and each room attendant supervises the other. We also have three supervisors who check the rooms daily, and I randomly select a few rooms to check myself.

When room attendants finish their own work, they check to see who has fallen behind during the day, and help each other catch up the workload. The fact that they supervise and help each other gives staff more independence and keeps them from feeling that management is breathing down their necks.

I also encourage our three supervisors to attend housekeeping seminars, and other employees to take classes to learn a second language. These courses give them an idea of what my job is about and help us communicate more effectively. All this pays off when we have to make decisions about how the department should be run.

Not only is our staff willing to work together, but they've demonstrated their willingness to work as a team with other departments. Recently, for example, our maintenance department, which consists of only three men, planned to spray paint the building. But this plan fell through because of high winds. With only a week's

(continued)

Housekeeping Success Tip *(continued)*

time, I was asked for help by the maintenance department. When I approached the staff, to my amazement the whole staff volunteered and we took out brushes and rollers and painted the building in a week. We also saved the property thousands of dollars.

It's important for staff to feel the hotel's management recognizes what they do. We have a comments/suggestion box in our lobby for guests and employees. At an employee's birthday party the box is opened and the notes read. The general manager then gives medals to the employee according to his or her achievements. The whole staff works very hard to earn these medals because once a certain number are accumulated, a paid day off is awarded to that individual.

There are a lot of things an executive housekeeper can do to achieve his or her goals. First and foremost, trust your employees to do the best job they can without a lot of interference. Next, be committed to your employees; let them have a say in some decisions. This will help them to grow and to achieve their own goals. Finally, if you really care about your employees, it will show—and they will shine."

---

supervisors will find discontent everywhere as supervisors, in turn, play favorites with employees under their direction.

Controlling refers to the executive housekeeper's responsibility to devise and implement procedures which protect the hotel's assets. Assets are anything the hotel owns which has commercial or exchange value. An executive housekeeper helps safeguard the hotel's assets by implementing control procedures for keys, linens, supplies, equipment, and other items.

Evaluating

Evaluating is the management function of assessing the extent to which planned goals are, in fact, attained. Important evaluation tools for all managers in a hotel are monthly budget reports prepared by the hotel's accounting staff. These reports provide timely information for evaluating housekeeping operations, especially the department's monthly labor expense. The executive housekeeper uses these reports to compare actual departmental expenses to amounts estimated by the budget plan. Significant differences between actual amounts and budgeted amounts are called variances. Significant variances may require further analysis and action by the executive housekeeper.

In addition, the executive housekeeper needs information on a daily and weekly basis in order to closely evaluate the performance of the staff and the overall productivity of the department. Evaluation in these areas begins with performance and productivity standards developed by earlier plans. Daily inspection reports and quarterly performance evaluations are used to monitor how well the actual performance of employees compares with performance and productivity standards.

Key Terms

area inventory list
frequency schedule
general (deep) cleaning
job breakdown
job list
job description
maximum quantity

minimum quantity
non-recycled inventories
organization chart
par number
performance standards
productivity standards
recycled inventories

Discussion Questions

1. What resources can the executive housekeeper use to attain the objectives set by top executives in a hospitality operation?

2. What areas are most housekeeping departments responsible for cleaning in a hotel?

3. What additional areas may housekeeping be responsible for cleaning depending on the property's service level?

4. Why is it important for an executive housekeeper to take a systematic, step-by-step approach to planning?

5. What is the purpose of an area inventory list? What is an ideal way to sequence such a list?

6. What is a frequency schedule? How is it used in conjunction with a property's general (or deep) cleaning program?

7. What is the difference between a performance standard and a productivity standard?

8. What is the most important aspect of developing performance standards? How are standards best communicated once they are developed?

9. What two important principles should guide the organization of the housekeeping department?

10. What are the fundamental management functions that should be carried out by every hotel manager?

REVIEW QUIZ

When you feel you have covered all of the material in this chapter, answer these questions. Choose the *best* answer.

True (T) or False (F)

T F 1. Area inventory lists can be used as a basis for developing inspection checklists.

T F 2. General cleaning of guestrooms should be scheduled for periods of high occupancy.

T F 3. Productivity standards for housekeeping positions are uniform throughout the lodging industry.

T F 4. A concern for productivity will lower performance standards.

T F 5. A purchase unit is equal to one par.

T F 6. Each employee should have only one supervisor.

T F 7. The executive housekeeper can delegate authority to department supervisors.

T F 8. A job list shows tasks that must be performed by an employee occupying a specific position in the department.

T F 9. The approved budget provides a month-by-month plan for operating the housekeeping department.

T F 10. The management function of controlling involves supervising and motivating employees.

Alternate/Multiple Choice

11. Daily inspections and periodic performance evaluations should be followed up with:

 a. specific on-the-job coaching and retraining.
 b. a revision of the department's performance standards.

12. Storage space, shelf life, and cash resources limit the _____ quantity of non-recycled items in inventory.

 a. minimum
 b. maximum

13. The housekeeping department's operating budget shows:

 a. actual expenses.
 b. estimated expenses.

14. Areas of the property to be cleaned by the housekeeping department are ultimately determined by the:

   a. executive housekeeper.
   b. general manager.
   c. rooms division manager.
   d. owner.

15. If the time available for guestroom cleaning during a shift is 6 hours and 48 minutes and each room attendant is expected to clean 17 guestrooms per shift, the time it takes for one room attendant to clean one guestroom is:

   a. 18 minutes.
   b. 20 minutes.
   c. 24 minutes.
   d. 28 minutes.

14/15

Part II

Managing Human Resources

Chapter Outline

Non-Traditional Labor Markets
Making Jobs Easier to Fill
Recruiting Employees
 Internal Recruiting
 External Recruiting
Employee Turnover
The Selection Process
 Evaluating Applicants
 Interviewing
 Conducting the Interview
 Techniques for Questioning
 Interview Evaluation
 Interview Follow-Up
The Hiring Period
 Making and Closing Offers
 Preparing Other Employees
 Processing Personnel Records
The Orientation Process
 The Employee's First Day
 The Executive Housekeeper's Role
 The Trainer's Role

Learning Objectives

1. Identify sources of labor from non-traditional labor markets.

2. Describe two major employment benefits the housekeeping department and the property can offer to attract qualified employees.

3. Distinguish between internal and external recruiting activities.

4. Identify five internal recruiting methods.

5. Identify five external recruiting methods.

6. Describe the direct and indirect costs of employee turnover.

7. List some of the causes of employee turnover and factors that may signal potential turnover.

8. Describe the function of a job specification.

9. Explain the steps of an effective selection process.

10. List the five primary objectives to achieve when interviewing applicants.

11. Summarize the steps that the executive housekeeper should take during the hiring period.

12. Define the role of the executive housekeeper in a new employee's job orientation.

13. Define the role of the trainer in a new employee's job orientation.

3

Recruiting, Selecting, Hiring, and Orienting

RECRUITING AND RETAINING QUALIFIED EMPLOYEES is one of the most critical issues that will face the hospitality industry in coming years. Lodging industry employment needs are expected to increase 25% to 39% (or 525,000 to 800,000 workers) by the year 2000. However, the population from which the industry draws most of its new employees—people 16 to 24 years old—is expected to decline 26%. And the total number of older workers is expected to decrease as well—almost 10% from current levels by 1995.[1]

The projected worker shortage will be worse for unskilled jobs that are not automated—which include many housekeeping positions. The candidate pool for managers is likely to be sufficient due to high numbers of students in hotel, restaurant, and institutional management programs and schools. But in general, the lodging industry will experience serious worker shortages since most entry-level employees have traditionally been young, inexperienced workers.

To alleviate the expected worker shortage, the hospitality industry will need to fill its employee ranks with members of groups it may not have considered previously. This chapter discusses such so-called non-traditional employee sources and suggests where and how to find them. The chapter also explains how to make housekeeping positions easier to fill by offering such employee benefits as transportation assistance or alternative schedules. Then, the chapter focuses on such ever-important topics as internal and external recruiting and managing employee turnover. Finally, the chapter discusses employee selecting, hiring, and orienting.

Non-Traditional Labor Markets

In the face of the declining youth employee market, the hospitality industry will have to turn to new labor sources. Doing so could well be a matter of the industry's survival. Additional labor sources include:

- Young mothers
- Displaced homemakers
- Student employees
- Dislocated workers over age 50
- Recent immigrants
- Handicapped persons

- Retirees

Young Mothers. Women will be forming an increasingly large part of the industry's work force: 56% by 1995, up from 54% in 1984.[2] Astute managers will consider hiring young mothers who wish to balance the simultaneous demands of work and family life.

Displaced Homemakers. Displaced homemakers are widows or divorcees who have spent their adult lives as homemakers. Thus, they typically lack work experience and the skills necessary to bring them ready employment. Many communities sponsor training programs which help displaced homemakers with career planning and acquiring such job-related skills as resume writing, job hunting, interviewing skills, and assertiveness. Such programs also teach displaced homemakers how to balance the concurrent demands of work, school, and family.

Student Employees. College students have long been work force mainstays of summer resorts. In addition, they are often interested in working during winter and spring breaks when hospitality business peaks. Students also provide other benefits. For instance, they are often willing to do heavy physical labor that older employees cannot or would rather not do. Furthermore, students may enter the industry as part of an internship program required by some hotel, restaurant, and institutional management programs. Hotels might also hire high school students through distributive education programs.

Dislocated Workers over Age 50. Such workers may have been laid off from other jobs, or idled because of such instances as factory shutdowns. Members of this group are often willing to work at lower rates of pay in exchange for the opportunity to learn new skills. Like retirees, workers over 50 are usually steadier than younger people since they often have ties in the community and families to support.

Recent Immigrants. Recent immigrants are frequently good candidates for entry-level positions. They tend to be hard-working, and speak English well enough to fit in. They may know friends and family members whom they can refer for employment. Communities frequently sponsor programs which help immigrants with education, finding work and housing, and adapting to life in the United States. What's more, because hotels attract increasing numbers of foreign guests, they value employees who speak more than one language.

Under the Immigration Reform and Control Act of 1986, all employees hired after November 6, 1986, must provide verification of their legal right to work. It is illegal for any employer to knowingly hire someone who is not authorized to work in the United States, or to continue employing someone who is or becomes unauthorized. It is also unlawful to hire or recruit an employee without complying with the Act's verification requirements. To avoid discrimination against applicants who appear to be foreign-born, employers cannot legally ask for verification until the time of hiring.[3]

The Handicapped. "Special" employees include those either mentally or physically impaired. Experience shows that many hospitality operations find such individuals to be cheerful, enthusiastic employees who enjoy work—and the guest.

Work for mentally impaired employees may include mopping floors or washing walls or dishes. A wheelchair-user could perform clerical duties in the housekeeping department.

Managers may want to investigate the availability of supported employment programs in the state and local area. The federal government provides grants to develop and expand supported employment, which offers physically and mentally impaired individuals opportunities to obtain jobs in regular work environments. The benefits to host companies include partial public funding, program administration by a non-profit agency, affirmative action assistance, possible turnover reduction, and improved public relations. A "job coach" or training specialist provides intensive initial and ongoing training at the work site.

Retirees. These days, retirees are enjoying increasingly longer, healthier lives. Growing numbers of retirees are looking for part-time and even full-time jobs to fight boredom, fill extra hours, and to extend limited retirement pensions and Social Security benefits. Such employees offer advantages that younger employees do not. For example, retirees are often available to work not only during the peak summer tourist season, but also in the spring and fall when college students are not available. And retirees often prove to be among the most reliable of all employees.

Of course, retirees will not be able to do the physical labor required in many housekeeping jobs, but could serve in such positions as linen and uniform attendants or inspectors/inspectresses.

Making Jobs Easier to Fill

In order to hire and retain a highly qualified team of employees, the housekeeping department and the property as a whole will want to do what they can to make the jobs, the department, and the property appealing. They can do this by offering special employment benefits which go beyond the typical benefit packages. Possibilities include assistance with transportation and child care.

Transportation Assistance. Hospitality operations may decide to offer their employees transportation assistance. There are several advantages to instituting such a policy:

- The property won't have to set aside as much employee parking space.
- Property-supplied transportation cuts down on employee tardiness and absences.
- Employees who share rides will get to know each other, often building team spirit.
- Employees who do not have to fight traffic may arrive in better spirits.
- It may decrease employees' vehicle maintenance and operation expenses.
- Such a program may benefit the environment by cutting down on pollution, reducing traffic, and conserving energy.

Housekeeping Success Tip

Jon M. Owens, CHHE
Director of Housekeeping
Clarion Hotel & Conference Center
Lansing, Michigan

❝In recent years, the diminishing labor market has become a growing concern for the hospitality industry. Traditionally, the industry has relied on young and first-time workers as the mainstay of their labor supply. Yet, as our work force ages, and the number of young people entering the labor market slows down, hospitality will have to look elsewhere for staff. Too, the industry itself is experiencing tremendous growth as the economy becomes more and more service-oriented. Given these two forces, hotels must begin to consider alternative sources of labor to fill positions.

Among the various groups the industry can tap are people with disabilities. These workers can often be brought into a property through a supported employment program with a local or regional rehabilitation agency. Such programs provide the disabled with opportunities to learn and perform task-oriented skills on-the-job in such areas as maintenance, food service, and housekeeping.

The Clarion Hotel & Conference Center in Lansing, Michigan, has established a supported employment program with Peckham Vocational Industries. Under this arrangement, Clarion contracts a portion of its guestroom cleaning and servicing to Peckham. Peckham, in turn, provides the labor in the form of its clients. Peckham also provides an on-site supervisor who directly supervises and inspects the contracted cleaning activities. This supervisor is trained by Clarion to ensure that the property's cleaning standards are upheld. The disabled individuals work Monday through Friday and provide basic housekeeping services in a minimum of 30 guestrooms.

The Peckham individuals are assigned to one floor and work as a team to clean all designated guestrooms. Each client is trained to perform a particular task. This training includes all the basic skills necessary to properly clean a guestroom such as making the bed, cleaning the bathroom, vacuuming the carpet, spot-cleaning walls, removing trash, and placing amenities and supplies. When a client has demonstrated the ability to perform a particular task, he/she is trained in another.

This program has been very successful for Clarion and for Peckham. Although Peckham's clients have limited learning skills, they receive extensive training and undergo the same quality inspection program as Clarion housekeeping employees.

Establishing such a program could prove very beneficial to your hotel. It can give the executive housekeeper greater flexibility in staffing and scheduling—and, at the same time, provide employment opportunities for individuals with special needs. I would encourage all executive housekeepers who are experiencing shortages in housekeeping personnel to explore the possibilities of installing this type of program. It's a viable alternative and will make a positive contribution to your community. Information is generally available from your state's supported employment office—which is usually part of the human services department.❞

Helping employees meet their child care needs pays off in higher morale, less absenteeism and lateness, and lower turnover. Opryland Hotel operates its own state-licensed child care facility on-site. Child care is available from 5:00 A.M. to midnight, seven days a week, including holidays. (Courtesy of Opryland Hotel, Nashville, Tennessee)

Properties in areas offering good mass transportation systems may provide employees with discounted or free transit passes. Remember, though, public transportation routes are often limited; for example, they may not service all geographic areas, and may not operate evenings and weekends.

Alternatively, the property may wish to organize or even subsidize ride-sharing programs. A small property may develop a car-pooling program, while a larger property might donate the use of company vehicles.

Child Care Assistance. Single working parents and families with two working adults are becoming the norm in today's work force. Because of this, hospitality operations may wish to consider implementing a child care assistance program. The federal government and some states offer tax incentives to persuade businesses to ease their employees' child care needs. Child care assistance programs could go a long way toward attracting young mothers and displaced homemakers, who constitute two of the non-traditional labor sources discussed earlier.

So far, only the largest hospitality operations can afford on-site child care facilities. For example, the Opryland Hotel in Nashville, Tennessee, operates a child-care center that has been very popular with employees *and* hotel management. It has improved employee attendance and decreased tardiness and turnover.

Instead of on-site facilities, smaller properties may consider other possibilities. For example, they could pay child care allowances, such as adding $1.50 per hour to an employee's usual wage. Other possibilities include subsidizing parental leaves, or offering alternative scheduling.

Recruiting Employees

Employee recruitment is the process by which applicants are sought and screened concerning their suitability for positions in the operation. The process involves announcing job vacancies through proper sources, and interviewing and evaluating applicants to determine whom to consider for open positions.

In large properties, the human resources division assists the executive housekeeper in finding and hiring the most qualified individuals. However, many lodging properties do not have human resources divisions. Therefore, the executive housekeeper is often involved in such tasks as initial interviewing, contacting applicants' references, and related selection tasks. In all properties, the executive housekeeper should personally interview top candidates for open positions in the department. Depending on the property's organization, the executive housekeeper may either hire the applicant or make a recommendation to the manager at the next higher organizational level.

Internal Recruiting

Internal recruiting is a boon to managers at every property. It gives the executive housekeeper access to candidates that have acquired some skills, are familiar with the property, are a known quantity, and have already proven themselves. Internal recruiting is also attractive from the employees' point of view. It gives employees the opportunity to advance within the ranks of the property. Possibilities for promotion may enhance employee productivity and morale.

Internal recruiting methods include cross-training, succession planning, posting job openings on property premises, paying employees for job performance, and keeping a call-back list.

Cross-Training. Whenever possible, all employees should be trained to perform the duties of more than one position. Each potential employee should understand the role that cross-training plays in each housekeeping position. **Cross-training** makes it easier for the executive housekeeper to draw up a complete employee schedule at all times, and to plan around employee vacations and absences. Employees also find cross-training attractive. It gives them the opportunity to acquire diverse skills, brings variety to the job, and makes employees more valuable to the employer. It may also lead to promotions.

Succession Planning. In **succession planning,** the executive housekeeper identifies a key position and designates a housekeeping employee who will one day fill it. Housekeeping management also decides whether the employee will require further training, and sees that he/she gets it. The manager draws up a plan outlining when the training occurs, who will conduct it, and when the employee will step into the new position.

Paying for Performance. As employees gain more experience through cross-training and through their own efforts, they should be paid accordingly. Employees have more incentive to excel if they know that there is a wage increase program in place

Housekeeping Success Tip

Edith Sasser, CHHE
Executive Housekeeper
Altamonte Springs Hilton and Towers
Altamonte Springs, Florida

❝Cross-training is an established policy for every employee, supervisory or otherwise, in the housekeeping department at the Altamonte Springs Hilton and Towers. In fact, cross-training is a part of every housekeeping job description. It is explained to every new-hire and to every property employee transferring into the housekeeping department. Since every employee is aware of the policy from the beginning, there is never any confusion or surprise when changes in responsibility assignments are posted.

Everyone on the housekeeping team is career-conscious. As employees gain experience in various areas, departmental professionalism is enhanced. Employees are less reluctant to meet new challenges, and the whole team is supportive. Special cleaning chores—whether occasional, routine, supplemental, or emergency—become the responsibility of the entire team. Individual effort is appreciated and judgment is withheld. This atmosphere encourages initiative among team members, who are hourly and supervisory employees alike.

Because of cross-training, supervisors are also able to fill each position—thus becoming 'working' supervisors. Supervisory cross-training includes getting acquainted with everyone on each new crew assignment. All our employees know one another even though they may work different shifts.

A supervisor who is cross-trained may discover ready employment opportunities in competitive properties. A good way to keep a valued supervisor is to give him or her 'a day in the life of the executive housekeeper.' On such days, the executive housekeeper switches jobs with the supervisor. The supervisor thus gets a realistic view of the trials and tribulations at 'the top.' If this is done in a timely manner, the employee will often be willing to reconsider and stay. What's more, supervisors will recognize that they are allowed to handle whatever level of responsibility they are comfortable with (which may even be representing the executive housekeeper at training seminars), without actually having to make a job change.

As executive housekeeper, I strive to be consistently supportive of all team members, encouraging initiative on the part of everyone. In this way, many emergencies are promptly handled, and, in fact, may never arise. Personal emergencies in employees' lives are also realistically handled, even if it means that I fill in for them temporarily in the ultimate gesture of support.

Other advantages of cross-training follow.

- The employee's flexibility makes scheduling—and keeping a full crew—very simple.

- When employees are familiar with every position, they are able to make better decisions, especially under abnormal or catastrophic circumstances. They also have more empathy for co-workers in the same circumstances.

(continued)

Housekeeping Success Tip *(continued)*

- Cross-training makes management decisions regarding promotions easier. A manager knows a person's potential when he or she has already seen the employee filling many positions and shifting from one responsibility and situation to another.

- Cross-training is a very good tool for developing teamwork. Teamwork serves as a positive reinforcement. It can increase pride in one's work, reduce jealousy, increase empathy among peers and supervisors, and result in more profit for the property—which makes everybody's job more secure and eventually supports better pay scales.

Our property has a relatively low rate of turnover. I believe it is because of the housekeeping department's positive experience with cross-training. The staff know that we recognize them and their talents, that we appreciate them, and that we allow them to grow. They do not feel 'used.' And they stay with us even though they might be able to make more money elsewhere. **"**

---

which rewards hard work and productivity. It can be very discouraging when all employees are given the same wage increase regardless of performance.

Posting Job Openings. Internal posting can reduce property-wide turnover and provide a known applicant pool. Employees in other departments may be interested in transferring into the housekeeping department. Current housekeeping employees may desire promotion within the department. In these instances, managers must remember that good performance in one job does not always guarantee the same level of performance in another. When promoting from within, managers must be sure that the employee has the skills for the "new" job, as well as a good track record for the "old" job.

The executive housekeeper or the property recruiter should post each position as soon as it is officially available. Each position should be open to people on staff before it is opened to outside applicants. Job postings should always be prominently located, such as in the employee lounge or kitchen. It can also be effective to post entry-level positions, so that employees can inform qualified friends and acquaintances.

Keeping a Call-Back List. As every executive housekeeper knows, recruiting is an everlasting process. To aid future staffing efforts, managers should keep a **call-back list** of all employees and applicants who possess special skills and interests or who express an interest in filling certain housekeeping positions. Furthermore, it is often a good idea to maintain a backup list of former employees (who retired or left on friendly terms) who may be willing to step in during a pinch.

External Recruiting

Housekeeping managers often recruit outside applicants to fill open positions. New employees may contribute fresh ideas and new ways of doing things. External recruiting activities include networking as well as contact with temporary and leasing employment agencies. See Exhibit 1 for a list of recruitment strategies.

Networking. Networking involves developing personal connections with friends, acquaintances, colleagues, business associates, and teachers and counselors at local schools. Such networks may lead to employee referrals. Representatives of companies that service or supply the property may also have useful information. Other possible network sources include members of trade associations, religious leaders, or volunteer associations. Remember, it is important to communicate regularly with members in your network. Let them know that the housekeeping department and the property are always interested in hiring good employees.

Temporary Agencies. Temporary agencies can provide staff to fill a wide range of positions. They often train their employees to fill specific needs. The hourly rate charged for temporary employees will probably be higher than the rate paid to permanent hourly employees, but these costs are usually offset in other ways. For example, temporary agencies:

- Supply capable help quickly, helping to reduce overtime, recruitment, and hiring expenses

- Employ workers whom they have screened and trained

- Often give temporary workers benefits and full-time positions to promote employee commitment

- Can often supply an entire supervised crew of workers when necessary

On the downside, temporary employees are usually not trained in appropriate or specific procedures. This can render such workers less productive than a property's own staff. Temporary workers, too, often require more supervision. Most properties use such workers only as a "Band-Aid" solution to an emergency—not as a regular practice.

Leased Employees. When the housekeeping department requires a work force for a longer period, it may be a good idea to look into leasing employees. The leasing agency hires employees and leases them to businesses, billing the employing business for the costs of hiring or leasing the employees. Some businesses even enter into contracts with leasing companies whereby the businesses "sell" their employees to the leasing companies, who lease them back to the businesses. This benefits the property because the leasing company provides the same work services, but takes over the costs of paying employee benefits. The leasing agency also handles all recruiting, selecting, training, and payroll activities. Employees profit because leasing companies offer greater job security and better benefits than some small properties can afford. When using any leasing company as a source of help, executive housekeepers should be sure to make a thorough check of the company's credentials.

Tax Credits. Some government programs, such as the federal Targeted Jobs Tax Credits Program, provide tax incentives to private employers who hire members of certain categories. Any employee hired under this program *must* be certified as a

Exhibit 1   Recruitment Strategies

Radisson Hotels International
WORLDWIDE · WORLDCLASS℠

RECRUITMENT STRATEGIES

1. **Youth**
   Schools, Vo-Techs, Colleges
   — Meet with counselors
   — Speak to classes
   — Sponsor work study programs
   — Participate in career days
   — Invite classes to tour hotel

2. **Minorities**
   — Meet with representatives from minority community agencies and invite for lunch and tour of hotel
   — Advertise in minority newspapers
   — Visit schools in minority neighborhoods
   — Notices at churches in minority communities
   — Visit youth centers and place notices there

3. **Disabled Persons**
   — State Rehabilitation Agencies
   — National Alliance of Business
   — Private Industry Councils
   — National Association of Retarded Citizens
   — Goodwill Industries
   — Other local agencies

4. **Women**
   — Local organizations which assist women in transition
   — Community colleges, universities
   — Bulletin board notices in supermarkets, libraries, YWCAs, exercise centers
   — Flyers in parking lots
   — Displaced Homemakers organizations
   — Craft centers
   — Child care centers

5. **Older Workers**
   — AARP Senior Employment Services
   — Senior Citizen Centers
   — Synagogues and churches
   — Retirement communities and apartment complexes
   — Newspaper ads worded to attract
   — Retired military

6. **Individuals in Career Transition**
   — Newspaper ads
   — University evening programs
   — Referrals
   — Teachers
   — Unemployed actors
   — Laid off workers from other industries
   — Speak at community functions, i.e., Rotary, Toastmasters

7. **Lawfully Authorized Immigrants**
   — Ads in foreign language newspapers
   — Churches
   — English as a Second Language classes
   — Citizenship classes
   — Refugee resettlement centers
   — Employee referrals

Courtesy of Radisson Hotels International, Minneapolis, Minnesota

member of a targeted category by a local office of your state
sion *before* the property hires him/her. In order to claim the
the property must be certain that the hired person is not a rela
the property's owner, and that he/she has not worked for the

Employee Referral Programs. The housekeeping department or
adopt an employee referral program which influences its employ
their friends or acquaintances to apply. Satisfied employees can
best recruiters. Indeed, as a general rule, good employees tend to re
employees.

Employee referral programs usually reward employees who h
new-hires to the company. Management should establish the size of the
wards at the outset, stating which criteria must be met and how the rel
be credited to the proper employee.

Employee Turnover

Employee **turnover** occurs when a work unit loses a member who must be replaced. Within the housekeeping department, a work unit may be the department
itself, a group of workers within the department, or a specific workshift.

Employees leave jobs for many reasons, including dissatisfaction with job duties, the desire for higher wages, or the wish to enter a new career. If employees
leave for these reasons, executive housekeepers may want to reconsider the selection process by asking themselves these questions:

- Was the position accurately defined?

- Were the company and the position clearly and fully explained to the applicant before he/she was hired?

- Did the executive housekeeper obtain complete, accurate information about
  the applicant before hiring him/her?

Dissatisfied employees can disrupt the entire housekeeping staff. They do this
by becoming unproductive (thus adding to the workload of co-workers) and by
having a high rate of tardiness and absenteeism. Discontented employees can also
poison the work environment by spreading their unhappiness around. Just as
goodwill and laughter are contagious, so are discontent and discouragement.
Sometimes, however, executive housekeepers will find that disgruntled employees
do not leave. When this happens, department morale is often lowered enough to
drive valued employees away.

The costs of employee turnover can be extremely high. One report states that it
costs the hospitality industry as much as $1,500 to replace its average hourly em-
ployee.[5] The direct costs of turnover include those incurred while recruiting, selecting, hiring, and training new employees. In addition, costs involved when an
employee leaves a department or property may represent the greatest expense; one
report calculates the costs to be as high as 3 $1/2$ times the costs of employee selection
and training.[6] Exhibit 2 lists some of the direct costs of turnover.

Exhibit 2   Direct Costs of Employee Turnover

Search and selection costs:

- advertising
- recruiting
- agency fees
- screening and testing
- candidate and recruiter travel
- relocation
- the wages and overhead costs (including office space and supplies) of all personnel involved in these steps

Hiring and training costs:

- preparation of offer letters, hiring agreements, or employment contracts, including preparers' time and any legal costs
- preparation, development, and printing of publications such as processing forms, the employee handbook, and benefits booklets
- wages and overhead associated with training and orientation, including the preparation, development, and maintenance of all training programs

Separation costs:

- overtime
- severance pay
- wages and overhead related to termination recordkeeping, processing, and any other special services, such as exit interviews or review boards
- lost production of the position and associated equipment while the vacancy remains unfilled
- additional FICA and unemployment insurance payments
- legal costs for any employment litigation arising as a result of a termination

The indirect costs of turnover include lower productivity. New employees are less experienced and require the close training and attention of supervisors and co-workers, whose own productivity is then reduced. In addition, experienced employees often must carry a greater workload to compensate for a new worker's lack of experience. Experienced staff members may feel overwhelmed by the necessity to shoulder someone else's work as well as their own. Indeed, managers have noticed that employee resignations often follow closely on the heels of previous ones.[7]

It is important to determine the causes of employee turnover. Possible causes include inadequate training and supervision, a distribution of the workload that employees perceive to be unfair, unhappiness with work hours, poor employee attitude, and trouble fitting in. Indeed, most turnover results from hiring the wrong people. Only when the causes are identified and analyzed can executive housekeepers take steps to alleviate the problem. Some of the factors signaling potential turnover include:

- Increased absenteeism
- Increased tardiness
- Decreased productivity
- Lower quality of work performance
- Negative attitude
- Employee's tendency to avoid supervisors
- Increase in disciplinary problems

With the high costs incurred in employee separation, selection, and training, housekeeping managers would be wise to refine their selection procedures and skills.

The Selection Process

Important tools used in the selection process are job descriptions and job specifications. A job description is a listing of all the tasks and related information which make up a work position. A job description may also outline reporting relationships, responsibilities, working conditions, equipment and material to be used, and other important information specific to the requirements of the property. Job descriptions are helpful in recruiting and selecting employees because they clearly state the duties involved in a particular job. They may also explain how the position relates to other positions in the housekeeping department or other departments in the hotel. Formats and contents of job descriptions vary according to the needs of individual properties.

Job specifications list the personal qualities, skills, and traits needed to successfully perform the tasks outlined by a job description. Although job specifications are unique to a particular position (based on the job description for the position), some general statements can be made about the skills, educational background, and personal qualities helpful in performing many housekeeping tasks.

Evaluating Applicants

Generally, executive housekeepers evaluate job applicants by reviewing completed job application forms, checking applicant references, and interviewing selected applicants. In properties with a human resources division, applicants are initially screened on the basis of housekeeping job descriptions and job specifications. In properties without a human resources division, the executive housekeeper may be responsible for all aspects of evaluating applicants.

The completed job application form provides basic information to assess whether the applicant meets minimum job qualifications. Information contained on the completed form indicates how well the applicant meets qualifications listed in job specifications. The form should be simple and should require only information which is important in considering how suitable the applicant is for the open position.

Care must be taken in structuring questions included in application forms because there are federal, state, and some local laws that prohibit discriminatory hiring practices. Exhibit 3 is a guide that lists pre-employment questions which are lawful and unlawful. Such a guide may also be helpful in developing questions used in interviewing applicants. Since laws and their interpretations vary from state to state, an attorney should review the property's application forms, related personnel forms, and interview procedures to ensure that they do not violate current anti-discrimination laws.

Interviewing

The applicant will be very influenced by his/her first impressions of the interviewer, the property, and what it would be like to work there. Once new-hires are employed, their job satisfaction and productivity will be shaped by whether their initial expectations are being met. Thus, it is essential that the first impression—as given in the employment interview—be as realistic as possible. In other words, be very careful to give an honest impression of yourself, the job, and the department.

In large properties, the human resources division usually handles all aspects of advertising for, recruiting, and initial screening of all job candidates. Then, each department head conducts the main, in-depth interview, and decides who is ultimately hired. In large housekeeping departments, the executive housekeeper may delegate the interviewing and hiring tasks to an assistant. However, the executive housekeeper is the person most responsible for hiring and maintaining a qualified housekeeping staff.

Whoever the interviewer, he/she should be thoroughly familiar with the job and what it involves. Furthermore, the interviewer should be a good, objective judge of people and their differences; a good department role model; a skillful communicator who is optimistic about the job, the housekeeping department, and the property; and a good salesperson. The interview should be conducted in a comfortable, private setting that allows few, if any, interruptions. A business office setting with the interviewer seated behind the desk with the applicant seated before the desk may prove intimidating. Therefore, it may be a good idea to hold the interview in or near the area in which the new-hire will be working. If the work site is too distracting, consider giving promising candidates a tour as part of the interview.

Conducting the Interview

Five primary objectives to achieve when interviewing applicants are:

1. Establish a basis for a working relationship.
2. Get enough accurate information to make a decision.
3. Provide enough information to help the applicant make a decision.
4. Sell the company and the job to your preferred applicant.
5. Create a feeling of goodwill between the company and the applicant.

When conducting the interview, be conversational and speak at the applicant's level—but do not be condescending. Treat all applicants the way you treat guests. Observe the applicant's physical appearance. Generally, serious applicants

Exhibit 3 Pre-Employment Inquiry Guide

SUBJECT	LAWFUL PRE-EMPLOYMENT INQUIRIES	UNLAWFUL PRE-EMPLOYMENT INQUIRIES
NAME:	Applicant's full name. Have you ever worked for this company under a different name? Is any additional information relative to a different name necessary to check work record? If yes, explain.	Original name of an applicant whose name has been changed by court order or otherwise. Applicant's maiden name.
ADDRESS OR DURATION OF RESIDENCE:	How long a resident of this state or city?	
BIRTHPLACE:		Birthplace of applicant. Birthplace of applicant's parents, spouse or other close relatives. Requirement that applicant submit birth certificate, naturalization or baptismal record.
AGE:	*Are you 18 years old or older?	How old are you? What is your date of birth?
RELIGION OR CREED:		Inquiry into an applicant's religious denomination, religious affiliations, church, parish, pastor, or religious holidays observed. An applicant may not be told "This is a Catholic (Protestant or Jewish) organization."
RACE OR COLOR:		Complexion or color of skin.
PHOTOGRAPH:		Requirement that an applicant for employment affix a photograph to an employment application form. Request an applicant, at his or her option, to submit a photograph. Requirement for photograph after interview but before hiring.
HEIGHT:		Inquiry regarding applicant's height.
WEIGHT:		Inquiry regarding applicant's weight.
MARITAL STATUS:		Requirement that an applicant provide any information regarding marital status or children. Are you single or married? Do you have any children? Is your spouse employed? What is your spouse's name?
SEX:		Mr., Miss or Mrs. or an inquiry regarding sex. Inquiry as to the ability to reproduce or advocacy of any form of birth control.
CITIZENSHIP:	Are you a citizen of the United States? If not a citizen of the United States, does applicant intend to become a citizen of the United States? If you are not a United States citizen, have you the legal right to remain permanently in the United States? Do you intend to remain permanently in the United States?	Of what country are you a citizen? Whether an applicant is naturalized or a native-born citizen; the date when the applicant acquired citizenship. Requirement that an applicant produce naturalization papers or first papers. Whether applicant's parents or spouse are naturalized or native born citizens of the United States; the date when such parent or spouse acquired citizenship.
NATIONAL ORIGIN:	Inquiry into languages applicant speaks and writes fluently.	Inquiry into applicant's (a) lineage; (b) ancestry; (c) national origin; (d) descent; (e) parentage, or nationality. Nationality of applicant's parents or spouse. What is your mother tongue? Inquiry into how applicant acquired ability to read, write or speak a foreign language.
EDUCATION:	Inquiry into the academic vocational or professional education of an applicant and the public and private schools attended.	
EXPERIENCE:	Inquiry into work experience. Inquiry into countries applicant has visited.	
ARRESTS:	Have you ever been convicted of a crime? If so, when, where and nature of offense? Are there any felony charges pending against you?	Inquiry regarding arrests.

*This question may be asked only for the purpose of determining whether applicants are of legal age for employment.

(continued)

Exhibit 3   *(continued)*

SUBJECT	LAWFUL PRE-EMPLOYMENT INQUIRIES	UNLAWFUL PRE-EMPLOYMENT INQUIRIES
RELATIVES:	Name of applicant's relatives, other than a spouse, already employed by this company.	Address of any relative of applicant, other than address (within the United States) of applicant's father and mother, husband or wife and minor dependent children.
NOTICE IN CASE OF EMERGENCY:	Name and address of person to be notified in case of accident or emergency.	Name and address of nearest relative to be notified in case of accident or emergency.
MILITARY EXPERIENCE:	Inquiry into an applicant's military experience in the Armed Forces of the United States or in a State Militia. Inquiry into applicant's service in particular branch of United States Army, Navy, etc.	Inquiry into an applicant's general military experience.
ORGANIZATIONS:	Inquiry into the organizations of which an applicant is a member excluding organizations, the name or character of which indicates the race, color, religion, national origin or ancestry of its members.	List all clubs, societies and lodges to which you belong.
REFERENCES:	Who suggested that you apply for a position here?	

Source: Michigan Department of Civil Rights, Lansing, Michigan.

wish to make good impressions. Assume that the applicant is presenting his/her highest personal standard.

Let the applicant set the pace. If he/she is nervous or shy, be patient. Do not tell applicants exactly what it is you are looking for; if you do, applicants may modify their responses to fit your expectations.

After a period of small talk designed to put the applicant at ease, the interviewer should begin the body of the interview by asking questions about the applicant's job expectations. For example, ask the candidate what kind of work he/she is looking for. Encourage responses by using appropriate gestures and comments. Listen carefully, keeping an eye on the applicant's body language. Sudden changes in position or tone of voice, eye movement, facial expressions, and nervous mannerisms could betray the applicant's uneasiness with the discussion.

Watch carefully as the applicant responds to queries. If he/she hesitates before answering a question, probe further by following it up with related questions. Moreover, if the applicant responds vaguely or changes the subject, it may mean that he/she has something to hide.

When the applicant asks questions about the position or the property, be as honest and forthright as you can. Interviewers, too, can arouse suspicion if they try to conceal or avoid topics.

Focus your interview on one principal area at a time. For example, begin with the applicant's work experience and discuss it thoroughly before talking about work history, education, or other areas.

In a smaller property without a human resources division, the executive housekeeper or designated interviewer should determine early in the interview whether the applicant meets the position's fundamental requirements. This is also the time to mention other hiring prerequisites, such as the federal government's requirement that employees prove their legal right to work in the United States. Also, determine whether the employee's personal job requirements will be met, such as working conditions, hours, pay rate, type of work, and the property's benefit package.

These topics can generally be discussed quickly. If it looks as though employment will not prove satisfactory or even workable for both sides, end the interview quickly and cleanly. A job offer made or accepted in desperation is likely to lead to more turnover. Do not even think about extending an offer unless you are quite certain of your impression of the applicant's overall suitability.

Techniques for Questioning

The most common interviewing technique is a two-step questioning process. The first step asks for specific information, such as who, what, when, or where. Follow each specific question with a question that seeks more in-depth responses by answering why or how. For instance, ask the applicant, "What did you like most about working for XYZ Hotel?" Then ask, "Why was that your favorite?" Other questioning techniques include the following:

- Ask the applicant for a list rather than one single response, allowing the applicant more spontaneity. You can narrow your field of questions afterward.

- Use direct questions to verify facts and cover a lot of ground quickly.

- When seeking more than standard responses, ask indirect or open-ended questions, or ask the applicant to make comparisons.

- If responses seem unreasonable, pursue the subject.

- If the applicant gives a partial answer, probe further by restating the reply as a question, such as, "So that department was just too big, wasn't it?"

- Use short affirmative responses to encourage the applicant to continue talking, such as, "I see," or "Please go on." It may also be helpful to nod.

- The use of silence will also indicate that you expect the applicant to continue speaking.

- Suggest sample answers when the applicant does not understand a question.

- Vary your responses by making comments rather than always asking questions.

What to Ask. Remember that there must be a good business reason for asking any question. In addition, make sure that interview questions are appropriate to the open position. That is, the executive housekeeper would not ask the same questions of applicants for both an assistant housekeeper and an entry-level room attendant position. Exhibit 4 presents a selection of questions that may assist you in developing your own interviewing process.

What Not to Ask. Again, refer to Exhibit 3 for a sample of one state's guide to lawful and unlawful pre-employment questions. Generally, avoid asking questions that will yield information which legally cannot influence the hiring decision. Avoid discussions of birthplace, national origin or citizenship, age, sex, race, height and weight, marital status, religion or creed, arrest records, and memberships in lodges or religious or ethnic organizations. Avoid questions about an applicant's physical or mental health.

Exhibit 4   Sample Interview Questions

Relevant to Job Background

- Did you regularly work 40 hours a week? How much overtime did you work?
- What were your gross and take-home wages?
- What benefits did you receive? How much did you pay for them?
- What salary/wage do you desire? What is the lowest amount you are likely to accept?
- Which days of the week are best for working?
- Have you ever worked weekends before? Where? How often?
- Which shift do you enjoy working the most? Which shift can't you work? Why?
- How many hours a week would you like to work?
- How will you get to work?
- Is your means of transportation reliable for the shifts you may be working?
- When you started your last job, what position did you hold? What position do you hold now or did you hold when you left?
- What was the starting salary of your present job or the last job you held?
- How often did you get pay increases on your present job or the last job you held?
- What three things do you want to avoid on your next job?
- What qualities do you expect in a supervisor?
- Why did you choose this line of work?
- Why are you interested in working at this hotel?
- Which work experience most influenced your career decisions?

Education and Intelligence

- When you were in school, what subjects did you like the most? Why?
- When you were in school, what subjects did you like the least? Why?
- Do you think your grades are a good indicator of your overall abilities?
- If you had to make the same educational decisions over again, would you make the same choices? Why or why not?
- What is the most important thing you have learned in the past six months?
- What good qualities did you find in your best teachers? Can these apply to work as well?

Personal Traits

Some of the following may be more suitable for people without much work background:

- What do you like to do in your spare time?
- How many times were you absent or late for your present or last job? Is that normal? What were the reasons?
- What does your family think of your working at this hotel?
- On your last job, were the policies concerning being absent without cause or late clearly explained to you? Were these policies fair?
- What was your first supervisor like?
- How did you get your first job? Your most recent job?

For the following questions about personal traits, job titles may be changed to meet the needs of the interview:

- Who has greater responsibilities—a front desk agent or a reservation sales agent? Why?
- Have you ever had to deal with an angry guest who complains about everything? If so, how did you work with the guest to resolve the issues?

Exhibit 4 *(continued)*

- What do you consider the main reason people in the position you are applying for leave their jobs? What would you do to change this?
- What do you consider the most important responsibilities of a good front desk agent?
- Suppose your supervisor insisted you learn a task in a certain way, when you know there is a better way. What would you do?
- Have you ever had a supervisor show favoritism to certain employees? How did you feel about this?
- Of all your job experiences, what did you like the most? Why?
- Of all your job experiences, what did you like the least? Why?
- When you go to a store to purchase something, what qualities do you look for in the sales person?
- What was your biggest accomplishment on your last job?
- What would you have changed about your last job if you had the opportunity?
- If the opportunity was offered to you, would you return to your last employer? Why or why not?
- How much notice did you give your last employer when you decided to leave (or plan to give your current employer)?
- How would your former supervisor and fellow employees describe you?
- What strengths and weaknesses do you bring to this new position?
- What frustrates you on the job? How do you handle this frustration?
- On your last performance review, what areas did your former supervisor mention need to be improved? Why do you think the comment was made?
- What three areas would you most like to improve about yourself?
- What one thing have you done of which you are the proudest? Why?
- What is the funniest thing that has ever happened to you?
- What is important to you about the job you are applying for? Why?

Questions for Managers

- What type of training program did you have for your employees? Who set it up and who implemented it?
- What have you done on your last job to improve the performance of the department you supervised? How was this measured?
- What are the most important attributes of a manager?
- What hotels were your biggest competitors? What were their strengths and weaknesses?
- How would your employees describe you as a supervisor?
- How many people did you have to discipline on your last job? Describe the circumstances. How do you feel about terminating employees?
- What did you do to motivate your employees?

Further, it is illegal to ask questions of one sex and not the other. For example, do not ask female applicants whether they have small children or what their child care plans are. If such questions are provably employment-related, you must ask them of males as well as females.

You will need to obtain certain types of information *after* an applicant is hired, such as proof of age and proof of the new-hire's legal right to work in the United States. A good time to obtain such information is when the recruit is filling out employment papers.

Exhibit 5   Sample Interview Evaluation Form

INTERVIEW EVALUATION FORM

Applicant
Name _____

Position
Evaluated _____   Date _____

	Poor Match		Acceptable	Strong Match	
	−3	−1	0	+1	+3
RELEVANT JOB BACKGROUND					
General background					
Work experience					
Similar companies					
Interest in job					
Salary requirements					
Attendance					
Leadership experience					
EDUCATION/INTELLIGENCE					
Formal schooling					
Intellectual ability					
Additional training					
Social skills					
Verbal and listening skills					
Writing skills					
PHYSICAL FACTORS					
General health					
Physical ability					
Cleanliness, dress, and posture					
Energy level					
PERSONAL TRAITS					
First impression					
Interpersonal skills					
Personality					
Teamwork					
Motivation					
Outlook, humor, and optimism					
Values					
Creativity					
Stress tolerance					
Performing skills					
Service attitude					
Independence					
Planning and organizing					
Problem solving					
Maturity					
Decisiveness					
Self-knowledge					
Flexibility					
Work standards					
Sub-totals					

TOTAL POINTS _____

Interview Evaluation

Exhibit 5 presents a sample interview evaluation form that lists some key traits for employees in the hospitality industry. Portions of this form can be adapted to the job specifications for housekeeping positions. The executive housekeeper can use such a form to evaluate an applicant's strengths and weaknesses. After interviewing an applicant, the executive housekeeper completes the form by scoring the applicant according to the following criteria:

- Score zero if the applicant meets an acceptable level of skill in a given area, or if the skill is not needed.

- Score plus one or three according to the degree to which the applicant surpasses the acceptable level of skill in a given area.

- Score minus one or three according to the degree to which the applicant fails to meet the acceptable level of skill in a given area.

Every applicant has both strengths and weaknesses. An interview evaluation form helps ensure that a shortcoming in one area does not unduly lower an applicant's chances for further consideration. With this interview evaluation form, the applicant with the highest point total will probably make the best employee.

All employment interviews should be documented, especially those for people who were not hired. These records should reflect only job-related information. The interviewer's personal notes should never be written on job application forms.

Interview Follow-Up

Immediately after each interview, evaluate the information and make further investigations. Always verify the information given by an applicant before hiring him/her. The quickest and most effective method of verifying applicant information is the telephone reference check. Use this method to verify such information as employment dates, work attitudes, and ability to work with others. When conducting reference checks, ask the same types of questions about *all* applicants, regardless of race, sex, age, or other characteristics.

Be sure to check with more than one previous employer or more than one individual within a company. Keep a record of the results of all background checks in the appropriate personnel files.

Check references that applicants provide to verify that the applicants are who and what they say they are. Exhibit 6 presents a sample telephone reference form. Although the reference check form attempts to secure a great deal of information, past employers may feel obligated to provide only information about the applicant's past job title, dates of employment, and salary verification. Some employers may not even reveal whether the person is eligible for rehire because their statements may increase their potential liability if the person charges libel, slander, or defamation of character. Executive housekeepers must be familiar with their own property's policy regarding how to handle callers inquiring about the work record of current or past employees.

The Hiring Period

The final stage of the selection process is the **hiring period**, which begins when the employer makes an offer to a prospective employee. This phase involves all the

Housekeeping Success Tip

John Baker, CHHE
Director of Housekeeping
Quality Inn Sports Complex
Lyndhurst, New Jersey

❝When interviewing is complete, references are checked, and starting date and time are established, another employee enters the ranks of the housekeeping department. Does the housekeeping director have questions? Plenty. Did I make a wise decision? Will this new-hire last long? Will this person 'fit in'?

We have all hired a number of employees, with both successes and failures resulting. You can probably remember a fair number of both. You can derive a lot of pride from knowing that you have put together a good staff. Don't be fooled into thinking that luck has much to do with hiring success.

Traditionally, our industry has a high turnover rate. Keeping that rate at a minimum is tough. There are ways, however, and most have nothing to do with pay rates. Getting a new employee successfully through the first week is half the battle. Your close attention is a must in this early phase of employment!

A good practice when welcoming a new-hire is to put yourself in his or her shoes. We've all had 'first day on the job' experiences. It's a time of nervousness and varying degrees of insecurity, hope, and eagerness.

Before beginning actual training, get to know the recruit on a personal level. Filling out necessary employment papers together may help. Talking will help break the ice. Most people like to talk about themselves; listen to the new recruit. Be friendly and supportive. Explain company policies and regulations, making sure that he or she understands them. Don't sugarcoat anything. Finally, avoid discussing the personalities of other employees; it would just confuse your new-hire—and he or she will get to know co-workers soon enough anyway.

Introduce the new employee to your entire staff on the first day. Current employees will be eager to get to know their new co-worker. If possible, see that it happens gradually. Make sure the new employee feels comfortable with co-workers and with the hotel itself. He or she should not be left alone at all during the initial training period. If the size of your staff makes it impossible for you to spend lunch periods and breaks with the new employee, make sure that reliable co-workers do. If possible, introduce the new-hire to the general manager. Make sure the new employee knows who his or her supervisors are.

Naturally, training is very important. If you are not doing the training yourself, select a trainer who is a thorough worker and also a good teacher. It is sometimes wise to use different trainers to teach the same person.

The amount of time required to train will vary with each individual. Don't rush the new employee or allow him or her to get bored. Be patient. Some of my best employees were very slow and nervous during their training. In fact, I think that's a good sign, and I let the new-hire know that. It indicates the new employee's willingness and desire to please. Maintaining a new-hire's morale is not always easy, but

Housekeeping Success Tip (continued)

learning how to do it is very important for your career. If you can't succeed at this, you won't succeed at all. Check on a new employee's progress repeatedly. Ask if he or she is having fun. The new-hire will generally respond with smiles, and that's what you want to see.

Your relationship with your staff should be of the utmost importance. Gaining staff respect and confidence is essential. How you do this will depend on your personality, and also on how well your employees interact. Employee selection is very important in this regard.

Your hiring success rate begins with *your* choice of whom to hire. Naturally, you'll feel personally committed to making good choices. There are no guarantees, however, and the most promising candidate can become a hiring casualty for various reasons. You have to give each new employee your best effort. Don't let a good employee slip away—start on day one to build morale. **"**

Exhibit 6   Sample Telephone Reference Form

OPRYLAND HOTEL
TELEPHONE EMPLOYMENT VERIFICATION

Date_____

Applicant's Name_____

Company_____

Dates Employed: From:_____   To:_____

Position_____

Reason for Leaving_____

Would you Rehire?_____

Person Verifying_____

Telephone number of company_____

Comments_____

FORM 3755D   **OPRYLAND HOTEL** ®

Courtesy of Opryland Hotel, Nashville, Tennessee

arrangements necessary to prepare the new-hire and current employees for a successful working relationship, including the processing of personnel records. The hiring period lasts through the new-hire's initial adjustments to the job.

Making and Closing Offers

Extending an offer to a potential employee is the crucial first step in the hiring process. Because of the skill needed to do this effectively while complying with complex labor laws, most companies rely completely on the human resources division or someone else specifically designated by management. When only one or two people extend all offers, the property has more control over how the job is represented and what promises are made.

The three steps of making and closing offers are extending, negotiating, and concluding.

Extending the Offer. A carefully worded offer can give the potential employee a good name to live up to and may begin his/her commitment to the property's goals. Timing is a very important aspect of making offers successfully. The longer you wait to extend the offer, the less likely the candidate is to accept it. This is because the applicant will have started to talk him/herself out of wanting the job, or perhaps the applicant will have accepted another job offer.

Negotiating the Offer. If your research was thorough during the selection process, you should be very familiar with the applicant's background and expectations. Before authorizing an offer, you should have covered areas that might arise as obstacles to negotiations, such as pay requirements, starting dates, and special benefits. Authorize the human resources or hiring representative to extend an offer only if you're reasonably sure it will be accepted, and only after you have all the facts to put together a complete offer package.

Establishing a starting date lets new-hires know that your company does not like employees leaving without giving proper notice. It is inconsistent to expect employees to give proper notice at your property if you are not also willing to let them do so before leaving current positions.

Concluding the Offer. Once the offer is accepted, assure the applicant that he/she has made the right decision. Tell new-hires that the property does not expect them to know everything, but that the property believes that they will be able to do their jobs. Tell the new-hire that his/her new supervisor will begin immediately to prepare for his/her arrival. Such preparation includes notifying other applicants and the housekeeping staff, or obtaining the new-hire's uniform size, for example.

Preparing Other Employees

As executive housekeeper, you can head off potential problems and ensure faster start-up time for the new-hire by properly preparing your existing staff. The smaller your department or the higher the level of responsibility of the new employee, the more important this step is.

Inform current employees when you hire a new employee. Tell them the new person's name, previous jobs, and starting dates. Other employees may have worked at some of the same places; friendships and positive expectations of the new employee can begin to form. If your department has strong cliques, meet with informal leaders and enlist their cooperation. Make them accountable for seeing that new employees are accepted.

Explain who the new employee will report to and what his/her responsibilities are. If the new employee will have any authority, explain the extent of the authority and emphasize that he/she has your full support. Finally, remember to notify the person who will conduct the employee's job orientation.

Processing Personnel Records

Processing new employees before they actually start working will better prepare them for the first day of work. This is a good time to fit uniforms and order name tags. These items are needed on the employee's first day on the job, not two weeks later.

The tone of the processing period should be warm, caring, and professional. If it's too light or casual, new employees may conclude that the property is lax in its policies and procedures. Processing is a good time to convey what you expect from the employee in the way of service, and to talk about the goals of the housekeeping department and the property.

During processing, the executive housekeeper or the human resources division should cover certain points. These may include time cards, pay procedures, house rules, instructions on reporting for the workday, and uniforms. The use of a checklist can ensure that all important points are covered.

The Orientation Process

Job orientation involves two steps: first day activities and job skills training. In addition, the property's human resources division may hold a general company orientation, which is a formal program designed to introduce the company's mission and values to a group of employees. New employees usually attend a general orientation a few weeks after beginning work. Since general company orientation is conducted by the property and not the housekeeping department, only first day activities and skills training are discussed here.

First day activities acquaint the employee with basic elements of the job, the workplace, and co-workers. Job skills training takes place on the first and subsequent days and imparts basic skills and knowledge required for the job.

The Employee's First Day

The first day on a job can be confusing and anxiety-provoking to new employees. To help them make a smooth transition, consider using a checklist on the first day that covers each key area. It is important to strive for consistency in department processing. If employees do not know about a policy or procedure, they will not follow it.

The Executive Housekeeper's Role

When you greet the new employee, smile, make eye-contact, and shake hands. Establish a preferred-name basis for both of you. Welcome the new-hire and mention that the rest of the staff are looking forward to meeting him/her.

As executive housekeeper, you must decide how much processing to do yourself and how much to delegate to assistants. Many items can be turned over to a trainer, but you should cover the following areas yourself: overview, guest relations and security, work schedule, paycheck procedures, appearance standards, and work standards.

Overview. Give the new staff member a summary of what the first day will entail to help alleviate any initial anxiety. It is important to set standards high from the outset, since it's much harder to raise expectations and standards after the employee has been on the job awhile. Cover department goals and establish a clear direction regarding what needs to be accomplished. Detail actual job duties and job performance standards. Consider reviewing the organization chart.

Guest Relations and Security. Two areas that are part of every employee's job are guest relations and security. Executive housekeepers should emphasize the importance of the property's service philosophy. In addition, all employees should be aware of the need for good security procedures.

Work Schedule. Cover work hours and how to take specific meal and break periods. Explain expected days off, whether they are rotated or fixed, and how to handle requests for days off. Indicate where the schedule will be posted and stress the importance of punching or signing in correctly. Review time card procedures. Spell out your policies regarding absenteeism and lateness.

Paycheck Procedures. Review where and when to pick up paychecks.

Appearance Standards. Stress the importance of having a complete, clean uniform. Go over personal cleanliness and personal appearance standards, including makeup, jewelry, hair length, and so forth. Stress that you expect all employees to clean up after themselves in work areas, locker rooms, and break areas.

Work Standards. Explain the quality and quantity of work that you expect. Discuss and document specific standards so employees know and accept the way their job performance will be evaluated. Explain any quotas you have, such as a given number of rooms per housekeeper. Emphasize quality and the necessity of learning the procedures correctly, stressing that speed will come with experience. Encourage new employees to ask questions, indicating that questions show interest and a desire to do a job well—qualities you welcome.

Final Steps. If you are not handling the actual job training yourself, introduce the new employee to the trainer, who will assume the balance of the hiring process. Explain the trainer's position and authority. Praise the trainer—in his/her presence—as a particularly competent staff member and trainer. Make an extra effort to check on the new employee several times during the first day and during the first few weeks. This will let you see how training is progressing, and give the employee opportunities to ask any questions that arise.

The Trainer's Role

The trainer should be one of your better employees—someone that both you and fellow employees respect (see Exhibit 7 to help you choose a trainer). The trainer should also be a good communicator, committed to department goals, and a positive

Exhibit 7   Are You a Good Trainer?

1. Do you consider preparation to be the first step in instruc...

2. Do you spend at least as much time getting things ready f... in the actual training session?

3. Do you use the performance standards from job breakdowns... plans?

4. Do you list key points around which you will build the instruction...

5. Do you devote time to explaining to the employee how he or she... from the training session?

6. Do you determine what the employee already knows about the job... start training?

7. Do you set up a timetable showing the amount of time you plan to sp... training employees, and when you expect their training to be complete...

8. Do you expect that there will be periods in the course of the training dur... which no observable progress will be made?

9. Do you expect some employees to learn two or three times faster than oth...

10. Do you both tell and show the employee how to do the skill involved?

11. When an employee performs incorrectly during training, do you acknowledge correct performance before pointing out areas that need improvement?

12. Do you give instructions so clearly that no one can misunderstand them?

13. Do you ask the employee to try out the skill and to tell you how to do it?

14. Do you praise correct performance frequently?

15. Do you expect 100% conformity to standards?

(All of the questions should be answered "Yes.")

---

role model. This is especially important if there are strong cliques operating within the department. The trainer should embody all the standards you want the new employee to learn.

The trainer should put new employees at ease just as the executive housekeeper did. Beginning this part of the orientation with a tour allows the trainer to learn more about the new employee. This makes the new employee feel comfortable and helps the trainer determine how to approach individual training. The tour itself should include the work area, time clocks, and locations of posted schedules, supply areas, first aid kits, restrooms, and break and smoking areas. A tour of related departments could be accompanied by an explanation of the workflow and the need for teamwork. Show new employees how their jobs fit into the operation. During the tour, the trainer should introduce fellow housekeeping staff members and other department heads with whom contact is likely. A tour of the entire facility will come later with the general company orientation, if the property conducts one.

At this time, the trainer should just cover those areas the employee will need to know to get the job done. For example, room attendants will need to know where supplies are kept on every floor. When the tour is over, the trainer and new employee should return to the work area for a review of the department's chief policies and procedures. The trainer should also review equipment and supplies needed for the job. After this, skills training can begin.

Endnotes

1. Laventhol & Horwath, *U.S. Lodging Industry 1987* (Philadelphia: Laventhol & Horwath, 1987), p. 7.

2. George Leposky, "Tapping New Labor Sources," *Lodging* reprint, pp. 2, 6.

3. For more information on this subject, see Jack P. Jefferies, *Understanding Hospitality Law,* 3d ed. (East Lansing, Mich.: Educational Institute of the American Hotel & Motel Association, 1995).

4. Readers interested in further information on the federal Targeted Tax Credit Program should contact the local office of their state's employment security commission.

5. Robert Bové, "In Practice," *Training and Development Journal*, April 1987, p. 14.

6. Dean B. Peskin, *The Doomsday Job: The Behavioral Anatomy of Turnover* (New York: AMACOM, 1973), p. 69.

7. For more information on employee turnover, see David Wheelhouse, *Managing Human Resources in the Hospitality Industry* (East Lansing, Mich.: The Educational Institute of the American Hotel & Motel Association, 1989), Chapter 3.

Key Terms

call-back list
employee referral program
external recruiting
general company orientation
hiring period
internal recruiting

leased employees
networking
recruitment
supported employment program
turnover

Discussion Questions

1. What sources of labor will executive housekeepers be likely to draw upon to staff housekeeping departments in the late 1990s?

2. What methods of internal recruiting would most benefit your department or property?

3. How does the use of job descriptions help in recruiting and selecting employees?

4. What are the direct costs of employee turnover? The indirect costs?

5. What are some of the things an executive housekeeper should do to prepare for an interview?

6. What are some of the sources of information you might use to check an applicant's references?

7. What are some specific examples of the common errors of evaluation?

8. What are the three steps of making and closing employment offers, and what do they involve?

9. What key areas should an executive housekeeper cover with a new employee on the latter's first day on the job?

REVIEW QUIZ

When you feel you have covered all of the material in this chapter, answer these questions. Choose the *best* answer.

True (T) or False (F)

ⓣ F 1. Displaced homemakers lack formal work experience.

T Ⓕ 2. Supported employment programs seek job opportunities for recent immigrants.

ⓣ F 3. In the United States, employers can't legally ask applicants for verification of their legal right to work until the time of hiring.

ⓣ F 4. The executive housekeeper should personally interview the top candidates considered for open department positions.

T Ⓕ 5. Cross-training is an effective external recruiting method.

T Ⓕ 6. Employees have more incentive to excel when all are given the same wage increase regardless of performance.

T Ⓕ 7. Jobs should be opened to outside applicants before they are opened to current staff members.

T Ⓕ 8. An employee referral program recruits employees from temporary employment agencies.

ⓣ F 9. Lower productivity is considered a direct cost of employee turnover.

ⓣ F 10. The longer you wait to extend a job offer, the less likely the candidate is to accept it.

Alternate/Multiple Choice

11. By the year 2000, the number of older workers in the United States is expected to:

ⓐ decrease.
b. increase.

12. Which of the following questions would be preferred when interviewing job applicants?

ⓐ What is your full name?
b. What is your full name and maiden name?

13. Orienting new employees to work schedules, paycheck procedures, and appearance standards should be done by the:

ⓐ trainer.
b. executive housekeeper.

14. Which of the following is <u>not</u> a signal of potential employee turnover?

    a. increased absenteeism
    b. increased tardiness
    c. increased productivity
    d. decrease in job performance

15. Which of the following questions should *not* be asked when interviewing job applicants?

    a. Do you intend to become a U.S. citizen?
    b. What is your birthplace?
    c. What countries have you visited?
    d. What were your previous gross and take-home wages?

13/15

Chapter Outline

Skills Training
    Job Lists and Job Breakdowns
    Training to Standards
    The Four-Step Training Method
Scheduling
    The Staffing Guide
    Developing Employee Work Schedules
Motivation
    What Is Motivation?
    Methods of Motivating Employees
Employee Discipline
    Elements of an Effective Disciplinary
        Program
    Informal Counseling
    Progressive Discipline
    Guidelines for Disciplinary Action
    Conducting a Disciplinary Conference

Learning Objectives

1. Describe how job breakdowns can be developed for large and small hospitality properties.

2. Explain how job breakdowns can be used to develop lesson plans for specific training sessions.

3. Identify guidelines for implementing the four-step training method.

4. Distinguish between fixed and variable staff positions in the housekeeping department.

5. Explain the steps involved in developing a staffing guide for the housekeeping department.

6. Identify alternative scheduling techniques.

7. Identify techniques that executive housekeepers can use to motivate department employees.

8. Describe guidelines for developing incentive programs.

9. Define the four basic steps of a progressive discipline program.

10. Identify guidelines for administering formal disciplinary action.

Training, Scheduling, Motivating, and Disciplining

Much of this chapter was written and contributed by Jon M. Owens, CHHE, Director of Housekeeping, Clarion Hotel & Conference Center, Lansing, Michigan.

EMPLOYEES ARE THE LIFEBLOOD of any hospitality operation; without them, an operation stands still. It stands to reason, then, that management must do all it can to offer employees the training they need to do their jobs well. Training is especially important in today's busy housekeeping departments, where skills must meet or exceed department standards in order to get the work done in a guest-pleasing manner.

Beyond training, housekeeping employees need to be scheduled efficiently to meet organizational goals; and if employees are motivated—if they really *want* to do their best—those goals will be more readily and easily met. Efficient scheduling and employee motivation are more challenges for the executive housekeeper.

This chapter takes a close look at training—especially the four-step training method—and poses some practical scheduling/staffing situations. The chapter then explores motivational techniques and methods, and examines the role of discipline.

Skills Training

Ensuring that department employees receive proper training is one of the executive housekeeper's major responsibilities. This does not mean that the executive housekeeper must assume the duties and responsibilities of a trainer. The actual training functions may be delegated to supervisors or even to talented employees. However, the executive housekeeper should be responsible for ongoing training programs in the department.[1]

Job Lists and Job Breakdowns

Job lists and **job breakdowns** for each housekeeping position are the basic tools with which to build an effective skills training program. They also form an efficient system for evaluating employee performance. A job list for a housekeeping position is a list of tasks which must be performed, and a job breakdown specifies how to perform each task.

Exhibit 1   Sample Job List: Morning Shift Room Attendant

Date: xx/xx/xx

JOB LIST

Position: Housekeeping Room Attendant

Tasks: Employee must be able to:

1. PARK in designated area.
2. WEAR proper uniform.
3. PUNCH in.
4. PICK up clipboard and keys.
5. MEET with supervisor.
6. OBTAIN supplies.
7. PLAN your work.
8. ENTER the room.
9. PREPARE the room.
10. MAKE the beds.
11. GATHER cleaning supplies.
12. CLEAN the bathroom.
13. DUST the room.
14. CHECK/REPLACE paper supplies and amenities.
15. CLEAN windows.
16. INSPECT your work.
17. VACUUM the room.
18. LOCK the door and mark your report.
19. TAKE breaks at designated times.
20. RETURN and restock cart.
21. RETURN to housekeeping with clipboard and keys.
22. PUNCH out.

The tasks on the job list should reflect the total job responsibility of the employee. Exhibit 1 presents a sample job list for a morning shift room attendant. Note that each line on the sample job list begins with a verb typed in uppercase letters. This format stresses action and clearly indicates to an employee what he/she will be responsible for doing. Whenever possible, tasks should be listed in an order that reflects the logical sequence of daily responsibilities.

The job breakdown format can vary to suit the needs and requirements of individual properties. Exhibit 2 presents a sample job breakdown designed to serve not only as a training guide for a newly hired room attendant, but also as a tool for evaluating the performance of all room attendants on the morning shift.

The first column in Exhibit 2 shows a task from the job list presented in Exhibit 1. The second column breaks down the task by identifying the specific observable and measurable steps that an employee must take in order to accomplish the task shown in the first column. These steps are written as **performance standards.**

Exhibit 2   Sample Job Breakdown

POSITION: Housekeeping Room Attendant, morning shift

NAME:

SUPERVISOR:

JOB LIST	PERFORMANCE STANDARDS	ADDITIONAL INFORMATION	1st Qtr. Yes/No		2nd Qtr. Yes/No		3rd Qtr. Yes/No		4th Qtr. Yes/No	
7. PLAN YOUR WORK.	A. STUDY your assignment sheet.	Early service requests, rush rooms, check-outs, VIP's and no-service requests will be noted on your chart.								
	B. CLEAN check-outs first, whenever possible.	Cleaning check-outs first gives the front desk rooms to sell.								
	C. CLEAN early service requests as noted on your report.									
	D. CLEAN VIP rooms before lunch, whenever possible.	A VIP is our most important guest.								
	E. LOCK your cart room door and proceed to your section.									
	F. HONOR "do not disturb" signs.	We must honor the privacy of guests. Many guests like to sleep in. Never knock on a door that has a "do not disturb" sign.								
	G. CHECK rooms marked c/o and then check the rooms which are circled on your report. These are rooms due to check out.	Rooms marked c/o have already checked out at the front desk. Check-out time is noon.								
	H. PLAN your work around early service requests.	If you have early service requests, be sure to clean these rooms at the proper time.								

The third column presents additional information. Generally, this information explains why each step of the task is performed according to the standards listed in the second column. The additional information column can also be used to stress the hospitality aspects of the job, such as desired attitudes, and safety tips and pointers which may help the room attendant perform a task according to the housekeeping department's standards.

The fourth column is a checklist for recording quarterly performance evaluations. A performance evaluation identifies an employee's strengths and areas for improvement. The areas for improvement indicate specific training needs.

Employees should know the standards that will be used to measure their job performance. Therefore, it is important to break down job tasks and document the standards. In order to serve as a performance standard, each item in the second column of the job breakdown must be observable and measurable. The executive housekeeper (or the supervising housekeeping manager) conducting a quarterly performance evaluation should be able to simply check either "Yes" or "No" in the checklist column to indicate whether the employee performed correctly.

Job breakdowns are useless if performance standards cannot be observed or measured. For example, a performance standard such as "BE happy" is useless when it comes to evaluating a housekeeping employee's performance. One manager may think that an employee looks happy, while another manager may not. However, a performance standard can state that an employee should "SMILE" when responding to guests. A smile is an observable behavior; an employee is either smiling or not smiling, regardless of who is doing the observing.

Developing Job Breakdowns. If one person in housekeeping is assigned the responsibility of writing every job breakdown, the job may never get done, unless the department is very small with a limited number of tasks. Some of the best job breakdowns are written by those who actually perform the tasks. In properties with large housekeeping staffs, standards groups can be formed to handle the writing tasks. Group members should include department supervisors and several experienced room attendants and public area attendants. In smaller properties, experienced employees might be assigned to write the job breakdowns. Exhibit 3 summarizes the process of developing job breakdowns.

Most hospitality organizations have a policy and procedures manual. Although this manual rarely contains the detail necessary to set up effective training and evaluation programs, portions of it may be helpful to members of a housekeeping standards group as they write job breakdowns for each department position. For example, if the procedure sections of the manual include job descriptions and job specifications, they may help a standards group in writing job lists and performance standards. The policy sections may be helpful sources of additional information which can be included in the job breakdowns.

The job breakdowns for tasks which involve the use of equipment may already be written in the operating manuals supplied by vendors. Standards groups should not have to write performance standards for operating floor buffers, wet vacs, and other types of machinery, for example. Instead, the standards group may simply refer to (or even attach) appropriate pages from the operating manual supplied by the vendor for in-house training.

Housekeeping Success Tip

Patricia B. Schappert, CHHE
Director of Housekeeping
Opryland Hotel
Nashville, Tennessee

❝High turnover and low morale are two of the executive housekeeper's biggest headaches, and thinking of ways to combat these problems poses special management challenges. Here at the Opryland in Nashville, we realized that the behind-the-scenes nature of housekeeping work makes employees feel they are overlooked. So we started the World Series of Housekeeping in 1985 as a way to make the housekeeping staff more visible, to help them take pride in their work and their property, and to say thanks for a job well done. Since then, the event has become bigger and more popular every year, and other properties around the country have started their own series.

Exactly what *is* a World Series of Housekeeping? Every year we devise a number of events so that teams from housekeeping departments at properties in the Nashville area can pit their skills against one another. Area executive housekeepers judge the events. To add to the fun, we try to build each game around a theme. In 1988, we had events named after movie titles. In the 'Attack of the Killer Vacs' contest, for example, teams of two people from each property competed to see who could dismantle and reassemble a vacuum cleaner in the shortest time. Another event, 'The Big Bed Makeoff,' tested the speed and quality of a room attendant's bedmaking ability.

Participating housekeeping departments send a six-person team to the event. Not everybody in the department may be athletically inclined or want to participate. But going to the event to cheer your co-workers on helps build a team spirit among all staff members at the property.

Organizing the event in the first place is a way to say, 'Thanks for the good job you do.' Awarding cash prizes to members of first-, second-, and third-place teams helps make the thank you a little more tangible. All team members receive a small memento—we had T-shirts and bags in 1988. And the executive housekeeper at the winning property also receives a plaque.

Last year [1988], we opened the series to the general public and earmarked concession proceeds to the Tennessee Special Olympics. That gave competitors the extra satisfaction of helping a worthy cause as well as gaining some recognition for themselves. We also drew community support with good television and newspaper coverage of the event. Steve Phillips, sports director for Nashville's WKRN-TV, emceed the series.

Twelve teams—a record number—participated in the Nashville event in 1988, and we hope the event will continue to grow each year. It was great fun—even if Opryland did lose its title. (We promise to get it back again in 1989!) I think it's a mark of high standards that the top three teams were quite different properties.

(continued)

Housekeeping Success Tip *(continued)*

They ranged from our 1,891 rooms to the Marriott's 400 to the Pickett Suite's 128. Only a three-point spread separated the first-place Marriott and third-place Pickett.

We were especially pleased when the Nashville area world series sparked a similar event in Las Vegas under the sponsorship of the local National Executive Housekeeper's Association (NEHA) chapter. That event drew a crowd of 700 spectators. Roscoe Green, director of housekeeping at the Las Vegas Hilton, said, 'The biggest success is that the event has generated a lot of morale among all the employees and among all the participating properties.' And that's the whole idea! **"**

Exhibit 3   Developing Job Breakdowns

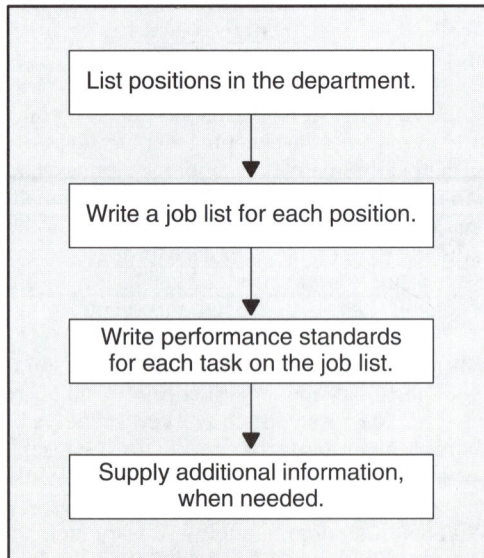

List positions in the department.

↓

Write a job list for each position.

↓

Write performance standards for each task on the job list.

↓

Supply additional information, when needed.

Developing job breakdowns involves breaking down each task on each housekeeping job list by writing the performance standards that state the specific observable and measurable steps an employee must take in order to accomplish the task. The executive housekeeper should assist the standards group in writing performance standards for at least two or three positions within the housekeeping department. While assisting the group, the executive housekeeper should stress that each performance standard must be observable and measurable. The value of each performance standard can be tested by asking whether a supervisor or manager can evaluate an employee's performance by simply checking a "Yes" or "No" in the quarterly performance review column.

Exhibit 4   Training with Job Breakdowns

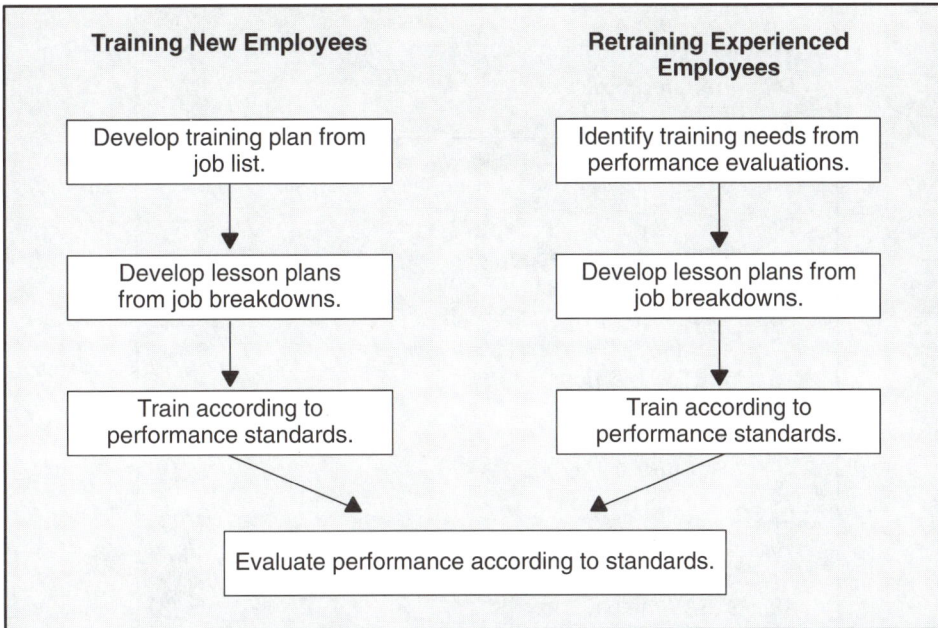

Training New Employees	Retraining Experienced Employees
Develop training plan from job list.	Identify training needs from performance evaluations.
↓	↓
Develop lesson plans from job breakdowns.	Develop lesson plans from job breakdowns.
↓	↓
Train according to performance standards.	Train according to performance standards.

Evaluate performance according to standards.

After the standards group has written job breakdowns for two or three tasks, the writing of job breakdowns for the other housekeeping tasks should be assigned to individual members of the group. Within a specified time, they should submit their work to the executive housekeeper or the executive housekeeper's assistant, who then assembles the breakdowns, has them typed in a single format (perhaps similar to that shown in Exhibit 2), and provides copies to all of the group's members. A final meeting can then be held, with the standards group carefully analyzing the breakdowns for each position within the department. After the job breakdowns have been finalized, they should be used immediately to train the department's staff.

Training to Standards

Exhibit 4 shows how job breakdowns can be used to train new employees and re-train experienced employees to perform tasks according to established standards. Housekeeping managers and supervisors can use performance evaluations based on job breakdowns to identify the training needs of experienced employees. Because of their detail, job breakdowns are also very useful in training newly hired employees who have little or no experience in hospitality housekeeping jobs. A comprehensive **training plan** can be developed from the job lists of the positions that the new employees will occupy. The lesson plans of specific training sessions can be based on the performance standards and additional information columns of the job breakdowns.

Exhibit 5   Sample Five-Day Training Plan

Position: Housekeeping Room Attendant

Date Prepared: xx/xx/xx

Employee: _____

Tasks: Employee must be able to:

Day 1
1. PARK in designated area.
2. WEAR proper uniform.
3. PUNCH in.
4. PICK up clipboard and keys.
5. MEET with supervisor.
6. OBTAIN supplies.
8. ENTER the room.
9. PREPARE the room.

Day 2
10. MAKE the beds.
11. GATHER cleaning supplies.
12. CLEAN the bathroom.

Day 3
13. DUST the room.
14. CHECK/REPLACE paper supplies/amenities.
15. CLEAN windows.

Day 4
16. INSPECT your work.
17. VACUUM the room.
18. LOCK the door and mark your report.

Day 5
7. PLAN your work.
19. TAKE breaks at designated times.
20. RETURN and restock cart.
21. RETURN to housekeeping with clipboard and keys.
22. PUNCH out.

For example, Exhibit 5 shows a sample five-day training plan developed from the sample job list in Exhibit 1. The trainee masters one group of related tasks at a time and must achieve 100% conformity to performance standards before progressing to the next group. Note that, except for task number seven, the training plan follows the same sequence of tasks which appear on the sample job list. However, it is not always possible or desirable to train employees in the same sequence in which tasks are actually performed on the job.

The Four-Step Training Method

The trainer's primary function is to communicate performance standards to employees. Trainers accomplish this by following a **four-step training method,** which can be used to train both newly hired and experienced employees. The four steps are:

1. Prepare to train
2. Conduct the training
3. Coach trial performances
4. Follow through

This on-the-job training method can be adapted to meet the special requirements of almost any housekeeping department. It can be used for either individualized instruction or for group training programs. The following sections briefly describe each step of the training method.

Prepare to Train. Some housekeeping managers think they know housekeeping skills so well that they can teach them to others without any thought or preparation. However, it is easy to forget important details. Trainers need a written format to guide them while conducting training sessions. The following sections discuss how department trainers should prepare for training.

Write training objectives. Training objectives describe the tasks which the trainee will be expected to demonstrate at the end of the training session. These tasks should already be listed in the job list column of the job breakdown. In addition, training objectives should clearly state that the only acceptable performance is 100% conformity to standards. Tell employees what the training objectives are at the beginning of the training session.

Develop lesson plans. Write step-by-step lesson plans for demonstrating the tasks which the employee is expected to learn. It should be easy to develop lesson plans directly from the performance standards shown in the second column of the job breakdown.

Decide on training methods. Determine which methods will be most appropriate for accomplishing the training objectives. Whenever possible, demonstrate the tasks to be learned and provide step-by-step visual aids.

Establish a timetable for instruction. Decide how long the training session will take. Then, determine when the volume of work will be such that training can be conducted without interruption or distraction.

Select the training location. Conduct job-related training at the work station(s) where the work will be performed. Determine when the training can best occur without interfering with daily housekeeping and hotel functions. Make sure that the employee's position at the training location will provide an unobstructed view of the demonstrated tasks. Also, the employee should be able to see the demonstration from the actual position from which he/she will be performing the task(s). If an employee stands facing the trainer, every movement will appear to be the reverse of the way it is actually performed. This may seem like a minor point, but such a situation can become so frustrating that an employee may develop resistance to learning the task(s).

Assemble training materials and equipment. Gather copies of the appropriate job list and job breakdowns. These are the most important materials needed for the training session. The job breakdowns should indicate other materials and/or equipment required for teaching a particular task. Set up all materials and necessary equipment in the area in which the training is scheduled.

Set up the work station. If a work station will be used as the training location, lay it out exactly the way it is usually stocked. Position each piece of equipment in the same way the employee must operate or maintain it.

Conduct the Training. After preparation, actual training can begin. The following sections suggest guidelines for presenting the training.

Prepare the trainees. Present an overview of the training session and help the trainees become motivated to learn. Explain the training objectives for the session and let employees know exactly what is expected. Tell them why the training is important, how it relates to the job, and how they will benefit from it. Help them understand how the objectives relate to the total responsibilities of that particular position in the housekeeping department. Emphasize how trainees can make immediate use of the skills they learn.

Begin the training session. Use the job breakdowns as a training guide. Encourage trainees to study the job breakdowns so that they will know the standards by which you or the department will evaluate their performance. It may be necessary to have job breakdowns translated in another language. Follow the sequence of each step in the performance standards column of the job breakdowns. Tell trainees how to do each step and why each is important.

Demonstrate the procedures. As you review each step, demonstrate how to do it. Trainees will understand and retain more by seeing a demonstration than by listening to a lecture. Encourage them to ask questions at any time they do not fully understand the demonstration or explanation.

Avoid jargon. Use words that trainees can understand. Jargon and technical terms that may be familiar to some may seem like a foreign language to a person new to the job.

Take adequate time. Proceed slowly. Remember that new employees are seeing and hearing many things for the first time. Carefully show and explain everything they should know about each step. Some trainers find it difficult to slow the pace of their instruction and maintain it at a level appropriate for new employees. Try not to become frustrated if trainees do not understand each step as well or as soon as you expect.

Repeat the sequence. Go through the entire sequence twice to ensure that trainees know the process thoroughly. When going through the process for the second time, ask questions to check on comprehension. Follow the job breakdowns and repeat the steps as often as necessary until trainees know the procedure.

Coach Trial Performances. When trainer and trainees agree that they are familiar with the job and able to complete the steps acceptably, trainees should try to perform the tasks alone.

Immediate practice results in good work habits. Have each trainee demonstrate each step of the task presented during the training session. Ask him/her to explain each step while performing it. This will help you check for comprehension.

Coaching will help the employee gain the skill and confidence necessary to perform the job. Compliment the employee immediately after correct performance; correct the employee immediately when you observe problems. Bad habits formed at this stage of training may be very difficult to break later. Be sure that the trainee understands and can explain not only how to perform each step, but also the purpose of each step.

Follow Through. A common training mistake is to ignore the need for follow-through until performance appraisal time. Then, the manager looks at each employee's performance and discovers that the procedures learned in training have not been followed since shortly after training ended. There was no follow-up on the training to coach employees to apply what they learned and to maintain performance standards. Follow-through involves coaching, reinforcement, feedback, and evaluation.

Continue positive reinforcement. Reinforcement reminds employees of what they have learned. Reinforcement may take the form of verbal feedback during or after training.

When an employee is on the job and strays from performance standards set by the job breakdowns, first compliment him/her for performing some of the tasks correctly. Then, guide the employee back to the correct procedures. This technique will improve employee performance and help the employee to develop a positive attitude.

Provide constant feedback. Tell employees what is correct as well as what is incorrect about their performance. An employee usually has questions about new tasks which he/she has learned. Encourage questions and discuss ways to improve performance and efficiency. A new employee also needs to know where to go for help when the trainer is not available.

Coach a few tasks each day. A new employee can retain only a limited amount of new information at a time before becoming tired and frustrated. In one training session, teach what the employee can reasonably master and allow time for practice. Teach additional material in subsequent sessions until all of the job responsibilities are covered.

Evaluate the employee's progress. Evaluate training efforts in terms of whether the trainee accomplished the training objectives. Did the employee actually demonstrate the behavior specified before training began? If not, provide additional guidance and practice.

Scheduling

Since labor is the greatest single housekeeping expense, one of the most important managerial functions of the executive housekeeper is to ensure that the right number of employees is scheduled to work each day. When too many employees are scheduled to work, the department is overstaffed. Overstaffing results in excessive labor costs that decrease hotel profits. When too few employees are scheduled to work, the department becomes understaffed. While understaffing decreases labor costs, it may also decrease hotel profits because performance standards will not be met, resulting in dissatisfied guests and lost business.

The first step toward efficient scheduling is to determine which positions within the housekeeping department are fixed and which are variable in relation to changes in occupancy levels at the hotel.

Fixed staff positions are those that must be filled regardless of the volume of business. These positions are generally managerial and administrative in nature and may include the following:

- Executive housekeeper
- Assistant executive housekeeper
- Supervisor (day shift)
- Department clerk (day shift)
- Department clerk (afternoon shift)

Employees occupying these positions are usually scheduled to work at least 40 hours a week, regardless of the occupancy level of the hotel.

The number of **variable staff positions** to be filled varies in relation to changes in hotel occupancy. These positions include:

- Room attendants (day and afternoon shift)
- Housepersons (day and afternoon shift)
- Inspectors
- Lobby attendants

The number of employees scheduled to work in these positions is determined primarily by the number of rooms occupied during the previous night. Generally, the higher the previous night's occupancy, the more employees must be scheduled to work the next day. The number of housepersons and lobby attendants needed for any given shift may also vary in relation to meeting room and banquet functions, convention business, and the volume of business at the hotel's restaurants.

In order to schedule the "right" number of employees occupying variable staff positions within the department, the executive housekeeper should develop a staffing guide.

The Staffing Guide

A **staffing guide** is a scheduling and control tool that enables the executive housekeeper to determine the total labor hours, the number of employees, and the estimated labor expense required to operate the housekeeping department when the hotel is at specific occupancy levels.

Developing a Staffing Guide for Room Attendants. The following sections present a step-by-step procedure for developing a staffing guide for day-shift room attendants at the fictional King James Hotel, a 250-room property providing mid-range services.

Housekeeping Success Tip

Ruth Ferguson
Executive Housekeeper
Holiday Inn South
St. Louis, Missouri

❝People traditionally like to plan family activities on weekends during the summer months, and that means summer weekends are the busiest times for hotels. That also means that hotel employees will be needed to work weekends, and that can hamper planning activities with their own families.

To improve morale during the summer months at the Holiday Inn South, we arrange our schedule so that all employees are given one weekend per month off during the summer months (May through September). Because our hotel runs at 100% occupancy during these months, weekdays are as busy as the weekends. Scheduling can therefore be rotated so that each staff member can have one weekend per month off.

We also allow staff to choose which weekend they want off. On the 20th of each month, a sign-up sheet for the following month is passed around. Each member signs up for a weekend.

Knowing in advance that they have a weekend off gives staff members a chance to make plans with their families. Besides being a morale booster, this program has eliminated last-minute absences as well as many requests for days off.

In fact, this scheduling system has been so successful that I continued it during the slower months of October and November so I could control the number of staff off on weekends.

Because this scheduling plan eliminates unexpected staff shortages due to call-offs, you can accurately predict how many room attendants you'll have to work on any given weekend by using the following equation:

Total number of room attendants ÷ Number of weekends in the month = Total number of housekeepers off each weekend

Say you have 40 room attendants, and there are four weekends in the month you are scheduling for. Divide 40 by four and you get 10 room attendants off each weekend. If there is an extra weekend in the month, it goes to senior staff.❞

Step 1. The first step is to determine the total labor hours for positions that must be scheduled when the hotel is at specific occupancy levels. This can be calculated by using **productivity standards.** Let's assume that the productivity standard for day-shift room attendants at the King James Hotel is approximately 30 minutes (.5 hours) to clean one guestroom. Given this information, we can calculate the total number of labor hours required for day-shift room attendants at various hotel occupancy levels.

For example, if the hotel is at 90% occupancy, there will be 225 rooms to clean the next day (250 rooms × .9 = 225). It will take a total of 113 labor hours to clean them (225 rooms × .5 hours = 112.5 labor hours, rounded to 113). At 80% occupancy, there will be 200 rooms to clean (250 rooms × .8 = 200). It will take 100 labor hours to clean them (200 rooms × .5 hours = 100 labor hours).

Step 2. Determine the number of employees that must be scheduled to work when the hotel is at specific occupancy levels. The staffing guide expresses this number only in relation to full-time employees.

Since the productivity standard is .5 hours to clean one guestroom, a day-shift room attendant at the King James Hotel is expected to clean 16 guestrooms during an 8-hour shift. Given this information, the number of full-time, day-shift room attendants that must be scheduled at different occupancy levels can be determined by dividing the number of occupied rooms by 16.

For example, when the hotel is at 90% occupancy, there will be 225 rooms to clean the next day. Dividing 225 rooms by 16 indicates that 14 full-time, day-shift room attendants will be needed to clean those rooms (225 rooms ÷ 16 = 14.06, rounded to 14). When the hotel is at 80% occupancy, there will be 200 rooms to clean. It will take 13 room attendants to clean them (200 rooms ÷ 16 = 12.5 room attendants, rounded to 13).

The actual number of room attendants scheduled to work on any given day will vary depending on the number of full-time and part-time employees the executive housekeeper schedules to work. For example, when the hotel is at 90% occupancy, the executive housekeeper could schedule 14 full-time room attendants, or 10 full-time room attendants (each working 8 hours) and 8 part-time room attendants (each working 4 hours). Either scheduling technique would approximate the required total of 113 labor hours.

Step 3. The third step in developing a staffing guide is to calculate the estimated labor expense required to operate the housekeeping department when the hotel is at specific occupancy levels. This can be done for day-shift room attendants at the King James Hotel simply by multiplying the total labor hours by the average hourly rate for room attendants. Assuming the average hourly rate to be $5.00, when the King James Hotel is at 90% occupancy, the next day's estimated labor expense for day-shift room attendants is $565 (113 total labor hours × $5.00/hour = $565). Regardless of the combination of full-time and part-time employees that are eventually scheduled to work, the total labor expense for day-shift room attendants should not exceed $565, when the hotel is at 90% occupancy.

Developing a Staffing Guide for Other Positions. Similar calculations must be made for other variable staff positions in the housekeeping department. Let's assume that the executive housekeeper at the King James Hotel has reviewed the productivity standards for other positions and has determined that during one 8-hour shift:

- One inspector is needed for every 80 occupied rooms, yielding a productivity standard of .1 (8 hours ÷ 80 occupied rooms = .1).

- One day-shift lobby attendant is needed to service public areas when 100 rooms are occupied, yielding a productivity standard of .08 (8 hours ÷ 100 occupied rooms = .08).

- One day-shift houseperson is needed for every 85 occupied rooms, yielding a productivity standard of .094 (8 hours ÷ 85 occupied rooms = .094).

- One afternoon-shift room attendant is needed for every 50 occupied rooms, yielding a productivity standard of .16 (8 hours ÷ 50 occupied rooms = .16).

- One afternoon-shift houseperson is needed for every 100 occupied rooms, yielding a productivity standard of .08 (8 hours ÷ 100 occupied rooms = .08).

These productivity standards are multiplied by the number of occupied rooms to determine the total labor hours required for each position when the hotel is at specific occupancy levels. Dividing the total labor hours for each position by 8 determines the number of full-time employees that must be scheduled to clean the hotel the next day. Multiplying the required labor hours by an average hourly rate determines the estimated labor expense for each position. Exhibit 6 presents the complete staffing guide for variable staff positions at the King James Hotel.

Developing Employee Work Schedules

Occupancy forecasts generated by the rooms division are used in conjunction with the staffing guide to determine the "right" number of employees to schedule each day for every position in the housekeeping department. Executive housekeepers have found the following tips helpful when developing employee work schedules:

- A schedule should cover a full work week, which is typically defined as Sunday through Saturday.

- Schedules should be posted at least three days before the beginning of the next work week.

- Days off, vacation time, and requested days off should all be indicated on the posted work schedule.

- The work schedule for the current week should be reviewed daily in relation to occupancy data. If necessary, changes to the schedule should be made.

- Any scheduling changes should be noted directly on the posted work schedule.

- A copy of the posted work schedule can be used to monitor the daily attendance of employees. This copy should be retained as part of the department's permanent records.

Alternative Scheduling Techniques. Alternative scheduling, as its name implies, involves a staffing schedule that varies from the typical 9:00 A.M. to 5:00 P.M. workday. Variations include part-time and flexible hours, compressed work schedules, and job sharing.

Part-time employees. Part-time employees may include students, young mothers, or retirees who for various reasons are unable to work full-time hours.

Exhibit 6   Sample Variable Staffing Guide

King James Hotel										
Occupancy %	100%	95%	90%	85%	80%	75%	70%	65%	60%	55%	50%
Rooms Occupied	250	238	225	213	200	188	175	163	150	138	125
Room Attendants (A.M.)											
(Productivity STD = .5)											
Labor Hours	125	119	113	107	100	94	88	82	75	69	63
Employees	18	17	16	15	14	13	12	12	11	10	9
Expense	$625	$595	$565	$535	$500	$470	$440	$410	$375	$345	$315
Housepersons (A.M.)											
(Productivity STD = .08)											
Labor Hours	20	19	18	17	16	15	14	13	12	11	10
Employees	3	3	3	2	2	2	2	2	2	2	1
Expense	$100	$95	$90	$85	$80	$75	$70	$65	$60	$55	$50
Lobby Attendants											
(Productivity STD = .07)											
Labor Hours	18	17	16	15	14	13	12	11	11	10	9
Employees	3	2	2	2	2	2	2	2	2	1	1
Expense	$90	$85	$80	$75	$70	$65	$60	$55	$55	$50	$45
Inspectors											
(Productivity STD = .09)											
Labor Hours	23	21	20	19	18	17	16	15	14	12	11
Employees	3	3	3	3	3	2	2	2	2	2	2
Expense	$115	$105	$100	$95	$90	$85	$80	$75	$70	$60	$55
Room Attendants (P.M.)											
(Productivity STD = .14)											
Labor Hours	35	33	32	30	28	26	25	23	21	19	18
Employees	5	5	5	4	4	4	4	3	3	3	3
Expense	$175	$165	$160	$150	$140	$130	$125	$115	$105	$95	$90
Housepersons (P.M.)											
(Productivity STD = .07)											
Labor Hours	18	17	16	15	14	13	12	11	11	10	9
Employees	3	2	2	2	2	2	2	2	2	1	1
Expense	$90	$85	$80	$75	$70	$65	$60	$55	$55	$50	$45
Total Labor Hours	239	226	215	203	190	178	167	155	144	131	120
Total Expense	$1,195	$1,130	$1,075	$1,015	$950	$890	$835	$775	$720	$655	$600

Employing part-time workers can give businesses extra scheduling flexibility. It can also reduce labor costs, because the benefits and overtime costs usually decrease.

Flexible work hours. Flexible work hours, or **flextime**, allows employees to vary the times at which they begin and end workshifts. Certain hours of each shift necessitate the presence of all employees. The rest of the shift can be flexible, allowing employees to determine for themselves when their shift begins and ends. Of course, the executive housekeeper must make sure that each hour of the housekeeping day is covered. The benefits of flextime include heightened staff morale, productivity, and job satisfaction. Moreover, the property will be more attractive to quality employees.

Compressed work schedules. Compressed work schedules offer housekeeping employees the opportunity to work the equivalent of a standard work week in fewer than the usual five days. One popular arrangement compresses a 40-hour work week into four 10-hour days. Compressed work schedules are usually inflexible, but many employees prefer inflexible hours for four days with an extra day off per week to flexible hours for five days. Primary benefits from the employer's viewpoint include enhanced recruiting appeal and employee morale, and reduced absenteeism.

Job sharing. In job sharing, the efforts of two or more part-time employees together fulfill the duties and responsibilities of one full-time job. The workers involved usually work at different times, but some overlap of work hours is desirable so that the workers can communicate. Job sharing may alleviate department turnover and absenteeism and increase employee morale. In addition, the department profits, because, even if one job-sharing partner leaves, the other usually stays and can train the incoming partner.

Motivation

Management in any organization is responsible for creating a work environment which fosters the professional growth and development of its employees. This includes providing training, guidance, instruction, discipline, evaluation, direction, and leadership. If management fails to perform these basic functions, employees may become passive, critical, and indifferent to organizational objectives. Such feelings manifest themselves in absenteeism, poor productivity, and high turnover.

A major challenge facing hospitality managers is motivating employees. Current changes in the labor market and the high cost of employee turnover demand that organizations seek ways to retain good employees. One way to accomplish this objective is to practice effective motivational techniques.

What Is Motivation?

Hundreds of definitions probably exist for the term **motivation**. For the purpose of this chapter, motivation can be described as the art of stimulating a person's interests in a particular job, project, or subject to the extent that the individual is challenged to be continually attentive, observant, concerned, and committed. Motivation is the end result of meeting and satisfying those human needs associated with feeling a sense of worth, value, and belonging to an organization or department.

Methods of Motivating Employees

Managers can select a number of ways to motivate employees. The end result should be that an employee's perception of his/her value and worth has increased from engaging in a particular activity. Employees who make a positive contribution to the success of an operation and receive recognition and praise for their efforts will more than likely be motivated, top performers. The following section suggests a number of motivational techniques.

Training. One surefire way to motivate employees is to get them involved in an effective training program. Training sends a strong message to employees. It tells them that management cares enough to provide the necessary instruction and

> ### Housekeeping Success Tip
>
> **Susan Hill**
> Rooms Division Manager
> Holiday Inn—Orange Park
> Orange Park, Florida
>
> ❝Employee recognition and motivation has always been a difficult challenge in the housekeeping department. Unless guests see an employee face-to-face, they don't realize who is the source of their satisfaction.
>
> Here at the Holiday Inn—Orange Park, we have devised a reward program so that room attendants have the opportunity to be recognized for their performance by guests. Each room attendant has a business card with his or her name to leave in the room. The card is placed next to the comment card left for guests. The employee is then awarded five dollars for each time his or her name is mentioned on the guest comment card at the employee luncheon and awards ceremony.
>
> This is an opportunity for a behind-the-scenes employee to be recognized for a job well done.❞

direction to ensure their success with the organization. Training significantly reduces the frustration employees may feel when they do not have a clear-cut idea of what they are doing—or the proper tools and supplies to do the job well. Effective training educates employees about the job itself and about the use of any necessary tools and supplies. Managers should take the time to invest in employee training since it can result in employees who are more productive, efficient, and essentially, easier to manage.

Cross-Training. Cross-training simply means teaching an employee job functions other than those he/she was specifically hired to do. Cross-training has many advantages—both for the employee and the manager. From the employee's perspective, it prevents feeling locked-into a particular job and allows him/her to acquire additional work skills. From the manager's perspective, it increases flexibility in scheduling. Cross-trained employees become more valuable to an organization or department because they can perform many job functions instead of just a few. Cross-training can be a valuable motivational tool and can remove many of the obstacles associated with an employee's growth and advancement.

Recognition. Positive guest comments and repeat business reflect a staff that works together to satisfy guest needs. Managers should relay this type of information to the staff as recognition for a job well-done. Graphs and charts are also effective motivators, since they provide staff with visual cues of achievements and progress. Written room inspection reports are also powerful motivators. Room attendants with high scores may be recognized through a room attendant-of-the-month program or some type of financial reward.

Communication. Communication is a key to any motivational program. Keeping employees informed about goings-on in the department and property will reap positive results. Employees who are aware of events taking place feel a greater sense of belonging and value.

Developing a department newsletter is an excellent way to keep lines of communication open. Some properties allow housekeeping employees to develop the newsletter and provide their own articles. Write-ups might be job-related or personal, including such topics as:

- Promotions
- Transfers
- New hires
- Resignations
- Quality tips
- Special recognition
- Employee-of-the-month
- Birthdays
- Marriages
- Engagements
- Births
- Potlucks
- Upcoming parties

A bulletin board provides a place to post schedules, memorandums, and other pertinent information in a clear, easy-to-understand manner. Bulletin boards are most effective when they are in an area accessible to all employees and when employees are told to view the boards daily.

Incentive Programs. Employees in almost every organization need special appreciation for the work they perform. Sometimes a simple thank you demonstrates sincerity when an employee's performance meets or exceeds expectations. Other times, it's not enough. An incentive program is one of the most effective methods of rewarding and recognizing employees who excel in their jobs. Several basic guidelines should be considered when developing incentive programs. Managers should:

- Develop an incentive program which is appropriate for the department or organization.
- Outline the specific goals and objectives for the program.
- Define the conditions and requirements which employees must meet in order to receive the recognition and rewards.
- Brainstorm a variety of rewards and get the necessary approvals if financial expenditures are involved.

- Determine the date and time to begin the program. Make sure every employee participates and make it as much fun for them as possible.

Incentive programs offer special recognition and rewards to employees based on their ability to meet certain conditions. Among the rewards managers may consider offering are:

- Commendation letters
- Certificates of appreciation
- Cash bonuses
- Pictures taken with the general manager and the department head that will be posted in public and back-of-the-house areas
- Recognition dinners, potlucks, or picnics
- Dinners for two in the hotel's restaurant
- Gift certificates
- Complimentary suites for a weekend in company-owned or operated hotels in nearby cities or states
- Special parking privileges for 30 days
- Recognition plaques

Incentive programs vary in structure and design and are an excellent way to award exceptional performance beyond the paycheck. Properties should develop and establish incentive programs that result in a "win-win" situation for the employee, the guest, and the company. The program should be challenging and create a spirit of competition among the staff.

There is an element of surprise associated with incentive programs. A pre-shift meeting or departmental staff meeting presents the best opportunities to announce the award recipients. Such announcements should be planned and presented in such a way that the recipients feel very good about themselves and the work they do.

A good incentive program:

- Recognizes and rewards exceptional performance
- Motivates employees to be more productive
- Demonstrates the organization's commitment to guest satisfaction by providing a work environment which encourages employees to take care of the guest
- Says thanks for a job well done

Performance Appraisals. Employees need to know where they stand at all times so they can feel secure in their job and know that the boss is pleased with their performance. Consider what happens in the following scenario:

> Sam applied for the position of night cleaner in a large downtown convention hotel. During his interview with Mr. Doe, the executive housekeeper, he demonstrated enthusiasm about the job and promised that the hotel would be

Housekeeping Success Tip

Virginia Manso, CEH, CHHE
Executive Housekeeper
Registry Hotel
Charlotte, North Carolina

"I believe that every executive housekeeper is only as good as the people who work for him or her. A well-trained and motivated staff is the key to an efficient and well-organized housekeeping department. And, of course, what motivates a lot of us are rewards and sense of pride.

At the Registry Hotel, the general manager and I have developed an incentive program that has been used with great success. This incentive program for chambermaids and inspectresses accomplishes a two-fold purpose. First, it rewards and recognizes employees based on individual achievements. Second, the program greatly encourages professionalism—both in the way employees perceive themselves and in the way guests perceive our property.

The incentive program uses a comprehensive inspection procedure to measure the achievements of individual employees. Once a month, the executive housekeeper uses what we call the 'White Glove Inspection Form' to inspect the guestrooms cleaning jobs of each chambermaid. Up to 10 rooms are inspected per employee. This form consists of an itemized list divided between the bedroom and bathroom. The employee receives a certain number of points for each area that is cleaned correctly. For example, an employee would receive four points if the bed is made neatly with a clean spread, or three points if the tub is clean, dry, and polished—and void of soap scum or hair. The total possible points an employee can earn is 100—52 for the bedroom and 48 for the bath. After each inspection, points are tallied and recorded on a chart in the housekeeping department. This way, each employee can keep track of individual progress and see the overall progress of the entire department. This system is especially motivational as scores improve.

To ensure constant attention and effort on the part of all employees, incentives revolve around the scores achieved on the inspection forms. The incentive and corresponding scores are as follows:

Total Cumulative Score of 90 to 100

First Month Incentive
    Full day off with pay

Quarterly Incentive
    Cash bonus

Mid-Year Incentive
    A payroll bonus per hour which would remain in effect as long as the cumulative inspection scores are 90 or more.

(continued)

Housekeeping Success Tip *(continued)*

Total Cumulative Score of 85 to 89

First Month Incentive
A half day off with pay

Quarterly Incentive
Cash bonus

Mid-Year Incentive
A payroll bonus per hour which would remain in effect as long as the cumulative inspection scores are 85 to 89.

Overall, this incentive program has encouraged employees to strive for a truly higher level of cleanliness and to pay more attention to detail in their day-to-day cleaning activities. **"**

more than pleased with his services, Mr. Doe hired Sam to work specifically on tile floors during the night shift.

Sam worked very hard to please his boss. But even though the floors were beautiful and shined, there was a small problem: in the three months' time since he'd been hired, Sam hadn't heard one comment from his immediate supervisor. One day he decided to disguise his voice and call the hotel. When connected to housekeeping, he asked to speak with Mr. Doe.

"Mr. Doe," he said when the executive housekeeper answered, "My name is Jim. I noticed several weeks ago that you ran an ad in the newspaper for a night shift cleaner."

"Yes we did," Mr. Doe replied.

When "Jim" asked if the position had been filled, the executive housekeeper said that it had.

"Tell me," Sam asked, "how is the employee doing?"

"The young man is doing a wonderful job," Mr. Doe said enthusiastically. "He's improved the appearance of the hotel 110%. He has excellent floor-care skills and everyone likes him. He's productive and gets the job done effectively every night he works."

"That's wonderful," Sam said.

"Why do you ask?" inquired Mr. Doe.

Pausing, Sam continued with a friendly laugh, "Because this is Sam—the guy you hired. I just wanted to see how I was doing!"

Interaction between an employee and a manager can affect an employee's perception of the job and an employee's self-image. A **performance appraisal** is one of the best tools a manager can use to increase employee motivation and morale. The reason this particular technique is so effective is because it:

- Provides the employee with formal written feedback on his/her job performance

- Identifies strengths and weaknesses in performance and provides plans and actions for improvement

- Gives the manager and the employee the opportunity to develop specific goals and due dates to accomplish the desired results

- Recognizes and rewards outstanding performance through possible promotions, wage increases, and additional responsibilities

- Reveals—in some cases—whether that employee is actually suitable for the position

An effective performance appraisal focuses on an employee's job performance and the steps the employee can take to improve job skills and performance. Appraisals should be fair, objective, and informative. While it is important to point out weaknesses, it is not necessary to dwell on them. The experience should be positive for the employee. When the process is completed, the employee should clearly understand the areas in which he/she is doing very well and those in which he/she needs to improve.

There are numerous methods and techniques for evaluating employee performance. Every organization must tailor its own appraisal program to meet particular needs and goals.[2]

Employee Discipline

Managers and supervisors sometimes find themselves in hot water because disciplinary action was administered to an employee with unjust cause. Employers often find themselves in an unfavorable position because of their inability to justify and defend their actions when challenged in court. There has been a tremendous increase in the number of lawsuits being filed against employers for sexual harassment, racial and age discrimination, unfair labor practices, and unjust terminations. Government agencies such as the Equal Employment Opportunity Commission (EEOC) and the Civil Rights Commission have been established to protect the rights of employees in the workplace.

People have the right to work in an environment free of harassment and discrimination. When the courts find employers guilty of such charges, stiff penalties and fines are often imposed. Many companies have been ordered to pay back wages and reinstate employees who were terminated for unjust cause. Others have paid thousands of dollars in punitive and compensatory damages. While an employer cannot stop an individual from seeking legal redress for what he/she considers unfair treatment, an employer can provide a peaceful, fair, and non-hostile work environment for all employees.

At some point, most managers are faced with the unpleasant task of disciplining an employee. Many managers and supervisors avoid such confrontations, hoping that the problem will disappear or correct itself. In some instances, problems can be corrected without a formal disciplinary process. Others cannot. When a manager fails to address improper behavior, the behavior is reinforced and encouraged. This sets the stage for accusations of favoritism, discrimination, and unfair practices.

Elements of an Effective Disciplinary Program

Every organization should have a formal disciplinary action program that addresses and corrects undesirable employee behavior. The goal of any disciplinary program should be to turn a wayward employee in the right direction. The purpose is not to "nail the employee to the wall," but to continue to help the employee correct performance problems.

Managers should ask themselves the following questions when assessing the effectiveness of a property's disciplinary program:

- Is the disciplinary program fair, firm, and consistent?

- Do all employees know and understand the policies and procedures? Do they know the rules and regulations and the consequences for violations?

- If the courts or some other government agency investigated the property's personnel files, would they find evidence of unfair hiring practices, discriminatory comments, or inconsistent application of disciplinary action for the same violations? Would they find female and minority employees being paid less for the same workload as non-minority and male employees?

- Do managers assist all employees in correcting negative behavior?

- Do the employees feel that the punishment fits the infraction? Are some employees fired for certain infractions while others simply have their hands slapped?

The greatest safeguard against discrimination, or any other type of charge, is to make sure that disciplinary action is fair, firm, and consistent—and that all employees know and understand the rules and regulations.

The goal of an effective disciplinary action program should be to provide a system that addresses and corrects improper behavior and helps employees become productive members of the organization. Many organizations have a formal written disciplinary program which outlines specific rules and regulations and the consequences for violating them. A written policy is beneficial to employers if they are challenged in court to prove that every means was exhausted to salvage an employee. A standard policy also demonstrates that the organization extends the same opportunity to every employee—not just a chosen few. Before involving any employee in the formal disciplinary process, a manager should consider the benefits of counseling.

Informal Counseling

Counseling is a process of one-on-one problem-solving during which a manager helps an employee seek solutions to his/her own problems. Counseling requires skill. It means permitting an employee to develop insight into his/her own behavior, attitudes, and perceptions. Counseling gives a manager and an employee an opportunity to sit down together and discuss whatever problems are present. It gives an employee the opportunity to "lay all the cards on the table." After problems have been identified, the manager and the employee agree on a solution and schedule a follow-up conference to discuss the employee's progress.

While lengthy written documentation is not necessary during the first counseling session, it is a good idea to document that a discussion did take place by noting the date, time, and a brief description of the content. The manager should keep this record for reference. Essentially, counseling is designed to correct a problem without formal disciplinary measures. In many cases, this informal conversation is all that is necessary to get an employee on the right track.

Progressive Discipline

The process of **progressive discipline** addresses undesirable behavior by administering progressively sterner measures for repeated infractions. Progressive discipline starts by correcting improper behavior immediately, using a positive tone and attitude. Initial action should take place in an informal setting and manner. If undesirable behavior continues, the disciplinary approach becomes more structured and the techniques more specific.

For the most part, progressive discipline involves four basic steps: the spoken warning, the written warning, suspension, and termination.

Spoken Warning. When attempts at counseling fail, and the undesirable behavior still exists, managers must use a stronger method to get the message across. A spoken warning is the first step in the formal disciplinary action process. The primary difference between counseling and a spoken warning is that, with a warning, a more aggressive presentation is made of the facts—and of the corrective actions. The employee should be told how his/her behavior is affecting the other employees and the operation. The employee should also be told that if the infraction occurs again, more severe disciplinary action may take place. The manager should offer help, support, and assistance. The manager should also document the occurrence of the conversation and arrange a follow-up date to discuss any progress.

Written Warning. When counseling and a spoken warning prove ineffective, a manager must move to the next step—the written warning. In this step, everything relative to the problem should be written down on a disciplinary action form. This form should describe dates, times, and violations of the rules. It should also state the dates on which the problem was previously discussed with the employee.

The employee should be asked to sign the written warning. This acknowledges that the problem has been addressed for the third time—once informally and twice formally. If the employee refuses to sign, another manager or the human resources director should note the employee's refusal on the warning notice. Copies of all documentation should be placed in the employee's personnel file. A sample employee warning notice is shown in Exhibit 7.

Suspension. Suspension—the temporary layoff of an employee without pay—has traditionally been used as a method of punishment. Today, it is more commonly used as a step toward termination, pending investigation. In many organizations, suspensions serve as a "cooling off period" and give management the opportunity

Exhibit 7   Sample Employee Warning Notice

EMPLOYEE WARNING NOTICE

EMPLOYEE NAME_____   DEPARTMENT_____

DATE OF PROBLEM_____   TIME OF PROBLEM_____

NATURE OF PROBLEM
(Please Check)

[   ] Absence / Lateness   [   ] Carelessness   [   ] Work Performance

[   ] Violation of Safety Rules   [   ] Conduct   [   ] Breaking Hotel Rules

[   ] Attitude   [   ] Insubordination   [   ] Other

EXPLANATION REGARDING THIS WARNING (BE SPECIFIC: STATE EXACTLY WHERE, WHEN, WHAT, INCLUDING DATES, ETC.)

_____

_____

_____

_____

_____

ACTION TAKEN:_____

_____

_____

PREVIOUS WARNING? [   ]No [   ]Yes   If Yes, Date_____

[   ] FINAL WARNING: The employee has been told that any further violation of this type, or any other infractions of company policy may result in termination.

NOTE TO EMPLOYEE: Employees who have a problem affecting job performance are encouraged to seek help on a confidential basis by contacting the Employee Assistance Center at _____, or the Personnel Office at _____ , for additional information about the program.

I HAVE DISCUSSED THIS MATTER WITH THE ABOVE EMPLOYEE ON

_____ at _____ Signature:_____
   Date   Time   Supervisor

YOUR SIGNATURE ON THIS FORM INDICATES ONLY THAT YOU HAVE BEEN ADVISED OF THIS WARNING NOTICE. IF YOU FEEL THAT THE WARNING IS UNFAIR, YOU ARE ENCOURAGED TO TALK TO THE PERSONNEL OFFICE.

_____
                       Employee Signature

Source: David Wheelhouse, *Managing Human Resources in the Hospitality Industry* (East Lansing, Mich.: Educational Institute of the American Hotel & Motel Association, 1989), p. 355.

to conduct an investigation. If the suspended employee is deemed innocent after the investigation is completed, he/she is brought back with full compensation for those days he/she was suspended. If, however, the investigation reveals that the employee did violate company rules, and that allowing the employee to return would severely hamper day-to-day business, the employee may be terminated immediately without compensation.

Termination. Termination is the final step in the progressive discipline process. This action permanently severs an employee from the organization and should only be used when every other method to assist the employee has failed. Under normal circumstances, termination should occur if the following situations and steps are documented in the employee's personnel file:

1. Informal counseling session

2. Spoken warning

3. Written warning

4. Suspension (and investigation)

In most organizations, an employee can be terminated immediately without any prior disciplinary action for such offenses as fighting, stealing, damaging company property, immoral conduct, and other such infractions. **House rules** governing such infractions should be outlined in the employee handbook. Employees should know the consequences of violating house rules. Exhibit 8 shows a sample set of house rules.

Guidelines for Disciplinary Action

Managers should follow several basic principles—and consider a number of variables—before, during, and after the administration of any formal disciplinary action. Some of these guidelines follow.

Conduct a thorough investigation of the incident. All facts should be gathered and analyzed. Determine an appropriate action to address and correct the situation with the employee(s) involved.

Document the action. Documentation is the single most important element in the disciplinary process. Documentation should be made of every disciplinary action and placed in the employee's personnel file. This is part of building a defense which will stand up in court should it come to that point. Failure to do so can place the manager and the organization in an unfavorable position.

Be firm, fair, and consistent. Disciplinary action should be applied with the same strength and severity to *everyone* who breaks the rules. This does not mean disciplining an employee every time he/she makes a mistake. It simply means identifying problems and moving swiftly to make corrections.

There is a big difference between making mistakes and deliberately violating the rules and regulations. Employees should perceive managers as being consistent and fair in all dealings. Managers must guard against being perceived as having favorites. Managers should never be accused of writing up one employee and not another. Managers should also treat everyone with dignity and respect. Yelling,

Exhibit 8   Sample Set of House Rules

Strict enforcement of these policies will help protect our employees and ensure that our hotel runs in an efficient manner. Listed below are some of the violations which may result in immediate suspension or termination, at the option of the hotel.

- Being discourteous, rude, insubordinate, or using abusive language to a guest or fellow employee.

- Fighting, stealing, unauthorized possession of hotel property, or gambling on hotel premises.

- Unauthorized use of alcohol, possession, use, or appearance of being under the influence of alcohol, narcotics, intoxicants, or other substances prohibited by law, or the abuse of medication whether obtained legally or illegally, while on hotel premises.

- Possession of lethal weapons or other items prohibited by law while on hotel premises.

- Indecent, immoral, or disorderly conduct in the hotel, including willful destruction of property and failure to follow safety procedures.

- Falsification of work or time records, reports, or guest checks.

- Being in an unauthorized area of the hotel while working or in a non-public area of the building after hours without prior permission from your department head.

- Socializing with guests on hotel premises.

- Removing anything from the hotel without permission.

- Sleeping while on duty.

Source: David Wheelhouse, *Managing Human Resources in the Hospitality Industry* (East Lansing, Mich.: Educational Institute of the American Hotel & Motel Association, 1989), p. 353.

screaming, and using abusive language is not appropriate or indicative of professional behavior.

Stick to the decision. Stick to whatever disciplinary action is determined to be appropriate. Do not be insensitive to the employee, but do not hesitate to lay all the cards on the table. If in doubt about a disciplinary decision, arrange a conference with the appropriate superior.

Be careful of unwritten rules. A property should never have any unwritten rules. At best, such rules involve personal preference rather than the cold hard facts. Unwritten rules do not hold water in court. If an employee is penalized for violating an unwritten rule, it will be extremely difficult to ever prove to an EEOC judge or in court that the employee knows which rule he/she violated. If a situation is important enough to require discipline, it is important enough to put in writing.

Support the employee, not the behavior. Keep in mind that the employee as an individual is separate from the behavior. The goal of behavior modification is to assist individuals in changing, altering, or adjusting specific behaviors. When this is accomplished, the employee will not feel attacked personally. Avoid making statements about an employee's "bad attitude," because such terminology may be perceived as suggesting a character or personality flaw.

Conducting a Disciplinary Conference

When it is necessary to hold a disciplinary conference, a manager should prepare for the meeting by reviewing the personnel file of the employee being disciplined. For purposes of documentation, the manager should have at hand the appropriate form to record dates, times, and violations, if the property has such forms. In conducting the conference, the manager should remember to:

- Greet the employee, and create a warm, receptive atmosphere. Explain the purpose of the conference.

- Review the problem. Explain in detail *exactly* what the problem is in a specific and objective manner. The employee should know how this problem affects the operation, the department, and the organization.

- Ask the employee for personal input, and give the employee a chance to explain the problem from his/her perspective. Be calm and understanding. Most of all, listen.

- Ask the employee to develop a plan of action to correct the undesirable behavior. Help the employee come up with ideas if the employee has difficulty doing so.

- Summarize. Address commitments step by step so that there is a clear understanding of who will be responsible for what. Establish a follow-up date to assess the progress that is or is not being made.

- Explain that there must be written documentation to support the discussion and that formal written procedures are necessary. Allow the employee to read and sign the disciplinary action report. The signature acknowledges that the conversation took place as well as the promises and commitments that were made.

- Explain what the next course of action will be if the undesirable behavior doesn't change. This could include another written warning, suspension, or termination—depending on the organization's policies and procedures.

Concluding the conference is a critical part of the process. The employee may feel emotionally wounded. He/she may be upset or even angry. This is an excellent time to express confidence in the employee, and to reassure him/her that you will be available to help in any way possible.

After the conference, written documentation should be sent to the personnel office and placed in the employee's personnel file. It is important that the documentation be comprehensive and easy to understand, since it demonstrates that steps are being taken to help the employee change his/her behavior.

Endnotes

1. A complete discussion of the training function is beyond the scope of this chapter. For a detailed treatment of this topic, see Lewis C. Forrest, *Training for the Hospitality Industry*, 2nd ed. (East Lansing, Mich.: Educational Institute of the American Hotel & Motel Association, 1989).

2. For a more complete discussion on developing a performance appraisal program in hospitality operations, see David Wheelhouse, *Managing Human Resources in the Hospitality Industry* (East Lansing, Mich.: Educational Institute of American Hotel & Motel Association, 1989), pp. 295–314.

Key Terms

alternative scheduling

coaching

counseling

cross-training

fixed staff positions

four-step training method

house rules

incentive program

job breakdowns

job lists

motivation

performance appraisal

performance standards

productivity standards

progressive discipline

reinforcement

scheduling

spoken warning

staffing guide

suspension

termination

training plan

variable staff positions

written warning

Discussion Questions

1. How does a job list differ from a job breakdown?

2. What is the relationship of performance standards to job lists and job breakdowns?

3. What is the last step of the four-step training method, and why is it important?

4. What is the difference between fixed staff and variable staff positions?

5. What purpose does a staffing guide serve?

6. What are four alternative scheduling techniques?

7. Why do both properties and employees welcome cross-training? What are the benefits of cross-training to both properties and employees?

8. How could an incentive program be used as a motivational technique for housekeeping employees?

9. The most effective performance appraisals are conducted for what reason(s)?

10. What are some of the primary considerations in establishing an effective disciplinary program?

11. What are the four steps in administering progressive discipline?

REVIEW QUIZ

When you feel you have covered all of the material in this chapter, answer these questions. Choose the *best* answer.

True (T) or False (F)

T F 1. The executive housekeeper is responsible for ongoing training programs in the department.

T F 2. The order of tasks on a job list should reflect priority work assignments for the day.

T F 3. Observable and measurable steps that an employee must take in order to accomplish specific tasks are shown in a job list.

T F 4. The necessary detail for setting up effective training and evaluation programs is found in policy and procedures manuals.

T F 5. Lesson plans for specific training sessions can be developed from job breakdowns.

T F 6. When demonstrating a task, the trainer should stand facing the employee.

T F 7. The best location for training is a classroom setting away from the work environment.

T F 8. Compressed work schedules offer housekeeping employees the opportunity to work the equivalent of a standard work week in fewer than the usual five days.

T F 9. Job sharing programs allow two or more part-time employees to fulfill the duties and responsibilities of one full-time position.

T F 10. Counseling is designed to correct a problem without implementing formal disciplinary measures.

Alternate/Multiple Choice

11. Tasks which a trainee will be expected to demonstrate at the end of a training session are called:

   a. policies and procedures.
   b. training objectives.

12. The first step in the formal disciplinary action process is to issue a _____ warning.

   a. spoken
   b. written

13.   The number of variable staff positions to be filled on any given day will largely depend on hotel occupancy for:

a.   the previous night.
b.   the next day.

14.   An example of a fixed staff position in a housekeeping department is:

a.   lobby attendant.
b.   room attendant.
c.   housekeeping supervisor.
d.   house person.

15.   By dividing the number of occupied rooms by the number of rooms cleaned during a work shift, the executive housekeeper can determine the:

a.   estimated labor expenses.
b.   productivity standard for room attendants.
c.   total labor hours.
d.   number of room attendants to schedule.

10/15

Part III

Management Responsibilities of the Executive Housekeeper

Chapter Outline

Par Levels
Linens
    Types of Linens
    Establishing Par Levels for Linens
    Inventory Control of Linens
    Taking a Physical Inventory of Linens
Uniforms
    Establishing Par Levels for Uniforms
    Inventory Control of Uniforms
Guest Loan Items
    Types of Guest Loan Items
    Establishing Par Levels for Guest Loan
    Items
    Inventory Control of Guest Loan Items
Machines and Equipment
    Types of Machines and Equipment
    Establishing Par Levels for Machines
    and Equipment
    Inventory Control of Machines and
    Equipment
Cleaning Supplies
    Types of Cleaning Supplies
    Establishing Inventory Levels for
    Cleaning Supplies
    Inventory Control of Cleaning Supplies
Guest Supplies
    Types of Guest Supplies
    Establishing Inventory Levels for Guest
    Supplies
    Inventory Control of Guest Supplies
    Printed Materials and Stationery

Learning Objectives

1. Define the term par in relation to inventory levels.

2. Identify three factors that must be considered when establishing a par number for linens.

3. Explain how the occupancy report generated by the front desk can be useful in inventory control of linens.

4. Describe the function of a linen discard record.

5. Explain the importance of taking a physical inventory of linens.

6. Identify the function of a master inventory control chart for linens.

7. Identify the factors to consider when establishing par levels for uniforms.

8. Identify the importance of a uniform inventory control form.

9. Identify the factors to consider when establishing par levels for guest loan items.

10. Explain the function of inventory cards and repair logs used in controlling inventories of machines and equipment.

11. Explain how par levels for non-recycled inventory items are established.

12. Explain how a perpetual inventory system is used to control inventories of cleaning supplies.

Managing Inventories

THE EXECUTIVE HOUSEKEEPER is responsible for two major types of inventories. **Recycled inventories** are those items that have relatively limited useful lives but that are used over and over again in housekeeping operations. Recycled inventories include linens, uniforms, guest loan items, and some machines and equipment. **Non-recycled inventories** are those items that are consumed or used up during the course of routine housekeeping operations. Non-recycled inventories include cleaning supplies, small equipment items, and guest supplies and amenities.

This chapter describes the types of inventories maintained by the housekeeping department and explains how par stock levels are established for each type of inventory item. This chapter also discusses important inventory control measures.

Par Levels

One of the first and most important tasks in effectively managing inventories is determining the par level for each inventory item. **Par** refers to the standard number of inventoried items that must be on hand to support daily, routine housekeeping operations.

Par levels are determined differently for recycled and non-recycled inventories. The number of recycled inventory items needed for housekeeping functions is related to the operation of other hotel functions. For example, the par level of linen depends upon the hotel's laundry cycle. The number of non-recycled inventory items a property needs is related to the usage rates of different items during daily operations. For example, par levels of particular cleaning supplies depend upon how fast they are consumed through routine cleaning tasks.

Inventory levels for recycled items are measured in terms of a **par number**—or a multiple of what is required to support day-to-day functions. Inventory levels for non-recycled items are measured in terms of a range between minimum and maximum requirements. When quantities of a non-recycled inventory item reach the minimum level established for that item, supplies must be reordered in the amounts needed to bring the inventory back to the maximum level established for the item.

Linens

Linens are the most important recycled inventory item under the executive housekeeper's responsibility. Next to personnel, linen costs are the highest expense in the housekeeping department. Careful policies and procedures are needed to control the hotel's inventory of linen supplies. The executive housekeeper is responsible for developing and maintaining control procedures for the storage, issuing, use, and replacement of types of linen inventories.

Types of Linen

The executive housekeeper is generally responsible for three main types of linen: bed, bath, and table. Bed linens include sheets (of various sizes and colors), matching pillowcases, and mattress pads or covers. Bath linens include bath towels, hand towels, specialty towels, washcloths, and fabric bath mats. The housekeeping department may also be responsible for storing and issuing table linens for the hotel's food and beverage outlets. Table linens include tablecloths and napkins. Banquet linens are a special type of table linen. Due to the variety of sizes, shapes, and colors, banquet linens may need to be kept separate from other restaurant linens in the inventory control system. The basic principles and procedures for managing linen inventories also apply to blankets and bedspreads.

Establishing Par Levels for Linens

The first task in effectively managing linens is to determine the appropriate inventory level for *all* types of linen used in the hotel. It is important that the inventory level for linens is sufficient to ensure smooth operations in the housekeeping department.

Shortages occur when the inventory level for linens is set too low. Shortages disrupt the work of the housekeeping department, irritate guests who have to wait for cleaned rooms, reduce the number of readied rooms, and shorten the useful life of linens as a result of intensified laundering. Although housekeeping operations run smoothly when inventory levels are set too high, management will object to the inefficient use of linen and to the excessive amount of cash resources tied up in an overstock of supplies.

The par number established for linen inventories is the standard stock level needed to accommodate typical housekeeping operations. One par of linens equals the total number of each type of linen that is needed to outfit all guestrooms one time. One par of linen is also referred to as a **house setup**.

Clearly, one par of linen is not enough for an efficient operation. Linen supplies should be several times above what is needed to outfit all guestrooms just once. Two par of linens is the total number of each type of linen needed to outfit all guestrooms two times; three par is the total number needed to outfit all guestrooms three times; and so on. The executive housekeeper must determine how many par of linens are needed to support efficient operations in the housekeeping department. When establishing a par number for linens, the executive housekeeper needs to consider three things: the laundry cycle, replacement linens, and emergency situations.

The hotel's laundry cycle is the most important factor in determining linen pars. Quality hotels change and launder linens daily. At any given time, large amounts of linen are in movement between guestrooms and the laundry. When setting an appropriate linen inventory level, the executive housekeeper must think through the laundry cycle in terms of the hotel's busiest days—when the hotel is at 100% occupancy for several days in a row. If housekeeping manages an efficient on-premises laundry operation, the laundry cycle indicates that housekeeping should maintain three par of linens: one par—linens laundered, stored, and ready for use today; a second par—yesterday's linens which are laundered today; and a third par—linens to be stripped from the rooms today and laundered tomorrow. Executive housekeepers also need to figure in guest requests for extra linens, and linens for roll-away beds, sofa beds, and cribs.

The laundry cycle in properties that use an outside commercial laundry service will be somewhat longer than the cycle in properties with their own in-house laundry operation. The frequency of collection and delivery services from the commercial laundry will affect the quantities of linen the property needs to stock. The more frequent the service, the less stock is needed to cover the times when the hotel's linen is being transported to and from the laundry service. A typical turnaround time for a commercial laundry is 48 hours. In this situation, the executive housekeeper may need to add another par of linen to cover linens that are in transit between the hotel and the outside linen service. In addition, some commercial laundries will not collect and deliver on weekends. This means that extra stock will be required to cover those days.

The second factor to consider when establishing linen par levels is the replacement of worn, damaged, lost, or stolen linen. Since linen losses vary from property to property, executive housekeepers will need to determine a reasonable par level for linen replacement based on the property's history. The need for replacement stock can be determined by studying monthly, quarterly, or annual inventory reports where losses and replacement needs are documented. A general rule of thumb is to store one full par of new linens as replacement stock on an annual basis.

Finally, the executive housekeeper must be prepared for any emergency situation. A power failure or equipment damage may shut down a hotel's laundry operation and interrupt the continuous movement of linens through the laundry cycle. The executive housekeeper may decide to hold one full par of linens in reserve so that housekeeping operations can continue to run smoothly during an emergency.

Therefore, the hotel's laundry cycle, linen replacement needs, and reserve stock for emergencies suggest that a minimum of five par of linen should be maintained on an annual basis. Properties using an outside commercial laundry service will need to add a sixth par to cover linens in transit.

Exhibit 1 illustrates a sample par calculation for the number of king-size sheets required for a hotel with 300 king-size beds. In this example, 3,000 king-size sheets should be in the hotel linen inventory at all times. Similar calculations need to be performed for every type of linen used in the hotel.

Inventory Control of Linens

To effectively manage linen inventories, the executive housekeeper needs to develop standard policies and procedures that govern how and where linens will be stored, when and to whom linens will be issued, and how to monitor and control the movement of linens through the laundry cycle.

The executive housekeeper needs to cooperate with the laundry manager to maintain an accurate daily count of all linens sent to and received from the laundry.

Exhibit 1   Sample Par Calculation

This is a sample calculation of how to establish a par stock level for king-size sheets for a hotel that uses an in-house laundry operation and supplies two sheets for each of the property's 300 king-size beds.

300 king-size beds × 2 sheets per bed = 600 per par number

One par in guestrooms	1	×	600	=	600
One par in floor linen closets	1	×	600	=	600
One par soiled in the laundry	1	×	600	=	600
One par replacement stock	1	×	600	=	600
One par for emergencies	1	×	600	=	600
Total number					3,000

3,000 sheets ÷ 600 sheets/par = 5 par

Effective communication with the laundry room manager can help the executive housekeeper spot shortages or excessive amounts of linen.

Storage. Much of a hotel's linen supply is in constant movement between guestrooms and laundry facilities. Laundered linens should rest in storage for at least 24 hours before being used. This helps increase the useful life of linens and provides an opportunity for wrinkles to smooth out in permanent press fabrics. Linen is stored in the department's main linen room, in distribution rooms near the laundry, and also in floor linen closets for easy access by room attendants.

Linen storage rooms need to be relatively humidity-free and have adequate ventilation. Shelves should be smooth and free of any obstructions that could damage fabric, and should be organized by linen type. Sufficient room is required to prevent linens from being crushed or crowded. Linen storage rooms should be kept locked, and all standard key control procedures should be followed. Special security measures should be taken with new linens that are stored in the main linen room but have not yet been introduced into service.

Issuing. An effective method for controlling linen is to maintain floor pars for all floor linen closets. A **floor par** equals the quantity of each type of linen that is required to outfit all rooms serviced from a particular floor linen closet. Linen pars should be established and posted in each floor linen closet. Issuing procedures ensure that each floor linen closet is stocked with its par amount at the start of each day.

The occupancy report generated by the front desk can be used to determine linen distribution requirements for each floor linen closet. With information from this report, the executive housekeeper can create a linen distribution list that indicates how much linen is needed to bring each floor linen closet back up to par for the next day. This list functions as a requisition form for replenishing floor linen closet pars. The list is delivered to the laundry manager who sets aside the required amount of clean linens and stores excess clean linens in the laundry distribution room.

Exhibit 2   Sample Linen Control Form

Guestroom Linen to Laundry		
Date _____		
Item	**Color**	**Number**
Pillowcases	Ivory	
King Sheets	Blue	
Queen Sheets	Blue	
Twin Sheets	Ivory	
Bath Mats		
Bath Towels		
Hand Towels		
Washcloths		

Room Attendant _____

Some hotels require room attendants to record the number of soiled linens, by type, that are removed from guestrooms and delivered to the laundry. Exhibit 2 illustrates a worksheet that can be used for this purpose. The total number of linens removed from guestrooms should be consistent with the occupancy report.

At the end of the day shift, a member of the housekeeping evening crew restocks the floor linen closets with the linen set aside by the laundry manager. This brings each floor linen closet back up to full par in preparation for the next day's work. Supervisors can spot-check floor linen closets to ensure that standard procedures have been followed. In this way, linens are issued daily only in the amounts needed to bring each floor linen closet up to its par level.

Special procedures are also required for linen that needs replacement. Any clean linen item that is judged unsuitable due to holes, tears, stains, or excessive wear should not be used in guestrooms. Nor should such damaged linens be placed in soiled linen hampers. Instead, room attendants should place damaged linens in a special discard container and hand-deliver them to the main linen room or housekeeping office. A special linen replacement request form should then be filled out detailing the type of linen involved, the nature of the damage, the linen closet in which it was stored, and the name of the room attendant who noted the damage. The laundry manager will increase the floor distribution count the next day to accommodate the need for replacement.

Clean but damaged linen should be held separately and delivered to the laundry manager (or other appropriate personnel) who determines whether it is unusable or whether it can be repaired. Careful records must be kept of all linen items that are condemned and discarded. Exhibit 3 shows a sample linen discard record that can serve as an important inventory control tool. The linen discard record should be kept in the laundry area and used by employees who sort damaged linens. The form provides columns for recording the specific types and numbers of discarded linen items. At the end of the accounting period, the form is dated and transferred to the executive housekeeper. The executive housekeeper reviews the record and transfers the totals to the master inventory control chart for linens at the time a physical inventory is taken.

Inventory control procedures for table linens should be designed in much the same way as for room linens. A par stock level of all table linens used in each food and beverage outlet should be established. Soiled linens should be counted nightly, and a list of items sent to the laundry should be prepared. Both the laundry manager and the executive housekeeper can use the list as a control and as an issue order for the next day. Each food and beverage outlet should be brought back up to its par level of table linens on a daily basis. Linen needs for special events can be noted on the nightly count sheet and included on the next day's delivery of table linen.

Taking a Physical Inventory of Linens

A physical inventory of all linen items in use and in storage is the most important part of managing linen inventories. A complete count should be conducted as often as once a month. At the very least, physical inventories should be taken quarterly. Typically, the physical inventory is taken at the end of each accounting month to provide the executive housekeeper with important cost control information needed to monitor the department's budget.

As a result of regular physical inventories, the executive housekeeper has accurate figures on the number of all items in use, as well as those considered discarded, lost, or in need of replacement. This control is vital to maintaining a careful budget and for ensuring that the housekeeping department has adequate supplies to meet the hotel's linen needs. The need to replenish the hotel's linen supply is determined on the basis of each physical inventory.

Typically, the physical inventory is conducted by the executive housekeeper and the laundry manager working together. In large hotels, other housekeeping staff may be recruited to help count linen supplies. In most cases, the inventory is taken by staff members working in teams. One person calls out the count for each type of linen while the other records the quantity on an inventory count sheet. It is not unusual for the hotel's controller or a representative of the accounting department to

Exhibit 3 · Sample Linen Discard Record

LINEN DISCARD RECORD

HOUSEKEEPER'S INITIALS: _____   GENERAL MANAGER'S INITIALS: _____   PERIOD ENDING: _____

DATE	Bath Towels	Hand Towels	Wash Cloths	Bath Mats	Shower Curtains	Double Sheets	King Sheets	Double Pillow Case	King Pillow Case	Double Pillow	King Pillow	Double Blanket	King Blanket	Double Matress Pad	King Matress Pad	Double Bed Spread	King Bed Spread	Crib Sheet		
Total Dis-carded																				

(continued)

Exhibit 3   Sample Linen Discard Record (Back of Page)—*(continued)*

DISCARD METHODS

ITEM	HOW DISCARDED	ITEM	HOW DISCARDED

COMMENTS: _____

_____

_____

_____

BY: _____

(EXECUTIVE HOUSEKEEPER)

Courtesy of Holiday Corporation, Memphis, Tennessee

be involved in spot-checking counts and verifying the accuracy of the final inventory report. When the inventory is complete, a final report is sent to the hotel's controller or general manager for final verification and entry.

It is necessary for *all* linens in *all* locations to be included in the count. The executive housekeeper should plan to take the inventory at a time when the movement of linen between guestrooms and the laundry can be halted. This typically means that the inventory is taken at the end of a day shift after the laundry has finished its work, after all guestrooms have been made up with clean linens, and after all floor linen closets have been brought back to their par levels. All soiled linen chutes should be sealed or locked to avoid any further movement of linen.

The next step is to determine all the locations in the hotel where linens may be found. The executive housekeeper needs to take all possible locations into consideration, including:

- Main linen room

- Guestrooms

- Floor linen closets

- Room attendant carts

- Soiled linen bins or chutes

- Soiled linen in laundry

- Laundry storage shelves

- Mobile linen trucks or carts

- Made-up roll-away beds, cots, sofa beds, cribs, etc.

The executive housekeeper should prepare a linen count sheet which can be used to record the counts for every type of linen in each location. Space should be allocated at the top of each count sheet for the date, location, and names of the staff members performing the count. Down the left side of each count sheet should be a listing of every type of linen item to be counted. In making up the inventory list for the count sheets, the executive housekeeper should be sure to differentiate among all types, sizes, colors, and other features. In addition, the counting process will be quicker and easier if the inventory listing is organized in the same way as linen items on storage shelves. Exhibit 4 shows a sample count sheet for recording linen quantities in a floor linen room and on corresponding room attendant carts.

Using the count sheets, two-person teams can conduct the physical inventory at each linen location. One person should count and call out the number corresponding to a particular kind of linen, while the other person records the quantities in the appropriate place on the standard count sheet. A third person might spot-check counts to ensure accuracy.

After the counting process is completed and all standard count sheets have been filled out, the executive housekeeper should collect the sheets and transfer the totals to a master inventory control chart. Once the totals are collected, the results of the inventory can be compared to the previous inventory count to determine actual usage and the need for replacement purchases.

Exhibit 4   Sample Linen Count Sheet

Inventory Count Sheet
Guestroom Linens

Name	Date		Floor	
Item	**Closet**	**Cart 1**	**Cart 2**	**Cart 3**
Pillowcases				
King-Size Sheets				
Queen-Size Sheets				
Twin Sheets				
Bath Mats				
Bath Towels				
Hand Towels				
Washcloths				

Exhibit 5 shows a sample master inventory control chart for linens. Across the top of this master control chart in the first part of the form is a line for listing all inventory items in the hotel's linen supply. This listing should correspond to the listing of linen items used on the standard count sheets.

The second line on the chart identifies the date that a physical inventory was last taken. The executive housekeeper should transfer linen counts from previous inventory records onto this line.

The third line on the chart is used to record the numbers of new linen items received since the last physical inventory. These figures should include both un-opened linen shipments received as well as new linen items that have already been put into use.

The fourth line totals the on-hand quantities from the previous physical inventory (line 2) and the quantity of newly received linen items (line 3).

Next, line five is used to record the number of linen items known to have been discarded since the last physical inventory. These totals can be obtained by examining the linen discard record (see Exhibit 3).

Exhibit 5   Sample Master Inventory Control Chart

HOUSEKEEPING LINEN INVENTORY

LOCATION NAME: _____

LOCATION NUMBER: _____

GENERAL MANAGER'S INITIALS: _____

PREPARED BY: _____

INVENTORY DATE: _____

PART I

1. ITEM
2. LAST INVENTORY DATE (   )
3. NEW RECORD
4. SUBTOTAL 2 + 3
5. RECORDED DISCARDS
6. TOTAL 4 − 5

PART II

7. STORAGE ROOM
8. STORAGE ROOM
9. STORAGE ROOM
10. LINEN ROOM
11. LAUNDRY
12. ON CARTS
13. IN ROOMS
14. ON ROLL AWAYS, CRIBS, ETC.
15. TOTAL ON HAND ADD 7 THROUGH 14

PART III

16. LOSSES 6 − 15
17. PAR STOCK _____ TURNS
18. AMOUNT NEEDED 17 − 15
19. ON ORDER
20. NEED TO ORDER 18 − 19

Courtesy of Holiday Corporation, Memphis, Tennessee

By subtracting the numbers of discarded linens (line 5) from the subtotals (line 4), the executive housekeeper knows the totals for each linen type that can be expected to be on hand. These expected totals are recorded on line 6.

The second part of the form provides spaces for recording the totals counted for every linen type at each of the linen locations. These totals are obtained by tallying and transferring the totals listed for each linen type on the standard count sheets. The quantities of each linen type counted at every location are totaled on line 15 of the form. These figures represent the actual on-hand quantities for every type of linen in the hotel's inventory.

The third part of the form helps the executive housekeeper analyze the results of the physical inventory. By subtracting the counted totals for each linen item (line 15) from the corresponding expected quantities (line 6), the executive housekeeper can determine an accurate number lost for each linen item. This figure is recorded on line 16. Linen loss is the variance between the totals from the previous inventory (plus new purchases received) and the results of the current inventory. While the physical inventory reveals the losses for linen items, it does not show *why* these losses occur. If the variance between expected and actual quantities is high, further investigation is needed.

After each physical inventory, the executive housekeeper should make sure that the par levels are brought back to the levels originally established for each linen item. The par numbers for each linen type are recorded on line 17. These figures represent the standard numbers of each linen type that should always be maintained in inventory. By subtracting the actual quantities of each linen type on hand (line 15) from the corresponding par levels (line 17), the executive housekeeper can determine the quantities of each linen type that are needed to bring inventories back up to par. These amounts are recorded on line 18. By subtracting quantities of linen items that are on order but not yet received (line 19), the executive housekeeper knows precisely how many of each linen type still need to be ordered to replenish the par stock. This figure is recorded on line 20. As a result of the physical inventory, the executive housekeeper can determine both the linens and amounts that are needed to replace lost stock and maintain established par levels.

The completed master inventory control chart should be submitted along with the linen discard record (Exhibit 3) to the hotel's general manager. The general manager will then verify and initial the report before transferring it to the accounting department. The hotel's accounting department will provide the executive housekeeper with valuable cost information related to usage, loss, and expense per occupied room. This information is useful in determining and monitoring the housekeeping department's budget.

Physical inventories of table linen used by the food and beverage department should be handled in much the same way as room linens. The same general rules and procedures should be followed—and the same general forms used. Inventory lists should be prepared for each food and beverage outlet—including banquet facilities—that itemize all types, sizes, and colors of table linens the hotel uses. The inventory should be taken when the movement of table linens to and from the laundry can be halted and each food and beverage outlet is fully stocked to its established par levels. By following the same procedures used for room linens, the total inventory of table linens can be calculated, and the executive housekeeper can determine the need for replacement stock.

Uniforms

Many hotel departments have uniformed staff members. Sometimes, each department is responsible for maintaining its own inventory of uniform types and sizes. More typically, the housekeeping department stores, issues, and controls uniforms used throughout the property. This can be a very complex responsibility—especially in a large hotel with many uniforms of varying types, quantities, and sizes.

Establishing Par Levels for Uniforms

Determining the number of the different types and sizes of uniforms to have on hand can be very difficult. Among the factors which can make the task a true challenge are varying department needs, uneven distribution of size requirements, unavoidable turnover, and unpredictable damage from accidents.

The executive housekeeper can ensure that a sufficient supply of all types of uniforms is placed into service based on information supplied by the various department heads. To establish par levels for all types of uniforms, the executive housekeeper needs to know how many uniformed personnel work in each department, what specific uniforms they require, and how often uniforms need cleaning. Sizing problems can be handled by having uniforms tailored when they are first issued to new employees.

Another factor to consider in establishing par levels is the turnaround time required by the laundry for processing uniforms. The par level for uniforms depends, in large part, on how frequently uniforms need laundering. If, for example, uniforms are laundered only once a week, each employee would have to be issued five uniforms weekly. In this unlikely situation, each employee would exchange five soiled uniforms for five cleaned uniforms at the beginning of each week. Counting the uniform worn on the day of the exchange, this would require maintaining a par of 11 uniforms for each employee.

A more likely scenario would involve laundering uniforms on a daily basis and exchanging a clean uniform for a soiled one each day. In this situation, a minimum of three par of uniforms would be required. One par is worn by employees, another is turned in to be cleaned, and a third is issued in exchange. Daily washing would be more practical and less costly than laundering uniforms on a weekly basis. Every employee would have a spare clean uniform (in addition to the one being worn) in case it was needed during the day. Uniform rooms could be staffed for daily uniform exchanges at the beginning of each shift.

The executive housekeeper may decide that five par is more reasonable. This would keep an adequate supply on hand for new employees and for replacing uniforms for existing personnel. Five par would also ensure that an adequate supply of spare uniforms was available during the day in case of accidents or unexpected damages.

The executive housekeeper may also want to consider the different needs of uniformed personnel across various departments. Since front-of-the-house employees are continuously in the public eye, their need to be neat and clean all day is particularly important. The executive housekeeper may want to maintain higher uniform pars for front office employees since they may need to change uniforms more frequently. Similarly, chefs, stewards, and other kitchen personnel may need two daily changes of uniforms since cleanliness is important for hotel staff dealing directly with food.

In many hotels, the employees themselves are responsible for the maintenance of the uniforms. However, since the law may require that employees be compensated for uniform cleaning, a hotel may be able to reduce costs by processing uniforms through the property's own laundry service.[1]

Inventory Control of Uniforms

All uniforms should be issued and controlled through the uniform room. Adequate storage space should be provided for stocking the different sizes and quantities of uniforms. In addition, the room should be well-organized. Uniforms should be categorized by department so as to save time and hassle when employees exchange uniforms at the start and end of each shift. The executive housekeeper needs to establish specific operating procedures for uniform control. A system should be in place for receiving uniforms for cleaning and for providing both the uniform room and laundry with a record of items received. The uniform room attendant should submit to the laundry a daily record of uniforms to be cleaned that day.

For control purposes, most hotels establish a policy whereby a clean uniform is issued only when exchanged for a soiled one. In some hotels, uniforms will be issued to an employee only with a special request form signed by a department manager.

Employees should receive a receipt when they turn in a uniform. If an exchange system is used, the employee receipt may simply be a clean uniform. When uniforms are first issued, all employees should acknowledge in writing the number and type of uniforms received. A card similar to Exhibit 6 can be used for this purpose and kept on file for each uniformed employee. Such uniform control cards can be kept by the uniform room itself, or by the employee's assigned department. Some properties also keep uniform records in the employee's personnel file. A master record of all uniforms issued to employees should be kept by the housekeeping department.

At the time of issue, the employee assumes full responsibility for the sub-custody, care, and control of the uniform. When an employee leaves the hotel's employ, he/she is expected to turn in all uniforms in his/her custody. The uniform room attendant should submit a statement to the accounting department that states whether the employee has properly returned all issued uniforms or if the employee's final paycheck should be adjusted to compensate for uniforms that have not been returned.

The executive housekeeper is responsible for seeing that all uniforms are kept in a state of good repair. Uniformed employees themselves are most aware of the repairs that their uniforms need. A simple repair request form could be used to notify the uniform room attendant. The employee should complete the form when exchanging a soiled uniform for a clean one, and the uniform room attendant should hang the repair tag where the clean uniform is usually hung. Information

Exhibit 6

Uniform Inventory Card

Name _____   Date _____

Position _____   Dept. _____

Uniform _____   No. _____

I understand that the uniform(s) issued me are my sole responsibility and that if I should change positions or leave the company, I will return the complete uniform(s). I authorize the company to deduct from my paycheck the cost of any missing items or the cost of repairing uniforms damaged from other than normal wear. I further understand that these uniforms will *NOT* be taken off hotel property at any time.

Employee
Signature _____

Housekeeping
Signature _____   Date _____

on the repair tag may include employee name, number, department, uniform item, repair request, date received, and the date the uniform is needed. When the soiled uniform is returned from the laundry, the repair tag alerts the uniform room attendant to send the uniform for repairs. If the damaged uniform is beyond repair, the executive housekeeper should condemn the uniform and put a replacement uniform into service. As in the case of discarded linens, records for discarded uniforms should be kept by the executive housekeeper.

All uniforms should be inventoried at least on a quarterly basis. The same general principles that pertain to taking physical inventories of linen supplies apply to counting uniforms. The uniform room should be closed to prevent movement of uniforms. Records should be consulted that account for uniforms issued to employees' sub-custody and for uniforms that have been taken out of service due to damages. All locations, including employee locker areas, need to be considered when counting uniforms. Uniform inventories are generally taken with the help of all hotel departments.

The executive housekeeper can collect accurate counts of each type and size of hotel uniform by using a uniform inventory control form such as the one shown in Exhibit 7. By comparing current on-hand quantities with those from previous inventories, the executive housekeeper can determine the number and kinds of uniforms that have been lost or discarded. In this way, quarterly uniform inventories can be used to establish an annual usage rate. The number and kinds of

Exhibit 7   Sample Uniform Inventory Control Form

QTY.	SIZE	DESCRIPTION	UNIT COST	QTY.	SIZE	DESCRIPTION	UNIT COST
		UNIFORM SERVICES				FRONT DESK	
		Bell Staff Jacket				F.O. Male Suit	
		Bell Staff Pant				F.O. Male Vest	
		Bell Staff Tie				F.O. Male Tie	
		Parking Att. Jacket				F.O. Female Suit	
		Parking Att. Pant				F.O. Female Scarf	
		Parking Att. Jac-Shirt					
		Parking Att. Tie					

replacement uniforms needed can be determined by comparing on-hand quantities to the par levels for each type of uniform.

Guest Loan Items

As a service to guests, hotels provide a variety of equipment that travelers commonly need. This equipment is loaned to guests upon request, and at no charge. The housekeeping department is typically responsible for maintaining the inventory of guest loan items, responding to loan requests, and tracking the items to make sure they are returned.

Types of Guest Loan Items

The types of items that a hotel makes available for guests to borrow vary from hotel to hotel. Generally, such items include irons, ironing boards, sewing kits, hair dryers, alarm clocks, cribs, bed boards, and voltage adapters. Other items include heating pads, hot water bottles, ice packs, razors, electric shavers, curling irons, non-allergenic pillows, heated blankets, feather down comforters, roll-away cots, and bridge tables and chairs. Canes, crutches, and wheelchairs may also be available for guests to use.

Establishing Par Levels for Guest Loan Items

The types of guest loan items maintained at the hotel generally depend upon the hotel's level of service and the typical needs of its clientele. The quantities maintained in inventory depend on the size of the hotel and the anticipated volume of guest requests. The frequency of guest requests for specific items varies according to the type of hotel, the hotel's occupancy level, the arrival/departure pattern of the day, and the kinds of guests staying in the hotel at any given time. The executive housekeeper needs to work with the hotel's general manager and marketing department to identify the kinds and quantities of guest loan items that need to be maintained. The executive housekeeper is responsible for maintaining an adequate supply to meet guest requests.

Inventory Control of Guest Loan Items

The executive housekeeper needs to develop procedures for maintaining accurate inventory records of guest loan items, responding to guest requests, tracking items on loan, and ensuring that borrowed items are returned.

The executive housekeeper needs to maintain a complete and accurate list of all guest loan items stored in the housekeeping department. For each item, the inventory record should reflect the item's name, manufacturer, supplier or vendor, date of purchase, purchase cost, warranty information, and storage location. The record should also note the par number for each item. This master inventory record of guest items should be kept up-to-date as worn or broken items are taken out of service and new items are put into use.

Specific policies and procedures for issuing guest loan items—and for tracking items in use—need to be developed. Procedures will be shaped by the nature of the property's usual clientele and the history of any lost or stolen loan items. Whatever method is used to track on-loan items, a balance must be reached between the need to control hotel losses and the need to provide good guest service.

The executive housekeeper can monitor requests by maintaining a log such as the example in Exhibit 8. This log records the type of item loaned; the guest's room number; and the times of the item's request, delivery, and return. The guest's expected check-out date is also noted to help track items such as special pillows and bed boards that are generally loaned for the duration of a guest's stay. By using this kind of log, the executive housekeeper can surmise what items guests request most, the times particular items are requested, and how long different items remain on loan. This log also helps the executive housekeeper track the locations of items in use and to ensure all items are returned.

Some hotels require that guests sign a receipt for loan items. In this situation, housekeeping employees who deliver loan items to guestrooms should record the type of item, guest name and room number, and the date and time of delivery on the receipt. The employee should also obtain the guest's signature. In addition, some hotels require that guests pay a deposit. The amount of the deposit will vary according to the type of item. In this situation, housekeeping employees who deliver loan items to guestrooms should explain that the amount of the deposit will be charged to the guest's folio in the event the loaned item is not returned. Some

Exhibit 8   Sample Log for Guest Loan Items

Date	Room No.	Loan Item Requested	Call Received		Delivered		Picked Up	
			Time	Who	Time	Who	Time	Who

hotels require that prepaying guests with no charge privileges pay a cash deposit for loaned items. In this situation, the guest should be required to come to the front desk to pay the deposit. Under no circumstances should housekeeping personnel or bell staff receive or handle cash deposits.

Receipts for deposits should be taken to the front desk for placement in the guest folio, but the amount should not be posted at this time. When guest items are returned, it is important that the receipt be removed from the guest's folio and immediately destroyed. If this is not done, guests may be charged for a loaned item even though it was returned.

Several other policies should be considered standard procedures for controlling guest loan items. Whenever possible, all guests requesting items should receive a follow-up call to confirm that he/she received the item and to see if any further assistance is needed. When items are delivered to guestrooms, the guest should be asked to call housekeeping later that same day to arrange for pickup. If housekeeping does not hear from the guest for several hours, the guest should be called to check on the status of the loan item. In most cases, items should not be loaned out overnight.

Each guest loan item should be checked regularly to see that it is in proper working condition and safe for guest use. An item should also be tested on the day it is loaned to ensure that the guest will be able to use it for its intended purpose. Worn, damaged, or broken items should be replaced on an as-needed basis.

Machines and Equipment

The executive housekeeper is responsible for seeing that members of the housekeeping department have the proper tools to carry out their assigned tasks. These tools include major pieces of machinery and equipment to clean guestrooms and public areas. All machines and equipment must be maintained in proper working order so employees may use them safely and effectively. The executive housekeeper needs to develop systems and procedures for controlling the hotel's inventory of machines and equipment.

Types of Machines and Equipment

A variety of machines and equipment is used by the housekeeping staff on a daily basis. Room attendant carts are among the more basic—and more visible—pieces an employee will use. Employees may also use a number of different types of vacuum cleaners for specific cleaning tasks. These include room vacuums, backpack vacuums, corridor vacuums, space vacuums, electric brooms, and wet vacuums. Carpet shampoo equipment, pile lifters, and rotary floor scrubbers are also essential to proper floor care. In addition, laundry equipment, sewing machines, and a variety of trash-handling equipment may also be maintained and controlled through the housekeeping department.

Establishing Par Levels for Machines and Equipment

The number and types of equipment that need to be maintained in-house will depend on the hotel's size and cleaning needs. The executive housekeeper may decide to rent—rather than purchase—equipment that is highly specialized or infrequently needed. Equipment needs are also affected by the number of guestrooms and their locations, the different kinds of floor and wall coverings, and the size of the laundry operation.

With the help and advice of the hotel's general manager, the executive housekeeper can determine the machines and equipment—and the number of each kind—that need to be maintained in inventory. The executive housekeeper should also keep a complete list of all machines and equipment stored in the housekeeping department.

Inventory Control of Machines and Equipment

Controlling the department's stock of major machines and equipment involves maintaining accurate inventory records, establishing issuing procedures, and ensuring that storage areas are secure.

An effective way to control inventories is to use an inventory card system. An inventory card should be prepared for each piece of major machinery or equipment used in housekeeping. The card should specify the item name, model and serial numbers, manufacturer, supplier from whom it was purchased, date of purchase, purchase cost, expected lifespan (usually measured in terms of work hours), warranty information, and local service contact information. These records help determine when a piece of equipment needs replacement. The card should list all the accessories that can be used with the machine and that are owned by the hotel. Any spare parts (hoses, belts, etc.) that are kept in inventory should also be listed. The proper storage or work areas where the equipment, accessories, and spare parts are kept should be indicated on the card.

Repair logs should be kept and filed along with the corresponding inventory card. Repair logs should record the date the item was sent for repair, a description of the problem, who performed the repairs, what repairs were made, which parts were replaced, the cost of the repairs, and the amount of time that the equipment was out of service. These records help pinpoint problems that should be addressed with the service representative. The executive housekeeper can also use these records to estimate repair costs and down-time during the useful life of the machine or equipment.

After issuing procedures have been established, equipment logs should be maintained to record all equipment that is issued and returned on a daily basis. The date, items issued, person to whom the equipment is issued, location in the hotel where the equipment will be used, and the time the items are returned should be noted on the log. Ideally, all equipment should be issued from a central location with one staff member assigned responsibility for issuing the equipment. Each employee should sign for receipt of the equipment and sign again when the equipment is returned.

Security is a major concern in determining storeroom requirements for major machines and equipment. When not in use, all equipment should be stored and locked. In-house equipment should never be permitted to leave the hotel. When machines or equipment are loaned to other departments, the executive housekeeper should keep careful records and follow up to ensure the equipment is returned.

Physical inventories of all major machines and equipment should be taken on a quarterly basis. To do so, a time should be established when all equipment and machinery will be stored and locked. Inventory cards should be consulted and the proper location of all pieces verified. All accessories and spare parts should be counted and recorded on the appropriate inventory card. Finally, all machines and equipment should be tested to ensure they are in good working order.

Cleaning Supplies

Cleaning supplies and small cleaning equipment items are part of the non-recycled inventory in the housekeeping department. These supplies are consumed or used up in the course of routine housekeeping operations. Controlling inventories of all cleaning supplies and ensuring their effective use is an important responsibility of the executive housekeeper. The executive housekeeper must work with all members of the housekeeping department to ensure the correct use of cleaning materials and adherence to cost-control procedures.

Types of Cleaning Supplies

A variety of cleaning supplies and small equipment items are needed to carry out the tasks of the housekeeping department. Basic cleaning supplies include all-purpose cleaners, disinfectants, germicides, bowl cleaners, window cleaners, metal polishes, furniture polishes, and scrubbing pads.

Small equipment items needed on a daily basis include applicators, brooms, dust mops, wet mops, mop wringers, cleaning buckets, spray bottles, rubber gloves, protective eye covering, and cleaning cloths and rags.

Establishing Inventory Levels for Cleaning Supplies

Since cleaning supplies and small equipment items are part of non-recycled inventories, par levels are closely tied to the rates at which these items are consumed in the day-to-day housekeeping operations. A par number for a cleaning supply item is actually a range that is based on two figures: a minimum inventory quantity and a maximum inventory quantity.

The **minimum quantity** refers to the fewest number of purchase units that should be in stock at any given time. Purchase units for cleaning supplies are counted in terms of the normal shipping containers used for the items such as cases, cartons, or drums. The on-hand quantity for a cleaning supply item should never fall below the minimum quantity established for that item.

Minimum quantities are established by considering the usage factor associated with each item. The usage factor refers to the quantity of a given non-recycled inventory item that is used up over a certain period. The rate at which cleaning supplies are consumed by housekeeping operations is the chief factor for determining inventory levels for these non-recycled items.

The minimum quantity for any given cleaning supply is determined by adding the lead-time quantity to the safety stock level for that particular item. The **lead-time quantity** refers to the number of purchase units that are used up between the time that a supply order is placed and the time that the order is actually received. Past purchasing records will show how long it takes to receive certain supplies. Executive housekeepers should keep in mind not only how long it takes suppliers to deliver orders, but also how long it takes the hotel to process purchase requests and place orders. The **safety stock** level for a given cleaning supply item refers to the number of purchase units that must always be on hand for the housekeeping department to operate smoothly in the event of emergencies, spoilage, unexpected delays in delivery, or other situations. By adding the number of purchase units needed for a safety stock to the number of purchase units used during the lead-time, the executive housekeeper can determine the minimum number of purchase units that always need to be stocked.

The **maximum quantity** established for each cleaning supply item refers to the greatest number of purchase units that should be in stock at any given time. An executive housekeeper needs to consider several important factors when determining maximum inventory quantities for cleaning supplies. First, he/she must consider the amount of available storage space in the housekeeping department and the willingness of suppliers to store items at their own warehouse facilities for regular shipments to the hotel. Second, the shelf life of certain items needs to be taken into account. The quality or effectiveness of some products deteriorates if they are stored too long before being used. Third, maximum quantities should not be set so high that large amounts of the hotel's cash resources are unnecessarily tied up in an overstocked inventory.

Inventory Control of Cleaning Supplies

Controlling the inventory of cleaning supplies involves establishing strict issuing procedures to regulate the flow of products from the main storeroom to the floor cleaning closets. It also involves maintaining accurate counts of the products on hand in the main storeroom.

The executive housekeeper can establish a system of par levels for floor cleaning closets from which room attendants supply their carts. (Properties without floor closets generally issue a day's worth of cleaning supplies to a room attendant at the beginning of the shift.) Based on usage rates for the various cleaning supplies under different occupancy levels, the executive housekeeper can determine par levels for every floor station so that each station has enough cleaning supplies to last for a week. Cleaning supplies can be issued from the main storeroom to replenish par levels established for each floor station. By tracking the amounts of cleaning supplies issued from the main storeroom to the various floor stations, the executive housekeeper can monitor usage rates and spot instances of under- or overuse. Floor cleaning closets can be inspected on a regular basis to ensure that par levels are maintained. Shortages of cleaning supplies in floor cleaning closets can result in inspection deficiencies, inconvenience to guests, and lost labor hours as room attendants search for supplies they need to do their job. Once par levels have been established, the executive housekeeper should regularly review and adjust them to accommodate changes in operation or occupancy.

The executive housekeeper needs to ensure that all storage facilities are secure and that all staff members strictly adhere to standard issuing procedures. The minimum and maximum quantities established for each cleaning supply item should be posted on the storeroom shelves where each item is kept. This enables the executive housekeeper to quickly determine whether an adequate amount of cleaning supplies is on hand.

A **perpetual inventory** of all cleaning supplies is often used in conjunction with the par stock system. The perpetual inventory provides a record of all materials requisitioned for supply closets. Together, these two systems enable the executive housekeeper to keep tight control over supplies used by housekeeping personnel in their cleaning assignments. As new purchases are received by the main storeroom and as quantities are issued to floor cleaning stations, the amounts of the different cleaning supplies are adjusted on the perpetual record. When the perpetual record shows that on-hand quantities for particular cleaning supplies have reached the reorder point, a requisition for sufficient quantities can be placed to bring the quantities back up to the maximum levels.

Regular physical inventories should be made of each property storeroom. A monthly physical inventory of all cleaning supplies will enable the executive housekeeper to determine order quantities. More frequent physical inventories need to be made for those items that are depleted more quickly. An inventory record, such as the one shown in Exhibit 9, can be used as a worksheet for taking the physical count of all cleaning supplies. By identifying the minimum and maximum

Exhibit 9 Sample Inventory Record

ITEM DESCRIPTION	STD. UNIT	MIN./ MAX	INVENTORY				TOTAL INV.	UNIT COST	TOTAL COST	
			STORE ROOM	1	2	3				

Holiday Inn — INVENTORY RECORD

HOTEL: _____ LOC. NO. _____ DEPT.: _____ DATE _____

CALLED BY _____ RECORDED BY _____ APPROVED BY _____ PAGE _____ OF _____

Courtesy of Holiday Corporation, Memphis, Tennessee

inventory levels for each item and tallying the on-hand totals from different store-room locations, the executive housekeeper can easily determine how much to order of each item to bring supplies back up to the established maximum quantities. Physical inventories are quicker and easier if the items listed on the inventory record are in the same order as they are arranged on storeroom shelves.

By recording both purchases and issues of cleaning supplies, the executive housekeeper can monitor the actual usage rates for each product kept in inventory. Exhibit 10 shows a form the executive housekeeper can use to determine the expected inventory for each cleaning supply and equipment item. The results of the previous physical inventory are listed in the beginning inventory column for the next month. Monthly purchases are added to these initial quantities, while the amounts of supplies issued are deducted. The total—or ending inventory column—estimates the quantity of each item that should be in stock at the end of the month. The results of the physical count can be compared to the expected ending inventory. The variance between the actual quantities on hand and the amounts expected to be on hand represent the loss of cleaning supplies and equipment during the month. If this variance is unacceptably high, the executive housekeeper should investigate whether proper storage, issuing, and recordkeeping controls are being followed.

Exhibit 10   Sample Form for Calculating Expected Inventories

Month _____

HOUSEKEEPING DEPARTMENT
SUPPLIES AND EQUIPMENT
INVENTORY CALCULATION SHEET

Item	Beginning Inventory +	Purchases −	Issues =	Ending Inventory
CLEANING SUPPLIES				
All-purpose cleaner				
Dusting solution				
Glass cleaner				
Wastebasket liners				
Carpet shampoo				
Stain remover				
Cleaning cloths				
Sponges				
Work gloves				
Eye covering				
EQUIPMENT				
Mops				
Brooms				
Dustpans				
Vacuums				
Buckets				
Bowl mops				
Carpet shampooers				
Window brushes				

Guest Supplies

Hotels provide a variety of guestroom supplies and amenities for the guest's needs and convenience. The executive housekeeper is typically responsible for storing, distributing, controlling, and maintaining adequate inventory levels of guest supply items and amenities.

Types of Guest Supplies

To a large measure, the types and quantities of guest supplies that a hotel routinely provides depends on the hotel's size, clientele, and service level. Guest supplies and amenities for which the housekeeping department is responsible typically include bath soaps, facial soaps, toilet seat bands, toilet tissue, facial tissue, and hangers. Other supplies may include glasses, plastic trays, water pitchers, ice buckets, matches, ashtrays, and wastebaskets. Some hotels may provide all guestrooms with lotions, shampoos, conditioners, bathfoam, shower caps, shower mats, sewing kits, shoe shine cloths, disposable slippers, and other items. Laundry bags, plastic utility bags, sanitary bags, emery boards, and candy mints may also be included on the list. Pens, stationery, and a variety of printed items, such as "do not disturb" signs, fire instructions, guest comment forms, and hotel or area marketing material may also be regularly distributed.

Establishing Inventory Levels for Guest Supplies

Each hotel will have its own standard room setup requirements for guest supplies. One par of guest supplies would be the quantity of each guest supply item needed to outfit all occupied rooms in the hotel one time. By knowing the forecasted number of occupied rooms, the executive housekeeper can determine the quantity of each guest supply item that will be needed to outfit guestrooms in the month ahead. However, since guest supplies are a part of the hotel's non-recycled inventory, usage rates are the most important factor for pinpointing inventory levels. Exhibit 11 shows a sample monthly par stock requirement for three guest supply items based on the number of occupied rooms forecasted for that month. Exhibit 11 also shows the actual usage of these guest supply items during the month. Notice that actual usage can far exceed the quantities expected to be used on the basis of par stocks required for room setups. If inventory levels for these guest supply items were based solely on the quantities needed to stock occupied rooms with the par amounts, severe shortages would result.

Like cleaning supplies, minimum and maximum inventory levels are used to establish and control the hotel's stock of guest supplies and amenities. Both occupancy levels and usage rates need to be considered in establishing minimum and maximum quantities for the hotel's inventory of guest supply items. As an example, consider how minimum and maximum inventory quantities are calculated for bath soap.

Since minimum and maximum quantities are calculated in terms of purchase units, the first step is to determine how many bars of soap are usually contained in a standard package. Suppose that one case of bath soap contains 1,000 bars.

Exhibit 11 Comparison of Par Stock and Actual Usage for Guest Supplies

Guest Supplies
Par Stock For One Month

Item	Potential Usage Per Occupied Room	×	Forecasted Number of Occupied Rooms	=	Par Stock Required
Shampoo	1.0	×	450	=	450
Bathfoam	1.0	×	450	=	450
Small Soap	1.0	×	450	=	450

Actual Usage For One Month

Item	Potential Usage Per Room		Occupied Rooms		Potential Consumed	Actual Consumed	Variance
Shampoo	1.0	×	450	=	450	370	<80>
Bathfoam	1.0	×	450	=	450	513	63
Small Soap	1.0	×	450	=	450	752	302

The second step is to calculate how many bars of soap will be used on an average day during the hotel's peak season. This, of course, depends on the number of occupied rooms and on the amount of the item that will be used in each room each day. Suppose that the average number of occupied rooms during the hotel's peak season is 200 and that one bar of soap is used in each occupied room each day. This means that the hotel's guests will use up 200 bars of soap each day.

The third step is to determine how many days it will take for the hotel's guests to use a standard purchase unit of soap. Since there are 1,000 bars of soap in each case, and 200 bars of soap will be used up every day, it will take five days to use up one case of soap (1,000 ÷ 200 = 5). This means that one purchase unit of bath soap will be used up every five days.

The fourth step is to determine the minimum number of purchase units of soap that should always be in stock at any time. The minimum quantity for any guest supply item is determined by adding the lead-time quantity to the safety stock level established for that particular item. Suppose that the executive housekeeper determines that an appropriate safety stock level for soap is one case—or enough for a five-day supply. The executive housekeeper knows that soap has a relatively short shelf life before deteriorating in quality and determines that a five-day supply is sufficient to cover any emergency, spoilage, or delivery delay. To determine the quantity of soap needed to cover the period between placing and receiving an order, the executive housekeeper has to consider how long it takes the hotel to process and approve a purchase request and how long the supplier needs to process and deliver the order. Suppose the executive housekeeper determines that it takes five days to process and deliver an order of soap. Since the amount of soap used in five days is one case, the executive housekeeper sets the minimum quantity at two cases (1 case safety stock + 1 case lead-time = 2 cases). Thus, the reorder point for soap is two cases.

The fifth step is to determine the maximum quantity of soap—or the greatest number of cases that should be in stock at any given time. In addition to concerns about storage space and conserving the hotel's cash outlay for inventoried products, the chief factor affecting the maximum inventory quantity for soap is the frequency of orders. A maximum inventory quantity can be calculated by dividing the number of days between orders by the number of days it takes to use one purchase unit, and then adding the minimum quantity. Suppose the executive housekeeper orders soap once a month. The amount of time from one order point to the next is 30 days. Since the number of days it takes to use one case of soap is five, the amount of soap that will be used in 30 days is six cases ($30 \div 5 = 6$). Adding the previously established minimum quantity of two cases, the maximum inventory quantity for soap can, in this case, be established as eight cases ($6 + 2 = 8$).

When the number of cases of soap in inventory reaches the minimum level, the executive housekeeper should place an order for enough soap to bring the inventory level back up to the maximum quantity. When the supply of soap reaches two cases, the executive housekeeper should place an order for six cases in order to bring the inventory level back up to the maximum eight cases ($8 - 2 = 6$).

Minimum and maximum inventory quantities can be determined in a similar way for all guest supplies and amenities. The key factors affecting the calculations are occupancy levels, usage rates, and the frequency with which supplies are to be reordered.

Inventory levels and usage rates for guest supply items should be carefully monitored and adjusted as necessary. A number of considerations may lead the executive housekeeper to adjust the minimum and maximum inventory levels for guest supplies and amenities. Seasonal changes in occupancy levels need to be considered because fewer supplies will be needed when occupancy decreases. Since occupancy levels may vary considerably from month to month, executive housekeepers may use occupancy forecasts to calculate par levels of guest supplies and amenities on a monthly basis.

The standard packaging for some items also affects the determination of minimum and maximum inventory levels. For example, shower caps may be sold in quantities as high as 1,000 caps per case. This item may be so inexpensive that the executive housekeeper may decide that an overstock is much less serious than the effects of any shortages.

Some guest supply items—such as drinking glasses—are actually recycled inventory items. Par levels for this kind of guest supply should be determined in a manner similar to room linens and other recycled inventory items. Considering the cleaning cycle alone, at least three pars of glasses may be needed to stock all guestrooms: one par of glasses will be clean, wrapped, and ready for distribution; a second par will be in guestrooms, soiled, and ready for collection; and a third par of glasses will be in the process of being cleaned and prepared for distribution on the following day. Considering theft, breakage, and potential guest requests for

Exhibit 12   Sample Par for Room Attendant's Cart

12 Small Soap	12 Matches	6 Large Soap	12 Pencils
12 Shampoo	3 Sewing Kits	12 Bathfoam	6 Postcards
6 Shower Caps	3 Security Cards	12 Laundry Bags	12 Envelopes
6 Plain Notepads		12 Letterheads	3 Magazines
3 Room Service Menus		3 Fire Instructions	12 Guest Comment Cards

additional glasses, the executive housekeeper may decide that four or five pars of glasses are required.

Inventory Control of Guest Supplies

Since guest supply items are among a hotel's non-recycled inventories, inventories are controlled in much the same way as cleaning supplies. Par levels are established, physical inventories taken, and records maintained. The principles of control and procedures for gathering accurate information about usage rates and inventory levels are substantially the same.

In addition to establishing par levels for the main storeroom and floor service closets, most hotels also establish par levels for room attendant carts. Exhibit 12 shows a sample par of guest supplies for a room attendant cart servicing 12 to 14 guestrooms. Inventory control procedures for issuing guest supplies and amenities to room attendants will depend on whether supplies are issued daily from the main storeroom in par levels established for room attendant carts, or whether supplies are issued on a weekly or other basis to replenish par levels established for floor service areas. In either case, the control procedure is the same: occupancy levels and usage rates determine the par levels needed, and supplies are issued only in quantities sufficient to replenish par stocks. Control forms, such as the one shown in Exhibit 13, can be used to monitor the quantities of guest supplies issued from the main storeroom.

Printed Materials and Stationery

Printed materials and stationery are distributed to guestrooms along with other guest supplies and amenities. Although the hotel's marketing department is usually directly involved in designing and producing these items, housekeeping is responsible for their distribution and for their inventory levels.

Writing paper, notepads, postcards, local and airmail letterheads, local and airmail envelopes, telephone message forms, telex forms, and even social diaries are examples of printed materials that a hotel may supply for the guest's use and convenience. "Do not disturb" signs, instructions for fire and other emergencies, room service menus, maps, brochures on area restaurants or attractions, TV listings, and forms for guest comments or evaluations of hotel services may also be provided.

Exhibit 13   Sample Control Form for Issuing Guest Supplies

Guestroom Supplies Requisition				
Item	Par Stock	Reorder Point	Requisition (same as par)	Cost of Item Requisition
Bar soap	1 case	$1/2$ case		
Tissue	1 case	$1/2$ case		
Toilet paper	1 case	$1/2$ case		
Shower caps	100	50		
Matches	6 boxes	3 boxes		
Pens	1 box	$1/2$ box		
Memo pads	2 pkgs	1 pkg		
Pencils	1 box	$1/2$ box		
Do Not Disturb signs	30	15		
Glasses	1 case	$1/2$ case		
Room folders	30	15		
Wastebaskets	6	2		

The frequency with which printed materials and stationery need to be changed or replaced will affect the minimum and maximum inventory levels for these items. Some printed items are relatively permanent features of the guestroom. These items do not need to be changed or redesigned very frequently, and rarely need to be replaced except in the case of damage or wear. Instruction sheets for emergencies and for equipment such as the telephone, TV, and heating and air conditioning system fall into this category. Inventory levels for these printed materials will likely depend upon printing costs.

Some printed materials such as TV listings or special event calendars may need to be replaced on a daily basis. Other materials, such as room service menus, may need to be changed less frequently. Excess inventories of such printed items will be sheer waste as they become outdated.

Other printed items with a relatively limited lifespan are marketing brochures and flyers related to hotel services. The marketing department generally has established schedules for the redesign and replacement of such promotional materials.

The executive housekeeper has to work closely with marketing personnel to determine adequate inventory levels for such items.

Par stock levels for stationery items such as notepads, letterheads, envelopes, and postcards are established in the usual manner for non-recycled inventory items. Occupancy levels, usage rates, safety levels, lead-time quantities, and purchasing schedules are considered when determining minimum and maximum inventory levels.

Endnotes

1. According to the Fair Labor Standards Act, if cleaning costs would reduce an employee's weekly earnings to under minimum wage, the employer must clean the uniforms or pay the employee to do so.

Key Terms

floor par	non-recycled inventories
house setup	par
inventory	par number
issuing	perpetual inventory system
lead-time quantity	receiving
minimum quantity	recycled inventories
maximum quantity	safety stock

Discussion Questions

1. Compare recycled and non-recycled inventories. What are some of the typical items in each category?

2. What is meant by par? by par number?

3. What is the basic premise for establishing par levels for recycled inventories? for non-recycled inventories?

4. What three factors should be considered when setting par for linen?

5. What are some of the typical ways the executive housekeeper and the laundry manager work together to control linen inventories?

6. What are the main benefits of conducting physical inventories? How often should physical inventories be taken?

7. What factors make it difficult to establish par levels for uniforms?

8. What type of information should be listed on an inventory card for a piece of major machinery or equipment?

9. What is meant by minimum quantity? by maximum quantity?

10. Describe how the concepts of minimum quantity and maximum quantity work together to control non-recycled inventories.

REVIEW QUIZ

When you feel you have covered all of the material in this chapter, answer these questions. Choose the *best* answer.

True (T) or False (F)

T   F   1. The same method is used to establish par levels for inventories of recycled and non-recycled items.

T   F   2. Linen costs are the largest expense item incurred by housekeeping operations.

T   F   3. It is not necessary to establish a par level for washcloths.

T   F   4. One par of guestroom linens equals the total number of each type of linen needed to outfit one guestroom.

T   F   5. Periodic inventory reports indicate the extent of linen losses at the property.

T   F   6. The housekeeping department should maintain an accurate daily count of all linens sent to and received from the laundry.

T   F   7. A physical inventory of all linen items should be conducted once a month.

T   F   8. The results of a physical inventory are recorded on a master inventory control chart.

T   F   9. A physical inventory of uniforms should be conducted once a year.

T   F   10. The on-hand quantity for a cleaning supply item should never fall below the minimum quantity established for that item.

Alternate/Multiple Choice

11. The most important recycled inventory items under the executive housekeeper's responsibility are:

    a. linens.
    b. cleaning supplies.

12. The executive housekeeper uses the occupancy report generated by the front desk to develop a:

    a. linen count sheet.
    b. linen distribution list.

13. The standard number of inventoried items that must be on hand to support daily, routine housekeeping operations is referred to as:

    a. safety stock level.
    b. lead-time quantity.
    c. par level.
    d. maximum quantity.

14.   Which of the following records is used to estimate repair costs and down-time during the useful life of machines and equipment used by the housekeeping department?

   a. inventory cards
   b. invoices
   c. purchase orders
   d. repair logs

15.   The minimum quantity of non-recycled items is determined by:

   a. adding lead-time quantity to the safety stock level.
   b. adding the maximum quantity to the usage rate.
   c. adding the safety stock level to the par level.
   d. subtracting lead-time quantity from the safety stock.

13/15

Chapter Outline

The Budget Process
    Types of Budgets
Planning the Operating Budget
Using the Operating Budget as a Control
    Tool
Operating Budgets and Income Statements
    The Hotel Income Statement
    The Rooms Division Monthly Income
    Statement
Budgeting Expenses
    Salaries and Wages
    Employee Benefits
    Outside Services
    In-House Laundry
    Linens
    Operating Supplies
    Uniforms
Controlling Expenses
Purchasing Systems
    Linen Replacement
    Uniform Replacement
    Purchasing Operating Supplies
Capital Budgets
Contract vs. In-House Cleaning

Learning Objectives

1. Identify the executive housekeeper's responsibilities in relation to the budget process.

2. Distinguish between operating budgets and capital budgets.

3. Explain how the concept of "cost per occupied room" serves as a valuable tool in the budget planning process.

4. Explain how the executive housekeeper uses the operating budget to monitor expenses in the housekeeping department.

5. Distinguish between operating budgets and income statements.

6. Identify the line items on a rooms division income statement that are affected by expenses incurred by the housekeeping department.

7. Distinguish between favorable and unfavorable variances as shown on a rooms budget report.

8. Explain how the executive housekeeper uses the rooms manager's forecast of rooms sales to estimate expenses for the housekeeping department.

9. Identify four basic methods that the executive housekeeper can use to control expenses.

10. Explain how an executive housekeeper determines the size of annual linen purchases.

11. Explain how the lifespan of linens is measured and how a cost per use is calculated.

12. Identify factors that an executive housekeeper should consider when arranging for outside contractors to perform cleaning services.

6

Controlling Expenses

Since housekeeping is not a revenue-generating department, the executive housekeeper's primary responsibility in achieving the property's financial goals is to control the department expenses. In addition to salaries and wages, inventoried items are a key area for the executive housekeeper's exercise of cost control measures.

This chapter describes the budgetary process and explains how budgets are determined for the operational expenses that fall within the executive housekeeper's responsibility. The chapter also examines the executive housekeeper's responsibility in controlling the costs associated with housekeeping operations. The role of the executive housekeeper in formulating capital budgets will also be discussed.

The Budget Process

The operating budget outlines the financial goals of a hotel. The purpose of the operating budget is to relate operational costs to the year's expected revenues. The yearly operating budget is broken down into budgets for each month of the fiscal year. In addition, each department prepares its own monthly budget. These budgets cover individual areas of responsibility and serve as a guide for how the department will achieve its expected contribution to the property's financial goals.

Essentially, a budget is a plan. It projects both the revenues the hotel anticipates during the period covered by the budget and the expenses required to generate the anticipated revenues. The executive housekeeper's responsibility in the budgetary process is twofold. First, the executive housekeeper is involved in the planning process that leads to the formulation of the budget. This entails informing the rooms division manager and general manager what expenses the housekeeping department will incur in light of forecasted room sales. Second, since the budget represents an operational plan for the year, the executive housekeeper ensures that the department's actual expenses are in line with budgeted costs and with the actual occupancy levels.

As a plan, a budget is not "set in stone." It may need to be adjusted in light of unforeseen or changing circumstances. If anticipated room sales do not materialize, then expenses allocated to different departments will need to be adjusted accordingly. If occupancy levels exceed expectations, then increased expenses need to be planned for and incorporated into a revised budget. If unexpected expenditures are required, their effect on the overall plan needs to be assessed. New ways may need to be determined for the property to meet its financial goals and objectives.

As a plan, a budget is also a guide. It provides managers with the standards by which they can measure the success of operations. By comparing actual expenses with allocated amounts, the executive housekeeper can track the efficiency of housekeeping operations and monitor the department's ability to control its expenses within the prescribed limits.

Types of Budgets

Two types of budgets are used in managing a hotel's financial resources: capital and operating budgets. The difference between the two essentially lies with the types of expenditures involved.

Usually, a **capital budget** plans for the expenditure of company assets for items costing $500 or more. Typically, these items are not used up in the normal course of operations; instead, they have a lifespan that exceeds a single year. Furniture, fixtures, and equipment are typical examples of **capital expenditures**. Capital expenditures in the housekeeping department may include room attendant carts, vacuum cleaners, carpet shampooers, pile lifters, rotary floor scrubbers, laundry equipment, sewing machines, and trash-handling equipment. In addition, major initial purchases of recycled inventory items—such as linen, towels, blankets, and uniforms—are capital budget items since they have a relatively long useful life and are not used up in the course of normal operations.

An **operating budget** forecasts revenues and expenses associated with the routine operations of the hotel during a certain period. **Operating expenditures** are those costs the hotel incurs in order to generate revenue in the normal course of doing business. In the housekeeping department, the most expensive operational cost involves salaries and wages. The costs of non-recycled inventory items, such as cleaning and guest supplies, are also considered operational costs.

Planning the Operating Budget

The budgeting process begins far in advance of the start of the period for which the budget is planned. The process of planning an annual operating budget generally takes several months. It involves gathering information, formulating initial plans, reconsidering goals and objectives, and making final adjustments. The budget planning process requires a closely coordinated effort of all management personnel.

Operating budgets are typically prepared for each fiscal year; the annual operating budget summarizes the anticipated year-end results. Monthly operating budgets are also prepared for the property's fiscal year. This enables managers to clearly outline seasonal variations in expected revenues and corresponding expenses. It also provides managers and department heads with valuable tools to monitor actual results.

In budgeting planning, the first step is always to forecast room sales. The reason for this is twofold. First, room sales generate the revenue for operating various departments. Second, and more important, most of the expenses that each department can expect—and the ones that departments are most able to control—are most directly related to room occupancy levels. This is especially true of the housekeeping department where salaries and wages, and the usage rates for both recycled and non-recycled inventory items, are a direct function of the number of occupied rooms. The concept of "cost per occupied room" is the major tool the

Housekeeping Success Tip

Thelma F. Alfelor, CHHE
Executive Housekeeper
UCLA Guest House
Los Angeles, California

❝Many people know that the housekeeping department is usually the largest department in any hotel operation and has the biggest budget. When I became executive housekeeper at the UCLA Guest House, I discovered that about 60% of the total budget was allocated for chemicals and other supplies.

Part of the problem was that chemicals were improperly ordered, but I realized immediately that cutting the cleaning chemicals and materials budget could save our department a great deal of money. To do this, I first made an inventory of all the chemicals in stock. I discovered many duplicate products, including five kinds of all-purpose cleaners, two disinfectant all-purpose cleaners, and three air fresheners.

Next, I decided which chemicals to eliminate by working with room attendants. Their input revealed that the inventory included two kinds of wax that did not work very well, two kinds of stripper which were difficult to remove because they weren't rinse-free, and a pesticide we weren't using at all.

After eliminating all these products, we were left with seven basic cleaners:

- All-purpose cleaner with disinfectant for tub, toilet, and sink
- Cleaner for glass doors, build-ups, and chrome
- Air freshener
- Glass cleaner
- Rinse-free stripper
- Wax
- Carpet cleaner

Then I returned the chemicals we decided to eliminate from the inventory to vendors, who replaced them with paper products. Reducing the chemicals inventory resulted in a 70% reduction in the chemicals allocation and saved us 50% of our budget overall. Moreover, trading the chemicals for paper products put me ahead of budget, since I did not have to purchase these items.

Reducing the chemical inventory, however, was only the first step. To keep our costs down, I implemented some control procedures. To help eliminate waste, I put all the chemicals (including samples given out free by suppliers) in a locked room and dispensed them to room attendants as needed. Previously, room attendants had carried large amounts of cleaners on their carts, which encouraged waste. Issuing smaller amounts makes them use cleaners more wisely. I buy cleaners in concentrated form and dilute them myself, which helps reduce waste.

(continued)

Housekeeping Success Tip *(continued)*

Besides waste, I suspected some theft problems. For example, the department had purchased expensive liquid laundry soap which room attendants carried on their carts "in case guests needed it." To eliminate this expense entirely, I stopped ordering the liquid soap and installed soap vending machines for guests and purchased cheaper powder soap for rugs and uniforms. I also made a weekly inventory of supplies kept in the locked room to discourage pilfering.

The savings we realized from the inventory reduction and control measures allowed us to reassign the money to in-house projects. For example, I was able to hire my own staff of housepersons for periodic projects—floor stripping, waxing, carpet cleaning, and window cleaning—previously performed by an outside contractor. I also bought some tools for gardening, a ladder which we formerly had to borrow from the maintenance department, etc.

My primary concern in my operation is to save money and cut my budget. That is how I impress my superiors.**"**

---

executive housekeeper uses to determine the levels of expense in the different categories. Once the executive housekeeper knows predicted occupancy levels, expected expenses for salaries and wages, cleaning supplies, guest supplies, laundry, and other areas can be determined on the basis of formulas that express costs in terms of cost per occupied room.

Occupancy forecasts are generally developed by the front office manager, working closely with the property's general manager. The forecast is based not only on historical data concerning past levels of occupancy (and their distribution among the budget periods), but also on information supplied by the marketing department concerning the anticipated effect on room sales of special events, advertising, and promotions. Some hotels generate forecasts for room sales that predict the level of occupancy for each day of the coming year.

Once occupancy levels are predicted, the departments whose costs fluctuate with occupancy levels can forecast expected costs and submit prepared budgets to the general manager and controller for review. Upper management analyzes and adjusts the departmental budget plans so they reflect the property's goals and objectives. Often, budgets are returned to department heads with comments and recommended adjustments. Such feedback primarily reflects the concern of upper management to maximize profits and control expenses while maintaining appropriate levels of service.

By specifying expense levels in relation to room sales, the budget actually expresses the level of service the hotel will be able to provide. In this regard, it is important for department heads to report how service levels will be affected by budget adjustments. This is especially important for the executive housekeeper. If upper management tones down the operating budget submitted by the executive housekeeper, the executive housekeeper should clearly indicate what services will be eliminated or downgraded in order to achieve the specified reductions.

The cycle of feedback and discussion continues as department heads revise their budgetary plans and provide additional input in response to recommended adjustments. It is through this back-and-forth process that agreement is ultimately reached. The final budget represents the forecasts, goals, and constraints that everyone adopts. Each department is then committed to operating under the limits expressed in the budget and achieving its contribution to the overall plan. Once approved, the operating budgets set a standard by which departmental performance can—and will—be evaluated.

Using the Operating Budget as a Control Tool

An operating budget is a valuable control tool to monitor the course of operations during a specified period. Each month, the hotel's accounting department produces statements reporting actual costs in each of the expense categories. The form of these statements is nearly identical to that of the operating budget; actual costs are listed alongside budgeted costs. Such reports enable the executive housekeeper to monitor how well the housekeeping department is doing in relation to the budgeted goals and constraints.

Controlling expenses in the housekeeping department means comparing actual costs with budgeted amounts and assessing the variances. When comparing actual and budgeted expenses, the executive housekeeper should first determine whether the forecasted occupancy levels were actually achieved. If the number of occupied rooms is lower than anticipated, a corresponding decrease in the department's actual expenses should be expected. Similarly, if occupancy levels are higher than forecasted, the executive housekeeper can expect a corresponding increase in housekeeping expenses. In either case, the decrease or increase in expenses should be proportional to the variation in occupancy levels. The executive housekeeper's ability to control housekeeping expenses will be evaluated in terms of his/her ability to maintain the cost per occupied room expected for each category.

Small deviations between actual and budgeted expenses can be expected and are not a cause for alarm. Serious deviations from the budgeted plan require investigation and explanation. If the actual costs far exceed the budgeted amounts while the predicted occupancy level remains the same, the executive housekeeper needs to find the source of the deviation. In addition to discovering why the department is "behind budget," the executive housekeeper needs to formulate a plan to correct the deviation and get the department back "on budget." For example, a re-examination of staff scheduling procedures or closer supervision of standard practices and procedures may be necessary. Other steps might include evaluating the efficiency and cost of products being used in the housekeeping department, and exploring the alternatives.

Even if the executive housekeeper finds that the department is far "ahead of budget," it is not necessarily a cause for celebration. It may indicate a deterioration of service levels that were built into the original budget plan. *Any* serious deviation from the plan is a cause for concern and requires explanation. Identifying and investigating such deviations on a timely basis is one of the most valuable functions an executive housekeeper can perform in terms of the operating budget.

Operating Budgets and Income Statements

An operating budget is identical in form to an **income statement.** An income statement—or statement of income—expresses the actual results of operations during an accounting period, identifying revenues earned and itemizing expenses incurred during that period. The difference between an income statement and an operating budget is that the first expresses the actual results of operations for a period that has ended, while the second expresses the expected results of operations for a current or coming period. The one is a report of what actually occurred, while the other is a forecast or plan for what is to come. The operating budget is a plan for the period in the sense that it predicts or anticipates what the income statement will actually show at the end of that period. The success of the hotel's plan as expressed in the budget is determined by how closely its forecasted numbers match the numbers on the end-of-the-period income statement.

In the budget planning process, upper management collects information from the various department heads to prepare a budget for the whole property. This budget takes the form of an income statement for the coming period. Income statements that predict the results of current or future operations, as opposed to reporting actual results, are often referred to as **pro forma income statements**.

The Hotel Income Statement

The statement of income provides important financial information about the results of hotel operations for a given period. The period may be one month or longer, but cannot exceed one business year. Since a statement of income reveals the bottom line—the net income for a given period—it is one of the most important financial statements used by top management to evaluate the success of operations. Although the executive housekeeper may never directly use the hotel's statement of income, this statement relies in part on detailed information supplied by the housekeeping department.

The sample statement of income shown in Exhibit 1 is often called a consolidated statement because it presents a composite picture of all the financial operations of the hotel. Rooms division information appears on the first line, under the category of operated departments. The amount of income generated by the rooms division is determined by subtracting payroll and related expenses and other expenses from the amount of net revenue produced by the rooms division over the period covered by the income statement. Payroll and related expenses charged to the rooms division include the wages, salaries, and benefits paid to housekeeping *and* front office staff, reservation agents, and uniformed service staff. Since the rooms division is not a merchandising facility, there is no cost of sales to subtract from the net revenue amount.

The revenue generated by the rooms division is often the largest single amount produced by revenue centers within a hotel. Using the figures shown in Exhibit 1, the amount of income earned by the rooms division during the year was $1,414,843, or 84.1% of the total income of $1,682,209. Since the rooms division is generally the hotel's major source of income, and since housekeeping is a major

Exhibit 1   Sample Consolidated Statement of Income

Holly Hotel
Statement of Income
For the year ended December 31, 19XX

Schedule A

	Schedule	Net Revenue	Cost of Sales	Payroll and Related Expenses	Other Expenses	Income (Loss)
Operated Departments						
Rooms	A1	$1,834,450		$ 292,495	$ 127,112	$1,414,843
Food and Beverage	A2	1,049,140	$ 356,620	408,360	109,406	174,754
Telephone	A3	102,280	120,088	34,264	3,174	(55,246)
Other Operated Departments	A4	126,000	20,694	66,552	13,462	25,292
Rentals and Other Income	A5	122,566				122,566
Total Operated Departments		3,234,436	497,402	801,671	253,154	1,682,209
Undistributed Expenses						
Administrative and General	A6			195,264	133,098	328,362
Marketing	A7			71,650	64,086	135,736
Property Operation and Maintenance	A8			73,834	49,274	123,108
Energy Costs	A9				94,624	94,624
Total Undistributed Expenses				340,748	341,082	681,830
Income Before Fixed Charges		$3,234,436	$ 497,402	$ 1,142,419	$ 594,236	$1,000,379
Fixed Charges						
Rent	A10					57,000
Property Taxes	A10					90,648
Insurance	A10					13,828
Interest	A10					384,306
Depreciation and Amortization	A10					292,000
Total Fixed Charges						837,782
Income Before Income Taxes and Gain on Sale of Property						162,597
Gain on Sale of Property						21,000
Income Before Income Taxes						183,597
Income Taxes						66,095
Net Income						$ 117,502

source of expense incurred by the rooms division, the executive housekeeper plays an important role in the hotel's overall financial performance.

The Rooms Division Income Statement

The hotel's statement of income shows only summary information. More detailed information is presented by the separate departmental income statements prepared by each revenue center. These departmental income statements are called schedules and are referenced on the hotel's statement of income.

Exhibit 2   Sample Rooms Division Statement of Income

	Holly Hotel		
	Rooms Division Income Statement		
	For the year ended December 31, 19XX		**Schedule A1**
Revenue			
Room Sales		$1,839,500	
Allowances		5,150	
Net Revenue			$1,834,450
Expenses			
Salaries and Wages	$245,218		
Employee Benefits	47,277		
Total Payroll and Related Expenses		292,495	
Other Expenses			
Commissions	5,100		
Contract Cleaning	10,853		
Guest Transportation	20,653		
Laundry and Dry Cleaning	14,348		
Linen	22,443		
Operating Supplies	27,226		
Reservation Expenses	20,419		
Other Operating Expenses	6,070		
Total Other Expenses		127,112	
Total Expenses			419,607
Departmental Income (Loss)			$1,414,843

Exhibit 1 references the rooms division schedule as *A1*. This rooms division income statement appears in Exhibit 2. The figures shown in Exhibit 2 for rooms division net revenue, payroll and related expenses, other expenses, and departmental income are the same amounts which appear on Exhibit 1 for the rooms division under the category of operated departments.

The format and specific line items used by the rooms division for its departmental income statement will vary with the needs and requirements of individual properties. The following sections briefly describe typical line items found on a rooms division income statement.

The first heading on the rooms division income statement records revenue from room sales during the period. The second heading, Allowances, identifies rebates, refunds, and overcharges of revenue. These are generally not known at the time that room sales are recorded. Instead, allowances are adjusted at a later date and may not appear as a budgeted line item on a pro forma income statement.

Net Revenue is arrived at by subtracting Allowances from Total Revenue. It is the Net Revenue figure that is transferred to the hotel's income statement as sales derived from the hotel's lodging operation.

The executive housekeeper is directly concerned with many of the line items listed in the expense sections of the rooms division's income statement. The largest single expense category listed is Salaries and Wages. The personnel costs associated with the housekeeping department are incorporated into this total which also includes payroll costs for all rooms division employees. Regular pay, overtime pay, vacation pay, severance pay, incentive pay, holiday pay, and employee bonuses are included in this expense category.

The expense item referred to as Employee Benefits is generally calculated by the personnel or accounting departments. It includes payroll taxes, payroll-related insurance expense, pension, and other related personnel costs. The share of Employee Benefit expense that belongs to the housekeeping department is included in this line item.

Many of the expense items listed under the heading Other Expenses fall under the direct responsibility of the executive housekeeper. These include:

- Contract Cleaning
- Laundry and Dry Cleaning
- Linen
- Operating Supplies
- Uniforms

Contract Cleaning includes the cost of contracting outside companies to clean lobbies and public areas, wash windows, and exterminate and disinfect areas of the rooms division. The pros and cons involved in the executive housekeeper's decision to employ contract cleaning services will be discussed at the end of this chapter.

The Laundry and Dry Cleaning expense item refers to the cost of both outside and in-house laundry and dry cleaning services. It also includes the cost of dry cleaning curtains and draperies as well as washing or cleaning awnings, carpets, and rugs in areas of the rooms division. All the expenses incurred by the property's in-house laundry facility (except salaries, wages, and benefits) are reflected in this expense item. The costs of supplies used to keep the house laundry in a clean and sanitary condition are included—plus the costs of all supplies used in the laundry operation itself. In addition, printing and stationery costs associated with laundry lists, printed forms, service manuals, and office supplies used by in-house laundry personnel are included. Finally, the cost of purchasing or renting uniforms for in-house laundry employees, along with the costs of uniform cleaning or repair, are incorporated in the total expense recorded in this category. Many hotels use a separate schedule to itemize all the costs within this expense category.

The Linen expense item includes the replacement costs or rental fees for sheets, pillowcases, towels, face cloths, bath mats, blankets, and other items included in the linen inventory.

The Operating Supplies expense item includes the cost of guest supplies, cleaning supplies, and printing and stationery items. All guest and cleaning supplies fall within the executive housekeeper's area of responsibility, and their inventories are maintained in the housekeeping department.

The Uniforms expense item includes the cost of purchasing or renting uniforms for all employees of the rooms division as well as other related costs.

Some of the expense categories listed under the heading Other Expenses fall outside the executive housekeeper's areas of responsibility. The Commissions expense refers to remunerations paid to outside sources, such as travel agents, who secure rooms business for the hotel. Guest Transportation includes the costs associated with transporting guests to and from the hotel. Reservations includes the cost of a reservations service and a central reservation system involving telephone, telegram, and teletype expenses.

In the budget planning process, the rooms manager will solicit information from the executive housekeeper concerning the expense categories that fall under the housekeeping department's areas of responsibility. In particular, the rooms manager will be interested in assessing expected expenses as a percentage of the revenue forecasted for room sales. Every controllable cost can be expressed as a percentage of revenue. For each expense category, the rooms manager will have a standard percentage that is considered to be an appropriate level of expense in relation to generated revenues. The rooms manager will expect that all projected expenses will fall within acceptable range of the standard cost percentage for each category. The rooms manager may also build improvements on past cost percentages into selected expense categories, assuming that greater efficiencies will be achieved through better training, closer supervision, and tighter controls. The rooms manager's goal is to maximize the department's income by minimizing its expenses while still preserving or enhancing the service levels. Crucial to achieving this goal are the executive housekeeper's calculations of anticipated expenses and comments on how budget adjustments may affect the quality of service.

The operating budgets under which the executive housekeeper operates take the form of monthly income statements for the rooms division. Projected revenues and expenses for each month of the budgeted period will represent the rooms division's operational plan. The executive housekeeper will be held accountable for controlling the expense areas that fall within the housekeeping department's areas of responsibility. As the budgeted period progresses, monthly income statements will be produced that show the actual amounts alongside the amounts originally budgeted.

Exhibit 3 shows a monthly budget report for rooms that indicates both budgeted forecasts and actual results. The last two columns in Exhibit 3 show dollar and percentage variances. The dollar variances indicate the differences between actual results and budgeted amounts. Dollar variances are generally considered either favorable or unfavorable based on the following situations:

	Favorable Variance	Unfavorable Variance
Revenue	Actual exceeds budget	Budget exceeds actual
Expense	Budget exceeds actual	Actual exceeds budget

For example, the actual amount of salaries and wages for rooms division personnel in the month of January was $20,826, while the budgeted amount for salaries and wages was $18,821, resulting in a variance of $2,005. The variance is

Exhibit 3 Sample Monthly Rooms Division Budget Report

Holly Hotel
Budget Report—Rooms Division
For January 19XX

			Variances	
	Actual	Budget	$	%
Revenue				
Room Sales	$156,240	$145,080	$11,160	7.69%
Allowances	437	300	(137)	(45.67)
Net Revenue	155,803	144,780	11,023	7.61
Expenses				
Salaries and Wages	20,826	18,821	(2,005)	(10.65)
Employee Benefits	4,015	5,791	1,776	30.67
Total Payroll and Related Expenses	24,841	24,612	(229)	(0.93)
Other Expenses				
Commissions	437	752	315	41.89
Contract Cleaning	921	873	(48)	(5.50)
Guest Transportation	1,750	1,200	(550)	(45.83)
Laundry and Dry Cleaning	1,218	975	(243)	(24.92)
Linen	1,906	1,875	(31)	(1.65)
Operating Supplies	1,937	1,348	(589)	(43.69)
Reservation Expenses	1,734	2,012	278	13.82
Uniforms	374	292	(82)	(28.08)
Other Operating Expenses	515	672	157	23.36
Total Other Expenses	10,792	9,999	(793)	(7.93)
Total Expenses	35,633	34,611	(1,022)	(2.95)
Departmental Income (Loss)	$120,170	$110,169	$10,001	9.08%

bracketed to indicate an unfavorable variance. However, if the revenue variance is favorable, an unfavorable variance in expenses (such as in payroll) is not necessarily negative. Rather, it may merely indicate the greater expense of serving more guests than were anticipated when the budget was created.

Percentage variances are determined by dividing the dollar variance by the budgeted amount. For example, the 7.61 percentage variance for net revenue shown in Exhibit 3 is the result of dividing the dollar variance figure of $11,023 by the budgeted net revenue amount of $144,780—and then multiplying by 100.

Virtually all actual results of rooms division operations will differ from budgeted amounts for revenue and expense items on a budget report. This is only to be expected because any budgeting process, no matter how sophisticated, is not perfect. Executive housekeepers should not analyze every variance. Only significant variances require management analysis and action. The general manager and

Exhibit 4   Summary of Forecasted Rooms Sales

	Budget Period	Occupancy Percentage	Number of Occupied Rooms	Average Price per Room	Total Rooms Sales
	1.				
	2.				
	3.				
	4.				
	5.				
Months of the Year	6.				
	7.				
	8.				
	9.				
	10.				
	11.				
	12.				

controller should provide the executive housekeeper with criteria for determining which variances are significant.

Budgeting Expenses

The budgeting process begins with a forecast of room sales. Since expense levels in all the expense categories on the departmental income statement vary with occupancy, everything in the operating budget depends upon how accurately occupancy levels are forecasted.

Early in the budget planning process, the rooms manager will give the executive housekeeper the yearly forecast of occupancy levels, broken down into monthly budget periods. This information may be delivered on a form such as the one in Exhibit 4. Using historical data, along with input from the hotel's marketing department, the rooms manager will predict the occupancy percentage for each budgeted period. The second column of the form translates the anticipated occupancy percentage into the actual number of rooms expected to be occupied. By multiplying the number of expected occupied rooms by the average rate per room, the rooms manager can forecast the amount of revenue anticipated from room sales. For the rooms manager, this projection of revenue is the most important part of the operating budget. The appropriateness of all expenses expected will be measured in terms of the percentage of revenue represented by each expense category.

For the executive housekeeper, the most important information in the rooms manager's forecast is not so much the total expected sales dollars, but the projected number of occupied rooms for each budget period. This is because nearly all the expense levels for which the executive housekeeper is responsible are directly dependent upon the number of occupied rooms the housekeeping department will have to service.

The executive housekeeper can predict a certain level of expense for each expense category when he/she knows: (1) the cost per occupied room for each category of expense and (2) the number of occupied rooms forecasted for each budget periods. At this point, the budgeting process simply involves relating costs per occupied room to the forecasted occupancy levels.

Salaries and Wages

Salaries and Wages expense for the housekeeping department is related to such positions as executive housekeeper, assistant housekeepers, inspectors, linen room attendants, room attendants, housepersons, lobby attendants, and others employed in the housekeeping operation.

By using a staffing guide, the executive housekeeper can determine how many employees of each job classification are needed to ensure smooth operations at varying levels of occupancy. When planning the Salaries and Wages expense for the operating budget, the executive housekeeper can use the staffing guide in conjunction with the occupancy forecasts to determine staffing needs for each budget period. After determining the number of labor hours needed for each job category, the executive housekeeper can multiply the number of hours by the position's average per-hour wage to calculate the expected cost for that job category. By summing the calculations for all positions, a total wage cost can be determined for each budget period. Costs associated with salaried positions in the housekeeping department can be averaged into each monthly budget period. In forecasting salary and wage costs, the executive housekeeper will also need to account for any scheduled salary and wage increases as well as any cost-of-living adjustments planned by the property.

Employee Benefits

Calculations related to employee benefits depend on the number of labor hours expected to be scheduled, the types of job classifications involved, and the property's policies regarding employee benefits. The kinds of benefits in this expense category may include charges for the cost of holiday or vacation pay, employee meals, payroll taxes, medical expenses or insurance, social insurance such as pensions, and staff parties or social events. With the help of human resources or accounting staff, the executive housekeeper can determine what levels of expense to budget for employee benefits.

Outside Services

If the hotel employs any outside contractors for major cleaning projects or for laundry and dry cleaning services, then the costs of these services are averaged throughout the budget periods. The executive housekeeper can consult current contracts or past invoices to determine the expense levels to budget.

In-House Laundry

The executive housekeeper needs to work closely with the laundry manager to budget laundry expenses. The forecasts of occupancy levels provided by the rooms division, along with the property's staffing guide, will be the basis for determining all expenses related to salaries, wages, and benefits for laundry personnel.

The cost of operating the hotel's on-premises laundry is directly related to the volume of soiled items to be processed. This, in turn, is a direct function of the hotel's occupancy levels. Therefore, the cost of laundering room linens and uniforms can be budgeted on the basis of historical information that shows the cost per occupied room of laundry operations. Multiplying the cost per occupied room for laundry operations by the number of occupied rooms forecast for each budget period will provide a figure for the expected laundry expense during the budget period.

Linens

Although linen supplies in the housekeeping department are a recycled inventory item, their lifespan is ultimately limited. New linens must be purchased throughout the year as older linens are removed from service due to loss, damage, or wear. Replacement cost for new linens is an expense that needs to be worked into the budget planning process.

Monthly physical inventories of linens show the executive housekeeper how long the existing stock of linens lasts and how much of each type of linen needs to be reordered to maintain appropriate par levels. The results of physical inventories of linens are submitted to the hotel's general manager who routinely transfers the information to the hotel's accounting department. In turn, the accounting department regularly processes the information and provides valuable statistical information related to usage rates, losses, and expenses per occupied room. The executive housekeeper can use the cost per occupied room for replacement linen to forecast linen expense for the periods covered by the operating budget. Multiplying the cost per occupied room for linen replacement by the number of occupied rooms forecasted for the budget period will yield the linen expense to be built into the operating budget.

Operating Supplies

The operating supplies expense category for the housekeeping department includes non-recycled inventory items, such as guest supplies and amenities, cleaning supplies, and small equipment items. As with the other housekeeping expense categories, the executive housekeeper can budget for the costs of these items on the basis of cost per occupied room.

Guest supplies include pens, stationery, matches, soap, shampoo, toilet and facial tissue, garment bags, and other amenities the hotel provides in each room for the convenience and use of its guests. The cost per occupied room for guest supplies is the same as the cost of one room par for these items. Budget amounts for guest supplies are determined by multiplying their cost per occupied room by the number of occupied rooms in the budget's forecast.

Cleaning supplies include not only chemical cleaners, polishes, and detergents, but also small equipment items needed on a daily basis such as applicators, brooms, brushes, mops, buckets, spray bottles, and a variety of cleaning cloths. By following inventory control procedures, the executive housekeeper has an effective system for tracking the usage rates for the various cleaning supply items at different levels of occupancy. By dividing the cost of the number of purchase units used each month by the number of occupied rooms that month, a cost per occupied room can be established for each item in the cleaning supply inventory. Summing the results for all inventoried items yields a cost per occupied room for cleaning supplies. Multiplying this figure by the number of occupied rooms forecasted for the budget period provides the cleaning supply expense for the operating budget.

Uniforms

Provisions must be made in the operating budget for the cost of new and replacement uniforms. In addition, the cost of washing or dry cleaning uniforms, as well as costs associated with repairing damaged uniforms, may need to be reflected in the operating budget.

Like linens, uniforms are a recycled inventory item. But unlike linens—whose usage rates and replacement needs are very predictable—the need for new uniforms during the budget period depends on additional factors such as personnel turnover and new hirings. To help organize information for the operating budget and for future purchasing, the executive housekeeper should maintain an itemized list of all types of uniforms maintained in the department's inventory. The cost information should be itemized for each part (e.g., shirt, blouse, pants, skirt, etc.) of each type of uniform. The number of people working in each uniformed position may be obtained from human resources. Since men and women in the same position may require different uniforms—sometimes at different costs—the executive housekeeper also needs to consider the number of men and women occupying each uniformed position.

There are some general rules of thumb that the executive housekeeper can use when budgeting for uniform purchases. While these rules of thumb may be helpful, executive housekeepers should keep in mind that uniform par levels vary from property to property.

The executive housekeeper should start by budgeting for one complete uniform for each person. Next, for uniforms that are dry cleaned, the executive housekeeper should budget for one additional uniform per person. For uniforms that are washed, the executive housekeeper should budget for two additional uniforms per person, since laundering greatly reduces a uniform's useful lifespan. As a final rule of thumb, the executive housekeeper should budget three additional sets of uniforms for cooks. Taking into consideration an annual plan for replacing uniforms, the executive housekeeper should divide the cost of new uniforms into those months during the budget period in which they will be purchased.

In determining the cost of repairing uniforms, the executive housekeeper needs to consider not only the materials needed for repairs, but also the cost of the time spent by the supervisor or seamstress to repair the uniforms. Records of past repairs and productivity standards for repair can provide information relevant to the executive housekeeper's estimate for budgeting the cost of repairs.

Controlling Expenses

Controlling housekeeping expenses means ensuring that actual expenses are consistent with the expected expenses forecast by the operating budget. There are basically four methods the executive housekeeper can use to control housekeeping expenses: accurate recordkeeping, effective scheduling, careful training and supervision, and efficient purchasing.

Maintaining accurate records is the first step in controlling expenses and identifying problems in relation to managing inventories. Accurate recordkeeping enables the executive housekeeper to monitor usage rates, inventory costs, and variances in relation to standard cleaning procedures.

Effective scheduling permits the executive housekeeper to control salaries and wages and the costs related to employee benefits. It is important to schedule all housekeeping employees according to the guidelines in the property's staffing guide. Since the staffing guide bases its guidelines on the level of room occupancy, it ensures that personnel costs stay in line with occupancy rates. At the same time, the need for adequate staffing to maintain the desired level of service leaves the executive housekeeper little room to "cut corners" by scheduling fewer employees than the staffing guide recommends. The executive housekeeper can ensure that the approved guidelines expressed in the property's staffing guide are consistently followed in all employee scheduling decisions. Adjusting weekly work schedules in light of anticipated occupancy levels is an ongoing responsibility of the executive housekeeper.

Training and supervision should not be overlooked as a cost control measure. The recommendations in the property's staffing guide are based on the assumption that certain performance and productivity standards are consistently achieved. Effective training programs that quickly bring new-hires "up to speed" can significantly reduce the time during which productivity is lower than the standards set for more experienced personnel. Close and diligent supervision, as well as refresher training, can ensure that performance and productivity standards are met—and may even bring about improvements. Finally, effective training and supervision are an important part of controlling the cost of inventoried items. For example, training employees in the proper use of cleaning supplies can improve usage rates and, over time, lower the cost of cleaning supplies per occupied room.

Efficient purchasing practices afford the executive housekeeper the greatest opportunity to control department expenses. The executive housekeeper bears an important responsibility in making sure that the hotel's money is well spent and the maximum value is received from products purchased for use.

Purchasing Systems

Efficient purchasing practices can make a significant contribution to the executive housekeeper's role in controlling housekeeping expenses. In fact, the most controllable expenses under the executive housekeeper's responsibility involve the various items whose inventories are maintained by the housekeeping department. Inventory control procedures enable the executive housekeeper to know when and how much to buy for each inventoried item. Deciding what to buy, whom to buy it from, and exactly how to purchase it requires careful consideration on the part of the executive housekeeper.

Although the actual purchasing may be done by the hotel's purchasing department, quantities and specifications are submitted to the purchasing department by department heads. When ordering items for the housekeeping department, the executive housekeeper will need to fill out and sign a purchase order form such as that shown in Exhibit 5. This order form then has to be approved by the controller and general manager. For all items purchased for the housekeeping department, the recommendation as to which product to purchase, in what quantities, and even from what vendor is made by the executive housekeeper. Although different properties have different procedures for processing and approving purchases, the evaluation of what's needed, when, how much, and from whom fall under the responsibility of department heads. The executive housekeeper needs to know how to obtain the best value when purchasing the variety of items needed by the housekeeping department.

Linen Replacement

Next to salaries and wages, linens are the highest expense item in the housekeeping budget. The initial purchase of linens for the hotel will greatly influence the costs of replacing linens that become lost or are taken out of service due to damage or excessive wear. The fabric type, size, and color will influence both initial purchases and replacement costs. Colored items are usually more expensive and have shorter lifespans since the colors fade through repeated washings.

The physical inventory records show the executive housekeeper how long the existing stock of linens lasts and how much of each type of linen needs to be reordered to maintain par levels. Typically, linen purchases are made annually with deliveries scheduled to be drop-shipped on a quarterly basis. This arrangement enables the executive housekeeper to conserve available storage space by using a supplier's warehouse facilities while periodically receiving replacement stock.

Planning linen purchases on a yearly basis can also result in considerable savings. Linen brokers provide a convenient and quick way to purchase linens, but they are expensive. Ordering larger quantities in bulk can often win lower per-unit prices. Planned annual linen purchases also enable large hospitality chains to order linen supplies directly from linen mills. Although these orders require considerable lead time to prepare, a property saves on the premiums charged by linen brokers to process orders and arrange deliveries. Unforeseen emergency needs could then be filled through a linen broker.

Quantities of linen to be purchased are determined by assessing the hotel's quarterly requirements to maintain linen at the proper par level. Physical inventories of linens can be used to calculate an annual consumption rate that shows how much linen is "used up" either by normal wear and tear, damage, loss, or theft.

Exhibit 5   Sample Purchase Order

Purchase Order					

Purchase Order Number: _____

Order Date: _____

Payment Terms: _____

To: _____
(supplier)

From/ Ship to: _____
(name of food service operation)

_____
(address)

_____

_____
(address)

Delivery Date: _____

Please Ship:

Quantity Ordered	Description	✓	Units Shipped	Unit Cost	Total Cost

Total Cost _____

Important: This Purchase Order expressly limits acceptance to the terms and conditions stated above, noted on the reverse side hereof, and any additional terms and conditions affixed hereto or otherwise referenced. Any additional terms and conditions proposed by seller are objected to and rejected.

_____
Authorized Signature

With this information, the executive housekeeper can use the following formula to determine the size of annual linen purchases:

$$\text{Annual Order} = \text{Par Stock Level} - \text{Linen On Hand}$$

The executive housekeeper is expected to carefully select suppliers and linen products to ensure that the hotel receives good value for money spent. The most important considerations are the suitability of the products for their intended uses and whether the products are economical. Regarding linen, the expected useful life of the linen is often more important than purchase price in determining whether alternative products are economical or not. The cost of laundering linens over their useful life is usually much greater and more important than their initial price.

The lifespan of linen is measured in terms of how many times it can be laundered before becoming too worn to be suitable for guestroom use. Linen that is purchased at bargain prices but that wears out after only moderate laundering will damage guests' perceptions of quality, increase annual linen usage rates, and increase costs in the long run. Durability, laundry considerations, and purchase price are the main criteria to use in selecting linen. A cost per use can be calculated in order to evaluate alternative linen purchases using the following formula:

$$\text{Cost per use} = \frac{\text{Purchase Cost + Lifespan Laundering Costs}}{\text{Number of Lifespan Launderings}}$$

The laundering costs over the lifespan of a linen product can be determined by multiplying the item's weight by the hotel's laundering cost per pound—and then multiplying again by the number of launderings the item can withstand before showing sufficient wear.

When orders of new linens are received, shipments should be checked against purchase orders and inspected to ensure that the linens meet all quality and quantity specifications. Newly received linen orders should be immediately moved to the main linen room for storage. In the main linen room, new linens that have not yet been put into service should be stored separately from linens that are already in use.

Inventories for all new linen received and issued at the hotel should be kept on a perpetual basis. This means that a running count should be kept for on-hand quantities of every type of new linen stored in the main linen room. The inventory record should show linen type, specific item, price, storage location, and dates of ordering and receiving. As linen items are put into service to replace worn, damaged, lost, or stolen linen, the quantity recorded on the perpetual inventory record should be adjusted accordingly.

The executive housekeeper is responsible for placing new linen in use on an as-needed basis to maintain the par level for each linen item. Issuing new linen to be used in daily operations typically occurs each month on the basis of shortages revealed by a physical inventory. New linen may also be issued between physical inventories to replace discarded linens. Some hotels inject a predetermined quantity of new linen into circulation at pre-established intervals based on past usage rates. New linens should be placed into service on a "first-in first-out" basis. New linen not in service should be under the control of the executive housekeeper or laundry manager in the main linen room or other secure place.

Uniform Replacement

Uniforms need to be replaced when they become damaged or worn. The executive housekeeper needs to establish a procedure for issuing new or replacement uniforms. A notation could be made on the employee's uniform card that a damaged uniform was received and discarded, and that a new uniform was issued. The date and the employee's signature should be recorded on the inventory card.

The executive housekeeper is generally responsible for receiving, storing, and controlling all new uniforms held in the hotel's custody but not placed into service. The executive housekeeper is also responsible for placing new uniforms into service to ensure that all uniform requirements are met, clean uniform replacements are available, and the laundry is not unduly burdened with clean uniform production.

As with linens, the main criteria for purchasing replacement uniforms are durability, lifespan, and the quality of materials. The purchase price of uniforms is a secondary consideration. Comfort, practicality, and ease of maintenance are also important considerations. New uniforms should be purchased to maintain the par levels established for the different kinds of uniforms. Comparing the on-hand quantities with the established par levels will show the executive housekeeper how many replacement uniforms should be ordered.

Purchasing Operating Supplies

Some hotel chains have centralized national purchasing systems for major housekeeping items in order to achieve quantity discounts. Other hotels may join together in purchasing groups to achieve savings on bulk purchases of commonly used items. But, for the most part, the large number and variety of operating supplies are purchased by the individual property and through the direct involvement of the executive housekeeper.

Inventory tracking forms can be used to create an exhaustive list of operating supplies that the executive housekeeper will need to purchase on a regular basis. Inventory control procedures will show how often and in what quantities supply items will need to be purchased to maintain par levels. Usage rates and cost per occupied room figures can be determined from the inventory records. This information can form the basis for an effective purchasing system. By following careful purchasing procedures, the executive housekeeper can help the hotel control costs while ensuring that adequate supply levels are maintained.

Before buying any product, the executive housekeeper should obtain samples in order to test the product and determine whether it meets specifications. Suitability for the intended task, quality, ease of handling, and storage requirements are just as important as the price in determining whether a product is economical.

Value—not price—should be the leading consideration in making purchase decisions. An inexpensive cleaning agent that has to be used in much larger quantities than a more expensive one may actually cost more in the long run. The crucial concern is to obtain the best value for the money.

Selecting the right vendors can often make the executive housekeeper's purchasing systems more efficient. The executive housekeeper needs to competitively shop suppliers and vendors for the products to be purchased on a regular basis. When asking for price quotations, the executive housekeeper needs to be as precise as possible regarding such specifications as weight, quality, packaging, size, concentration, quantities, and delivery times.

In evaluating alternative suppliers, the executive housekeeper needs to be concerned with how well the supplier will service the hotel's account. It is important that selected vendors appreciate the operations of a hotel's housekeeping department, fully understand the products they sell, and are able to provide demonstrations and even training in how to use the products. It is not unusual for the executive housekeeper to select one vendor for all guest supply items, another for cleaning products, and still another for all paper products. By limiting the number of suppliers with whom the housekeeping department has to deal, the executive housekeeper can streamline the purchasing process, reduce paperwork, and use time more efficiently. In addition, concentrating business with a limited number of suppliers often achieves greater purchasing—and thereby bargaining—power, resulting in improved quantity discounts and better service.

Another consideration in selecting vendors is whether they will be able to stock the products the hotel purchases at their own warehouse facilities and drop-ship the products to the hotel on an as-needed basis. This enables the executive housekeeper to achieve savings by purchasing products in bulk whenever possible and, at the same time, solve the problem of limited storage space.

In the process of reordering operating supplies, the executive housekeeper needs to periodically re-evaluate the suitability of existing products for their intended purposes. Meeting with housekeeping staff who use the product can help determine any problems that may lead to a reconsideration of quality or functionality. The functionality of the product should be tested, and the executive housekeeper should determine whether the existing specifications for the product should remain the same. Alternative products should be investigated and compared to existing products in terms of performance, durability, price, and value.

Worksheets can be used to monitor usage rates and costs for the different types of operating supplies kept in inventory. Exhibit 6 illustrates one such worksheet for tracking the use of various chemical cleaners. For each product, the Monthly Chemical Use Report identifies the vendor, the product name, and its intended use. Each month, physical inventories provide the executive housekeeper with information concerning how many purchase units of each chemical cleaner have been used. Multiplying the number of units used by the cost per unit yields the total cost of the product used during the month. Dividing the total cost by the number of occupied rooms yields a cost per occupied room figure for each product. By reducing the size of each purchase unit (e.g., gallons, cans, pints, quarts) to a common-sized unit (e.g., ounces) and multiplying the number of purchase units used by the common-sized amounts, the total amount used for each product can be determined in terms that render the different-sized products comparable. After using a common measure to calculate the actual amounts used, the executive housekeeper can divide by the number of occupied rooms to determine the usage of each product per occupied room. In this way, the Monthly Chemical Use Report enables the executive housekeeper to compare the relative efficiency of using different products for similar tasks. By comparing the costs per occupied room and the usage per occupied room achieved by alternative products, the executive housekeeper can evaluate which products yield greater cost savings and base purchasing decisions accordingly.

Exhibit 6   Monthly Chemical Use Report

						COST			**USAGE**
					COST	**PER**	**AMOUNT**		**PER**
			UNITS	**PER**	**TOTAL**	**OCC.**	**PER**	**TOTAL**	**OCC.**
VENDOR	**PRODUCT**	**INTENDED USE**	**USED**	**UNIT**	**COST**	**ROOM**	**UNIT**	**UNITS**	**ROOM**
Johnson	G.P. Forward	All-purpose cleaner	108 gal.	$3.60	$388.80	$.0131	128 oz.	13,284	.4669
Johnson	J-Shop 600	Degreaser	39 gal.	4.95	193.05	.0065	128 oz.	4,992	.1686
Johnson	Freedom	Floor stripper	48 gal.	8.24	395.52	.0136	128 oz.	6,144	.2075
3M	Trouble Shooter	Special area stripper	15 can	6.01	90.15	.0030	23 oz.	345	.0017
Johnson	Complete	Composition floor wax	142 gal.	6.60	937.20	.0317	128 oz.	18,176	.6139
Johnson	Fortify	Porous floor seal	67 gal.	7.70	515.90	.0174	128 oz.	8,576	.2897
Johnson	Snap Back	Spray buff	.5 gal.	7.45	3.73	.0001	128 oz.	64	.0022
Johnson	Conq-R-Dust	Dust mop treatment	2 gal.	10.95	21.90	.0007	128 oz.	256	.0086
SSS	Waterless Cleaner	Wood floor stripper	35 gal.	8.90	311.50	.0105	128 oz.	4,480	.1513
SSS	Traffic Wax	Wood floor wax	5 gal.	12.30	61.50	.0021	128 oz.	640	.0216
Johnson	Rugbee Dry Foam Shampoo	Carpet shampoo	17 gal.	11.99	203.83	.0069	128 oz.	2,176	.0735
SSS	SSS Defoamer	Steam cleaner	6 gal.	14.53	87.18	.0029	128 oz.	768	.0259
Johnson	Rugbee Carpet Deodorizer	Powdered deodorant	3 can	3.65	10.95	.0004	24 oz.	72	.0024
P-C	Odor-Out–Carpet	Powdered deodorant	24 can	7.12	170.88	.0058	32 oz.	768	.0259
Core	Unbelievable	Carpet spotter	70 pint	3.04	212.80	.0072	16 oz.	1,120	.0378
SSS	Gum Remover	Tar & gum remover	61 can	2.88	175.68	.0059	12 oz.	732	.0247
P-C	Kilroy	Graffiti remover	3 can	4.54	13.62	.0005	15 oz.	45	.0015
Zep	Once Over	Wall & vinyl cleaner	27 can	3.13	84.51	.0029	22.5 oz.	607.5	.0205
P-C	Purge	Bathroom bowl cleaner	1 quart	4.79	4.79	.0002	32 oz.	32	.0011
Butchers	Glad Hands	Dispenser soap	12 gal.	9.17	110.04	.0037	128 oz.	1,536	.0519
SSS	Sani-Fresh	Dispenser soap	67 pack	3.88	259.96	.0088	32 oz.	2,144	.0724
Johnson	Lemon Shine Up	Furniture polish	35 can	2.04	71.40	.0024	15 oz.	525	.0177
3M	Stainless Steel Polish	Metal cleaner	52 can	4.25	221.00	.0075	21.5 oz.	1,118	.0378
3M	Tarni-Shield	Brass cleaner	43 bottle	5.58	239.94	.0081	10 oz.	430	.0145
	Vinegar	Neutralizer	26 gal.	1.05	27.30	.0009	128 oz.	3,328	.1124

For April, 19XX
Number of Occupied Rooms—29,608

Courtesy of Opryland Hotel, Nashville, Tennessee

Capital Budgets

Purchases of most inventoried items in the housekeeping department occur monthly. These costs appear in the operating budget as expenses against the revenue generated over the same period. Major purchases of machines and equipment in the housekeeping department are not included on operating budgets. Instead, purchases for items with relatively high costs and long lifespans are planned as part of capital budgets because they involve additional capital investments by the hotel.

Capital budgets are prepared annually. The executive housekeeper will be asked to specify the need for funds to purchase machines and equipment for the housekeeping department. It is crucial that the executive housekeeper be prepared to justify any requests for capital expenditures. Although such requests may be part of an overall modernization or renovation program, they more typically involve a need to replace existing machines or equipment.

Typically, the need to replace major machines and equipment is discovered when a particular item cannot be repaired. However, the executive housekeeper can effectively predict the useful life of each machine in the housekeeping department based on how often it is used and the estimated number of working hours provided by the machine's manufacturer and supplier. Executive housekeepers should be aware, however, that machines and equipment that receive high usage will not live up to the guarantees and estimates of useful life provided by suppliers.

When purchasing housekeeping equipment, executive housekeepers need to focus on long-range considerations. Major purchases of machines and equipment represent a capital expense for the hotel, and planning is required. Whenever possible, it is important to choose a supplier who can service the machines in a quick and efficient manner. If such a supplier cannot be found, the executive housekeeper will need to order an adequate number of replacement parts so that the hotel itself will be able to service the machines.

The executive housekeeper is expected to be able to recommend the proper type, quality, and quantity of equipment needed to keep guestrooms and public areas clean and attractive. The housekeeping department needs equipment that will last through continuous use with a minimum of maintenance. Cost effectiveness is the most important consideration. As always, purchase price needs to be considered along with the quality and durability of the product.

Contract vs. In-House Cleaning

A number of outside contractors offer a variety of cleaning services to hotels. Outside contractors are available for nearly any cleaning task that needs to be done, including outside laundry and dry cleaning services, floor cleaning and care, outside window cleaning, overhead cleaning, masonry cleaning, and descaling and scouring of restroom fixture traps. Given these services, a hotel could even use outside contractors for its entire housekeeping operation.

An important decision that arises in an increasing number of contexts is whether to contract outside services for cleaning tasks or undertake them as in-house operations. The issue is often approached in terms of how to best control costs while ensuring that necessary tasks are accomplished and quality standards are maintained. In many situations, the issue is one that involves both capital and operating budget considerations.

While wages and materials are monthly expenses that can be budgeted, the equipment needed to start an in-house cleaning program is a capital expense that occurs all at once. Often, after the initial start-up cost for machines and equipment, the monthly expense that the hotel will incur with an in-house cleaning program is less than the monthly expense it would incur with an outside contractor. In addition, many executive housekeepers believe that an in-house staff will perform higher quality work than an outside contractor because of the opportunity for increased control.

The executive housekeeper may be asked to demonstrate how long it would take to recover the initial start-up costs for machines and equipment through the monthly savings achieved by an in-house cleaning program. By dividing the savings achieved each month into the total amount of capital expenditure for the needed equipment, the executive housekeeper can calculate how many months it would take to pay back the initial investment. In determining monthly expenses that an in-house operation would incur, the executive housekeeper needs to consider the costs of salaries and wages, employee benefits, materials and supplies, training, and supervision. The decision as to whether the initial investment is possible and worth the monthly savings is one that belongs to the hotel's upper management and ultimately to the owners.

In some situations, the initial start-up costs are deemed too high for the monthly savings achieved. In other cases, the nature of the cleaning task may be so specialized or infrequent that initiating an in-house cleaning program is not reasonable or cost-efficient. In still other situations, the monthly expense for outside contractors is lower than the monthly expenses an in-house operation would incur in performing the same tasks. There are always pros and cons to consider when assessing the need for contracted cleaning services.

For whatever reason, the executive housekeeper will sometimes be charged with the task of arranging for outside contractors to perform some cleaning service. The initial problem is choosing the appropriate contractor. The executive housekeeper should request cost estimates from at least three different contractors. For area cleaning tasks, contractors base quotations on the exact size of the area to be cleaned; the executive housekeeper cannot collect comparable cost quotations until exact measurements are obtained. For laundering tasks, quotations are based on dry weight or on a per piece basis. Quotations also specify the frequency of the service desired as well as collection and delivery times. For any kind of contract cleaning, the executive housekeeper should obtain cost quotations on the basis of carefully defined needs, precise descriptions of work, and clear indications of the frequency of service. The same specifications should be submitted to each contractor to ensure that cost estimates are fully comparable.

Both previous and current clients of each contractor should be checked for reports on the quality and efficiency of the services. The reputation and ability of the contractor's local organization, as opposed to the credentials of the home office, should be assessed. Visiting the contractor's place of business may provide insights into the kind of operation the contractor runs.

After selecting an appropriate contractor, it is important to establish the precise nature and frequency of the services desired in the written contract. The terminology describing the task, frequency, and expected performance needs to be as clear and precise as possible. All contracts should incorporate a cancellation clause. Certain contracts should also have penalty clauses to ensure compliance with all specifications.

After contracting an outside cleaning service, it is essential to monitor the quality of the contractor's work. Routine inspections and regular meetings with the contractor will enable the executive housekeeper to identify and discuss any problems or concerns in a timely manner. Assigned tasks and completion dates should be discussed clearly and documented in writing. It is also important to monitor invoices received from outside contractors; invoices should be checked for accuracy before being submitted to the accounting department for payment.

While the use of outside contractors for cleaning services appears to be increasing in the hospitality industry, the executive housekeeper should periodically assess whether replacing outside services with in-house operations can be justified as a cost control measure. After the initial capital investment in machines and equipment is recovered through monthly savings in operating expenses, the reduced costs and increased control that often accompanies in-house operations can be of significant value to the property.

Key Terms

capital budget
capital expenditures
contribution margin
income statement

operating budget
operating expenditures
pro forma income statement
schedules

Discussion Questions

1. What are the basic responsibilities of the executive housekeeper in the budget process?

2. What is the difference between a capital budget and an operating budget?

3. Why is forecasting occupancy levels such a critical part of the budget planning process?

4. How can the operating budget be used as a tool to control expenses?

5. What is the relationship of an operating budget to an income statement?

6. What are some of the typical expenses (or line items) an executive housekeeper might encounter when preparing the department's budget?

7. What two factors can help the executive housekeeper predict a certain level of expense for each expense category?

8. What are the four basic methods an executive housekeeper can use to control expenses?

9. What basic responsibilities does the executive housekeeper have in terms of purchasing operating supplies?

10. What types of situations should be assessed when deciding whether to use outside contractor for cleaning services?

REVIEW QUIZ

When you feel you have covered all of the material in this chapter, answer these questions. Choose the *best* answer.

True (T) or False (F)

T   F   1. The operating budget outlines the financial goals of a hotel.

T   F   2. A capital budget provides the executive housekeeper with financial standards by which to measure the success of the department.

T   F   3. The operating budget estimates expenses while the capital budget forecasts revenues.

T   F   4. The first step in the budget planning process is to forecast room sales.

T   F   5. Most expenses that the executive housekeeper can expect during the course of everyday operations are directly related to room occupancy levels.

T   F   6. If the number of occupied rooms is higher than anticipated, a corresponding decrease in the housekeeping expenses should be expected.

T   F   7. The hotel's income statement expresses the actual results of operations over a specified period.

T   F   8. The revenue generated by the rooms division is often the largest single amount produced by revenue centers within a hotel.

T   F   9. The Uniforms expense item on the rooms division income statement includes only the cost of purchasing or renting uniforms for housekeeping staff.

T   F   10. Dollar variances for expense items on a budget report are unfavorable when actual amounts exceed budgeted amounts.

Alternate/Multiple Choice

11. Which of the following is a capital expense item incurred by the housekeeping department?

    a. room attendant carts
    b. cleaning supplies

12. The executive housekeeper is directly concerned with many of the line items listed in the expense sections of the:

    a. hotel's income statement.
    b. rooms division income statement.

13. The most important information in the rooms manager's forecast in relation to budgeting expenses in the housekeeping department is the:

    a. total expected sales dollars.
    b. projected number of occupied rooms.

14. The largest single expense category on the rooms division income statement is typically:

  a. energy costs.
  b. allowances.
  c. salaries and wages.
  d. contract cleaning.

15. For items purchased by the housekeeping department, the recommendation as to which product to purchase, in what quantities, and from what vendor is generally made by the:

  a. general manager.
  b. purchasing agent.
  c. executive housekeeper.
  d. controller.

Chapter Outline

Safety
    Insurance and Liability Concerns
    Employee Morale and Management
Concerns
    Potentially Hazardous Conditions
    Job Safety Analysis
    Safety Training
OSHA Regulations
    Work Areas
    Means of Egress
    Hazardous Materials
    Sanitation
    Signs and Tags
    First Aid
    OSHA Inspection
Security
    Security Committees
    Suspicious Activities
    Theft
    Bomb Threats
    Fires
    Key Control
    Lost and Found
    Guestroom Cleaning

Learning Objectives

1. Discuss how an unsafe work environment can be costly from a medical, legal, and productivity standpoint.

2. Identify safety procedures that relate to tasks commonly performed by the housekeeping staff.

3. Describe the steps involved in a job safety analysis.

4. Identify the basic elements of a safety training program.

5. Explain how OSHA regulations pertain to housekeeping operations.

6. Describe the function of a hotel security committee.

7. Identify several methods for reducing the incidence of guest and employee theft.

8. Describe three different types of fire detection systems.

9. Describe the basic components of a fire safety training program.

10. Identify elements of a key control program.

11. Describe the lost and found function as it applies to housekeeping.

Safety and Security

This chapter was written and contributed by Sheryl Fried, Assistant Professor,
School of Hotel and Restaurant Management,
Widener University, Chester, Pennsylvania.

SAFETY AND SECURITY are two major responsibilities of hotel managers. Guests expect to sleep, meet, dine, and entertain in a facility that is safe and secure—and are entitled to reasonable care under law. Housekeeping personnel can help meet this guest expectation and, in some cases, make the difference in the property's safety and security system.

In a hospitality operation, **safety** refers to the actual conditions in a work environment. **Security** refers to the prevention of theft, fire, and other emergencies. This chapter will focus on safety and security needs of hospitality operations, the role that housekeeping plays in meeting those needs, and how safety and security issues affect housekeeping personnel.

Safety

The two hotel departments most likely to have the largest number of accidents and injuries are maintenance and housekeeping. One basis for this frequency is the sheer labor-intensity of these two departments. In many operations, housekeeping and maintenance employ more people than any other department. Another reason lies in the fact that working in housekeeping or maintenance involves physical activity and equipment use—both of which increase the risk of accident and injury.

To reduce safety risks, the executive housekeeper must be aware of potential safety hazards and develop procedures to prevent accidents. Safety should be a top priority. Ongoing safety training programs help ensure that safe conditions are maintained in all work areas. To develop such programs, management must be aware of the laws that regulate the work environment—and more specifically, how those laws affect housekeeping personnel.

Insurance and Liability Concerns

An unsafe work environment can be costly from a medical, legal, and productivity standpoint. Many work-related accidents result in loss of work. In addition, an employee injured on the job may require medical care. In some cases, the medical care can be extensive and expensive.

A chronic record of employee accidents will generally result in increasing costs for liability and medical insurance. A track record of unsafe work conditions can also result in the hotel being fined or sued. Liability insurance rates are high enough without being compounded by lawsuits from injured employees and guests. Overall **workers' compensation** rates may also increase if many workers' compensation claims are made over a period of time. All these charges add up. In other words, poor safety habits can cost a very substantial amount of money over time.

Employee Morale and Management Concerns

Unsafe working conditions have a negative effect on employee morale. If employees are preoccupied with hazardous conditions in the workplace, they will not be able to perform to the best of their ability. For the most part, it is difficult to motivate employees until the unsafe conditions are corrected.

One of management's top concerns should be for the health and welfare of employees. Employees are one of the most important assets a hotel has. If managers want employees to provide quality service, they must treat employees fairly and with respect. Respect for an employee's right to work in a safe and hazard-free environment is a good place to begin.

Potentially Hazardous Conditions

Managers and employees must work together to keep *all* job functions—no matter how routine or difficult—from becoming hazardous. The key is to identify a hazardous condition before it threatens employees, guests, and the property. Managers must train employees to recognize potentially hazardous conditions and to take appropriate corrective action. An alert and careful employee can be a property's best defense.

Wet floors and slippery walkways are accidents waiting to happen. Cluttered floors or cleaning equipment left out and in the way are invitations for injury. Improper lifting techniques and lifting or moving too much at once can threaten employee health. Such hazards result in the most common forms of employee injury in housekeeping departments: sprains, strains, and falls.

Accidents and injuries do not have to occur. By following three simple rules, employees can contribute to a safe, accident-free work environment.

1. **Take adequate time.** No job is so urgent that it is necessary to do it in an unsafe, hurried manner.

2. **Correct unsafe conditions immediately.** If you cannot correct an unsafe or hazardous condition yourself, report it at once to your supervisor.

3. **Do it safely the first time.** Every employee must do his/her job in a safe and correct manner. This is the best way to prevent accidents.

All lodging properties should have a list of safety rules. These guidelines should be part of a housekeeping safety plan to encourage employees to develop and practice safe work habits. Exhibit 1 shows a sample list of safety rules for housekeeping areas.

On any given day, housekeeping employees may lift heavy objects, climb ladders, operate machinery, and use dangerous cleaning chemicals—all of which

Exhibit 1   Sample Safety Rules List for a Lodging Operation

Rooms and Housekeeping Area: Safety Rules

- See General Safety Rules. (Section I)
- Keep glass out of linens.
- Be alert for cracked glasses that are wrapped.
- Empty razor containers before they become full.
- Keep cords out of pathways.
- Keep bed covers off the floors.
- Never smoke on the elevators.
- Place ashtrays on dressers, not beside beds where guests might be encouraged to smoke in bed.
- Do not overcrowd elevators.
- Use caution when pulling carts on and off elevators.
- Use the correct cleaning equipment for the job.
- Do not leave room service trays in guest hallways.
- Walk on the right side of corridors.
- Carry pointed objects with the sharp end down and away from yourself.
- Never use a chair or box as a substitute for a ladder.
- Put broken glass and metal waste in the proper containers.
- Correct tripping and slipping hazards immediately.
- Use handrails on stairways.
- Report defective wiring, plugs and uninspected appliances to your supervisor immediately.
- Check the cord and plug of any electrical appliance before plugging in. If there is a break or fray, or if sparks fly, *do not* try to plug in. Return appliance to Electric Shop and secure a replacement.
- Look for broken glass before kneeling on carpet or bathroom tile. If glass is present, sweep up with broom and then use portable vacuum. *Never* handle glass with bare hands.
- Dispose of broken glass in the "Broken Glass Containers" in the service halls.
- All ashtrays should be emptied in toilets, not in waste cans.
- Report any evidence of careless smoking in guestrooms, e.g., burnt carpets or bedspreads, burnt matches on floors, etc.
- Do not use bare hands to pull trash out of cans as there may be broken glass or razor blades present.
- Be careful in the placement of luggage in public areas and the baggage room.
- Never attempt to carry more luggage than you can handle safely.
- Pick up any foreign objects that guests may throw on stairs or floors.
- Wait until incoming cars have stopped at the curb before attempting to open the door.
- Be sure that passengers' hands and feet are clear of the door frame before closing automobile doors.
- Know the location of wheelchairs and stretcher.
- Know the procedures for dealing with guest injuries and illnesses.

Courtesy of the Hotel duPont, Wilmington, Delaware

Exhibit 2   Guidelines for Safe Lifting

1. Inspect the object before lifting. Do not lift any item that you cannot get your arms around or that you cannot see over when carrying.

2. Look for any protrusions, especially when lifting trash or bundles of linen. Quite often, these items can contain pointy objects or broken glass. Exercise special care to avoid injury.

3. When lifting, place one foot near the object and the other slightly back and apart. Keep well balanced.

4. Keep the back and head of your body straight. Because the back muscles are generally weaker than the leg muscles, do not use the back muscles to lift the object.

5. Bend slightly at the knees and hips but do not stoop.

6. Use both hands and grasp the object using the entire hand.

7. Lift with the leg muscles.

8. Keep the object close to the body. Avoid twisting your body.

9. If the object feels too heavy or awkward to hold, or if you do not have a clear view over the object, set it down.

10. When setting an object down do not use your back muscles. Use the leg muscles and follow the procedures used to lift objects.

represent potentially hazardous conditions. The following sections provide safety tips for completing these routine activities safely and efficiently.

Lifting. Housekeeping tasks often involve lifting heavy objects. Employees may also be required to move furniture in order to complete a thorough cleaning task.

Incorrectly lifting heavy objects such as bags, boxes, and containers may result in strained or pulled muscles and back injury. In turn, these injuries can result in loss of work and long-term pain and suffering. Employees can also incur cuts and scratches when lifting items such as trash or dirty linens which contain pointy objects or broken glass. In all instances, employees should know what conditions to look for and the special precautions to take. Exhibit 2 outlines safe lifting techniques for housekeeping personnel.

Employees should also be careful when doing related housekeeping tasks such as opening windows. An attendant should never hit or tug on the window frame if a window sticks or will not open easily. Improper pushing or pulling on a stubborn window may result in a back injury or a cut from broken glass. If a window is stuck, the attendant should call for assistance. Maintenance can usually open a window after applying a lubricant or fixing the window frame. Room attendants should also be cautioned against improperly pulling or tugging on heavy items such as room attendant carts and bundles of laundry. Like lifting, these actions can often result in injury.

Mattresses in guestrooms should be turned on a regular basis, depending upon the hotel's housekeeping rules or mattress manufacturer's specifications. Mattresses are often too large for one person to turn over without help. Two people can do the job in a much safer fashion.

Ladders. Ladders can be used when cleaning areas on or near the ceiling or for such tasks as changing light bulbs. When selecting a ladder for a particular cleaning job, its condition, height, and footing should be inspected. Check the ladder for stability and examine crosspieces for sturdiness. If the ladder is broken or defective, do not use it. Rather, tag the ladder, place it out of service, and report it to the appropriate housekeeping supervisor or the maintenance department.

Ladders come in many types. Ladders range from small step stools to 28-foot extension ladders and are usually constructed of wood, aluminum, or metal. An aluminum or metal ladder should never be used when working near or on electrical equipment. Ladders with rubber footings should be used on tile floors or in kitchen areas to prevent slipping. In all instances, the floor should be dry and clean.

A ladder must be high enough so that an attendant can stand on it and do the job without overreaching. Never stand on the top step of a ladder. If the area cannot be reached while standing on the step below the top step, the ladder is too short for the job. Ladders should be placed so the footing is at least one fourth of the ladder length away from the wall. For example, if the ladder is twelve feet tall, the footing should be three feet away from the wall. Never place a ladder against a window or an uneven surface.

Before climbing, test the ladder for stability; it should be well balanced and secure against the wall and floor. Always be sure to face a ladder when climbing and have clean and dry hands and feet. Do not hold any items or tools that may prevent the use of one or both hands. Mark the area underneath the ladder with caution signs so that guests or employees do not walk under the ladder. Walking under a ladder may be considered unlucky by the superstitious but it is also a very unsafe practice. Additional standards concerning ladders are presented later in this chapter.

Machinery. Employees should be authorized and trained in the use of machinery and equipment before operating such devices. Most equipment, machines, and power tools come with instructions. Some employees may need additional training and supervised practice before operating equipment and machinery on the job by themselves.

Many power tools and other machinery are equipped with protective guards or shields. These safety guards should not be removed. Employees may also be required to wear protective eye goggles or gloves. All protective gear should be worn per instructions.

Equipment and machinery should never be left unattended while in use. When not in use, all tools and equipment should be turned off and stored in the proper place. Never use a piece of equipment or machinery that is not operating correctly. Contact the appropriate supervisor or the maintenance department to have it repaired as soon as possible.

Electrical Equipment. Extra care must be taken when operating electrical equipment. Even one of the most common housekeeping appliances like a vacuum cleaner can be harmful or deadly if operated improperly or in unsafe conditions. Electrical equipment and machinery used at a lodging property should be approved by the Underwriters Laboratories. The **Underwriters Laboratories** is an independent non-profit organization that tests electrical equipment and devices. The purpose of such testing is to ensure that electrical equipment is free of defects which can cause fire or shocks. Approved equipment bears the initials "UL" in a circle on packaging, instructional material, or on a tag. Most equipment purchased in a retail store or through a wholesale purveyor will meet the Underwriters Laboratories standard.

An employee should never operate electrical equipment when standing in water or when hands or clothing are wet. It is also unsafe to operate electrical equipment near flammable liquids, chemicals, or vapors. Sparks from electrical equipment could start a fire.

Equipment that sparks, smokes, or flames should be turned off immediately. If it is possible and safe to do so, the equipment should be unplugged. In no instance should an attendant attempt to restart the equipment. The malfunction should be reported to the appropriate housekeeping supervisor or the maintenance department.

Equipment wires and connections should be checked periodically. Equipment with loose connections or exposed wires should not be used. An appliance should never be unplugged by pulling or yanking the cord. This will loosen the connection between the cord and plug and cause sparks and shorts. Equipment should be unplugged by grasping the plug and pulling it gently away from the outlet.

When using electrical equipment, the cord should be kept out of traffic areas such as the center of hallways or across doorways. This is not always possible, particularly with such tasks as vacuuming corridors. In such situations, keep the cord close to the wall and post caution signs in the work area. If the appliance will be stationary and in use for a lengthy period, tape the cord to the floor and place caution signs over the taped cord.

Extension cords are sometimes required—particularly when an electrical outlet is not located near the work area. Extension cords should be inspected for exposed wire before use just like any other electrical cord. There are many types of extension cords; not all are acceptable for use in a hospitality operation. The local fire department can pinpoint which types of cords meet the local, state, or federal fire codes and regulations.

When cleaning guestrooms, room attendants should check electrical lamps, appliances, and other fixtures for frayed wires, loose connections, and loose plugs. Exposed electrical wire may result in shock, injury, or even death when touched. Outlets and switch covers should be checked to ensure that they are covered properly and not cracked or broken. If any of these conditions are found, the room attendant should not attempt to fix them. Rather, potential problems should be reported to the appropriate housekeeping supervisor or to maintenance.

Chemicals. Many housekeeping employees are exposed to dangerous chemicals in their daily work routines. These chemicals are powerful cleaners, and, when used properly with proper protective gear, are relatively harmless. However, when used improperly, these same helpful chemicals can cause nausea, vomiting, skin rashes, cancer, blindness, and even death.

Chemicals are used to clean all areas of a lodging property including bathrooms, kitchens, and floors. Potentially hazardous chemicals are also used to kill insects and rodents. Some housekeeping situations require employees to handle toxic substances to unstop clogs in toilets and other plumbing fixtures. Often the use of such hazardous and toxic chemicals cannot be avoided.

Continual training in chemical safety is necessary for two reasons. First, misused chemicals can cause serious injury in a short period. Second, new employees—especially in properties with high employee turnover—need to be trained immediately.

Job Safety Analysis

Safety information is often best communicated through orientation and ongoing training. The design and use of a housekeeping safety manual is also an excellent communication vehicle. This manual should detail the different housekeeping jobs and instruct employees on the safe and proper way to perform each job. The first step in designing such a manual is to perform a job safety analysis.

A **job safety analysis** is a detailed report that lists every job function performed by all employees in a housekeeping department. The job list provides the basis for analyzing the potential hazards of a particular housekeeping position. The format for each analysis can parallel that of job breakdowns, with safety tips and potential hazards being cited in the additional information column.

For example, a job list for a room attendant may include the tasks of vacuuming, cleaning bathrooms, stocking room attendant carts, and emptying trash. Each task is then further broken down into a list of steps. The steps for vacuuming a guestroom may be:

1. Check the condition of the vacuum.

2. Take the vacuum to the farthest corner of the guestroom.

3. Plug the vacuum into the nearest outlet.

4. Vacuum your way back toward the door. Cover all exposed areas of the carpet.

5. Unplug the vacuum.

6. Reposition the vacuum on attendant cart.

For each step, a brief explanation should be written that describes a safe and standard method for performing that step. The how-to portion for step one, "Check the condition of the vacuum," could be:

1. Inspect the vacuum's wheels for dust or debris.

2. Check the cord for frays or missing insulation, especially near the plug.

3. Examine the handle, frame, and body for damage.

4. Check the vacuum bag for rips and tears.

Finally, the safety tips section of the analysis lists any possible hazards that an employee may encounter when performing a particular job step. For example, the safety points that could be listed when checking the vacuum might be:

1. Do not operate the vacuum if it needs mechanical attention.

2. By operating a vacuum in need of repair, a person increases his/her chances of suffering from slips, falls, strains, lacerations, bruises, burns, or electrical shock.

A complete job safety analysis should include all housekeeping jobs and the tasks involved. The analysis should be produced in booklet form; each employee should receive the sections that focus on the various jobs he/she will perform while working at the property.

It is not enough just to give the employee the analysis and hope that it gets read. Housekeeping managers should demonstrate and explain each task. At the end of the training period, the employee should sign a statement that he/she understands the job safety analysis and rules. The statement should include the safety and disciplinary consequences of not abiding by the job safety analysis rules and guidelines.

Safety Training

Safety training begins the first day of the job. Housekeeping employee orientation must include an introduction to property safety rules and regulations. Employees must know what is expected of them in the area of safety in order to perform their jobs more safely.

A good safety policy is written with the employee in mind; it states how safety and being safe benefits the employee and the company. This statement should be a general safety philosophy that covers the entire property—not just housekeeping. Exhibit 3 shows the safety philosophy presented by one property during employee orientation.

Specific safety rules and procedures should also be presented at orientation. Orientation is a good opportunity to hand out any written safety information. This information may be part of an orientation packet or a separate booklet. In some employee orientation booklets, a space is included next to each area for the employee to initial and date. This reinforces the importance of these rules and ensures that the employee receives and understands the information.

Safety training does not end at orientation. At least once a month every employee should participate in a safety education program. These sessions can be used to discuss new safety rules and the proper use of new equipment. Periodic safety education programs should also serve as refresher sessions for both new and seasoned employees. Familiar safety procedures should be reviewed and refined.

Early on in the training process—usually on the first or second day of the job—the employee should learn about specific safety conditions mandated by law—particularly those concerned with the use of hazardous chemicals. The following section provides an overview of regulations imposed on the workplace for employee safety.

OSHA Regulations

The federal government regulates work areas and businesses with respect to safety. The **Occupational Safety and Health Act** (OSHA) was enacted in 1970 to protect

Exhibit 3   Sample Safety Philosophy Statement

Hotel duPont Safety Philosophy

Safety is the number one priority of all employees at the Hotel duPont. We share the duPont Company's philosophy that all injuries and incidents can be eliminated through an effective safety program. At the Hotel, our safety program is very much employee oriented with maximum employee participation the key ingredient to its success. Our employees are taught that safety is a way of life, not something left at work, but carried home to their families.

Safety in the Hotel duPont does take on several meanings. When we talk about our safety programs, we may refer to any of the following:

- Employee Work Practices
- Off-the-Job Safety
- Life Safety
- Equipment Safety
- Safety Administration
- Food Sanitation
- Security
- Injury-Prevention

These areas combined form the safety philosophy of the Hotel duPont.

Courtesy of the Hotel duPont, Wilmington, Delaware

the worker at the work place. OSHA regulations are quite extensive and mandate safety regulations and practices for many industries—including hospitality.[1]

OSHA standards cover a variety of areas that concern housekeeping employees. Regulations focus on the areas where employees work, materials used on the job, and other safety issues. OSHA standards are primarily designed to protect the employee—not the guest. However, some OSHA standards cover areas where guest safety may be a concern.

The following sections discuss only a portion of the OSHA regulations that pertain to hospitality operations. For a more detailed review of the standards, managers should contact local OSHA or U.S. Department of Labor offices.

Work Areas

OSHA standards cover such areas as hallways, storerooms, and service areas. These standards require that work areas be kept clean, neat, and sanitary. Standards also require that hallways, passageways, and stairways have guard rails and railings. Any stairway with four or more steps must have at least one railing.

OSHA also covers portable ladders used in such areas. Regulations state that the ladders must be of good construction and cannot have inappropriate additions.

When in need of repair, ladders should be tagged properly. Regulations also cover scaffolds, mobile ladders, and towers.

Means of Egress

OSHA standards require exits to be clearly marked. The code details both the size and level of illumination of the sign. The exits and paths to the exits should not be visually or physically blocked. Exit doors should never be locked in such a way that employees or guests could not escape from a fire. The route to the exits must also be marked so that an escape path can be easily followed.

The standards dictate the occupancy level of individual rooms and the building based on the number of exits, available fire protection, and building construction. OSHA standards should be referenced for more specific information.

All properties must have written emergency escape plans. Emergency escape routes and procedures must be clearly indicated. The plan should specify employee duties and placement within the facility during an emergency. These plans should enable management to evacuate the property safely and account for all employees during and after an emergency. Written procedures should be specified for the preferred means of reporting emergencies and fires. Plans must also establish rescue and medical duties for employees or local services.

OSHA also requires a list of the job titles or people who can be contacted for further information or explanation of duties under the emergency plan. The executive housekeeper and housekeeping supervisory staff may be included on this list.

Hazardous Materials

OSHA standards cover the use of many chemicals, cleaning compounds, and flammable and combustible materials in the workplace. OSHA regulates the type of containers required for hazardous materials—including aerosol containers. In general, storage containers for hazardous liquids must be approved by the Underwriters Laboratories. These containers should have spring-loaded lids, spout covers, and should hold no more than five gallons. Cleaning supplies, compressed gases, and paints fall under this category.

Material Safety Data Sheets. The OSHA Hazard Communication Standard requires employers throughout the United States to tell their employees about hazardous materials that they may be required to handle in order to do their jobs. The standard is commonly referred to as **HazComm** or OSHA's right-to-know legislation. HazComm stipulates that material safety data sheets (MSDSs) must be collected for each chemical and filed where employees may read them at any time. Recent action concerning such right-to-know laws and regulations should be ascertained on a state-by-state basis because standards may vary (see Exhibit 4).

Material safety data sheets are forms with information concerning chemicals or cleaners used at a property. The MSDS form lists the hazardous ingredients, health hazard data, and spill or leak procedures for a product. The MSDS form also states any special precautions or protective gear required when using the product. These sheets may be obtained from the chemical supplier or a local purveyor.

Exhibit 4 Michigan Right To Know Law Poster

This Workplace Covered by the Michigan Right To Know Law

Employers must make available for employees in a readily accessible manner, Material Safety Data Sheets (MSDS) for those hazardous chemicals in their workplace.

Employees cannot be discharged or discriminated against for exercising their rights including the request for information on hazardous chemicals.

Employees must be notified and given direction (by employer posting) for locating Material Safety Data Sheets and the receipt of new or revised MSDS(s).

*Employees may also request MSDS from the Michigan Department of Public Health, Division of Occupational Health, 3500 North Logan, Lansing, Michigan, 48906, 517/373-1410.

SET #2105

MIOSHA
Michigan Occupational Safety and Health Act

MSDS(s) For This Workplace Are Located At

____
LOCATION(S)

____
LOCATION(S)

____
PERSON(S) responsible for MSDS(S)

____
PHONE

OSHA stipulates that MSDS forms must be shown to employees who use or are exposed to potentially hazardous cleaning products. A good place to store MSDS forms is in the executive housekeeper's office, housekeeping break rooms, and in housekeeping storage areas near the corresponding chemical compound and cleaner.

Sanitation

OSHA standards are quite particular concerning sanitation. Waste disposal, employee washing facilities, and food and beverage consumption are among the areas addressed.

Waste Disposal. OSHA specifies that all waste containers be leak-proof and have a tight-fitting cover. The waste containers must be maintained in a sanitary condition. The removal of all waste must be done without creating a menace or hazard to public health. Waste should be removed often or on a timely basis.

Washing Facilities and Showers. OSHA stipulates that washing facilities be kept clean and each bathroom be equipped with hot and cold running water. Hand soap or cleaner must be provided. Hand towels, paper towels, or warm air dryers should also be conveniently located for employee use. OSHA standards even dictate the number of toilets a lodging property must have based on the number and sex of employees.

If employees are required to shower before, during, or after their shift, the employer must provide the same items as required in the washing facilities. Individual towels must also be provided. OSHA standards require one shower for every ten employees of the same sex.

Food and Beverage Consumption. OSHA forbids the consumption of food and beverages in washroom areas. OSHA also stipulates that no food and beverage be consumed in areas such as storage rooms that contain hazardous or toxic chemicals.

If an operation provides employee meals, OSHA standards state that the food must be wholesome, and not spoiled. Food must be prepared, served, and stored in a sanitary way so as to avoid contamination.

Signs and Tags

OSHA standards require special signs for safety reasons. Three different types of signs are generally needed by the housekeeping department: danger, caution, and safety instruction. Accident prevention tags are also used to alert employees of potentially hazardous conditions.

Danger Signs. Danger signs are used only in areas where there is an immediate hazard. These signs indicate immediate danger and the necessity of special precautions. An appropriate use of danger signs would be in an area where a caustic cleaning liquid has been spilled. OSHA standards dictate that the colors of a danger sign be red, black, and white.

Caution Signs. Caution signs are used to warn against a potential hazard. A wet floor caused by water spillage or mopping would be a reason to use a caution sign. Caution signs should be stored near buckets and mops for easy access. Caution signs are yellow and black.

Safety Instruction Signs. Safety instruction signs are green and white, or black and white. They are used when general instructions need to be given in a certain area. For example, these signs may be used to instruct employees not to eat, drink, or smoke in a storage area.

Accident Prevention Tags. Accident prevention tags are a temporary means of alerting employees of a hazardous condition or defective equipment. One sample use of this type of tag would be to place it on a vacuum cleaner that has a frayed cord. The tag should read "Do not start," "Defective Equipment," or "Out-of-Service." This type of tag is not meant to be a complete warning, but only a temporary solution to a potential safety hazard. The tags should be red with white or grey letters. Accident prevention tags should be kept near electrical power equipment, ladders, and other housekeeping tools.

First Aid

OSHA stipulates that the employer should provide employees with readily accessible medical personnel. Some properties have a house doctor or a nurse's station.

In the absence of medical staff, the employer should indicate a local medical facility or doctor who can handle on-the-job injuries.

First aid supplies approved by a consulting physician should be stocked and readily available. If an employee works with corrosive materials, suitable facilities should be available to drench or flush eyes, face, and body in case of an accident. Caustic and corrosive cleaning solutions are generally used in kitchens to clean filters, exhaust hoods, and grills.

OSHA Inspection

Housekeeping managers should be aware that OSHA compliance officers have the authority to inspect a property. These inspections are often done without any advance notice. An inspection may be refused but the compliance officer may return with a court-approved warrant and inspect the property. It is usually best to allow a compliance officer to inspect the property once they have shown proper identification.

Compliance officers may wish to inspect the general property, equipment, and records. In addition to OSHA records, the officer may want to see that the OSHA poster is prominently displayed (See Exhibit 5). The inspection can also include safety committee reports, environmental sampling, and private discussion with individual employees. A management representative—perhaps the executive housekeeper—should accompany the officer during the inspection.

Security

Providing security in a hospitality operation is the broad task of protecting both people and assets. Security efforts may involve guestroom security, key control, perimeter control, and more. Each lodging property is different and has differing security needs.

Security should be recognized and used as a management tool. Whether the property requires a large security staff or one or several on-premises supervisory personnel, the security role should be clearly defined and implemented. In the development of security guidelines, all members of the property's management and supervisory team should be involved.

The information presented here is intended only as an introduction to security. Only those elements relevant to housekeeping are included. Hotel management should consult legal counsel to ensure that the property is in compliance with applicable laws.[2]

Security Committees

Some properties develop and refine security guidelines through a committee process. The **security committee** should consist of key management personnel—including department heads. Supervisors and selected hourly employees can also contribute important security information and add to the committee's effectiveness.

Among the major agenda items for the security committee are the development of a security handbook and the design of training and awareness programs. These materials and programs should cover guestroom security, key control, lighting,

Exhibit 5   OSHA Job Safety and Health Protection Poster

JOB SAFETY & HEALTH PROTECTION

The Occupational Safety and Health Act of 1970 provides job safety and health protection for workers by promoting safe and healthful working conditions throughout the Nation. Provisions of the Act include the following:

EMPLOYERS

All employers must furnish to employees employment and a place of employment free from recognized hazards that are causing or are likely to cause death or serious harm to employees. Employers must comply with occupational safety and health standards issued under the Act.

EMPLOYEES

Employees must comply with all occupational safety and health standards, rules, regulations and orders issued under the Act that apply to their own actions and conduct on the job.

The Occupational Safety and Health Administration (OSHA) of the U.S. Department of Labor has the primary responsibility for administering the Act. OSHA issues occupational safety and health standards, and its Compliance Safety and Health Officers conduct jobsite inspections to help ensure compliance with the Act.

INSPECTION

The Act requires that a representative of the employer and a representative authorized by the employees be given an opportunity to accompany the OSHA inspector for the purpose of aiding the inspection.

Where there is no authorized employee representative, the OSHA Compliance Officer must consult with a reasonable number of employees concerning safety and health conditions in the workplace.

COMPLAINT

Employees or their representatives have the right to file a complaint with the nearest OSHA office requesting an inspection if they believe unsafe or unhealthful conditions exist in their workplace. OSHA will withhold, on request, names of employees complaining.

The Act provides that employees may not be discharged or discriminated against in any way for filing safety and health complaints or for otherwise exercising their rights under the Act.

Employees who believe they have been discriminated against may file a complaint with the nearest OSHA office within 30 days of the alleged discriminatory action.

CITATION

If upon inspection OSHA believes an employer has violated the Act, a citation alleging such violations will be issued to the employer. Each citation will specify a time period within which the alleged violation must be corrected. The OSHA citation must be prominently displayed at or near the place of alleged violation for three days, or until it is corrected, whichever is later, to warn employees of dangers that may exist there.

PROPOSED PENALTY

The Act provides for mandatory civil penalties against employers of up to $7,000 for each serious violation and for optional penalties of up to $7,000 for each nonserious violation. Penalties of up to $7,000 per day may be proposed for failure to correct violations within the proposed time period and for each day the violation continues beyond the prescribed abatement date. Also, any employer who willfully or repeatedly violates the Act may be assessed penalties of up to $70,000 for each such violation. A minimum penalty of $5,000 may be imposed for each willful violation. A violation of posting requirements can bring a penalty of up to $7,000.

There are also provisions for criminal penalties. Any willful violation resulting in the death of any employee, upon conviction, is punishable by a fine of up to $250,000 (or $500,000 if the employer is a corporation), or by imprisonment for up to six months, or both. A second conviction of an employer doubles the possible term of imprisonment. Falsifying records, reports, or applications is punishable by a fine of $10,000 or up to six months in jail or both.

VOLUNTARY ACTIVITY

While providing penalties for violations, the Act also encourages efforts by labor and management, before an OSHA inspection, to reduce workplace hazards voluntarily and to develop and improve safety and health programs in all workplaces and industries. OSHA's Voluntary Protection Programs recognize outstanding efforts of this nature.

OSHA has published Safety and Health Program Management Guidelines to assist employers in establishing or perfecting programs to prevent or control employee exposure to workplace hazards. There are many public and private organizations that can provide information and assistance in this effort, if requested. Also, your local OSHA office can provide considerable help and advice on solving safety and health problems or can refer you to other sources for help such as training.

CONSULTATION

Free assistance in identifying and correcting hazards and in improving safety and health management is available to employers, without citation or penalty, through OSHA-supported programs in each State. These programs are usually administered by the State Labor or Health department or a State university.

POSTING INSTRUCTIONS

Employers in States operating OSHA approved State Plans should obtain and post the State's equivalent poster.

Under provisions of Title 29, Code of Federal Regulations, Part 1903.2 (a) (1) employers must post this notice (or facsimile) in a conspicuous place where notices to employees are customarily posted.

More Information

Additional information and copies of the Act, specific OSHA safety and health standards, and other applicable regulations may be obtained from your employer or from the nearest OSHA Regional Office in the following locations:

Atlanta, GA	(404) 347-3573
Boston, MA	(617) 565-7164
Chicago, IL	(312) 353-2220
Dallas, TX	(214) 767-4731
Denver, CO	(303) 844-3061
Kansas City, MO	(816) 426-5861
New York, NY	(212) 337-2378
Philadelphia, PA	(215) 596-1201
San Francisco, CA	(415) 744-6670
Seattle, WA	(206) 553-5930

Robert B. Reich,
Secretary of Labor

Washington, D.C.
1992 (Reprinted)
OSHA 2203

U.S. Department of Labor
Occupational Safety and Health Administration

To report suspected fire hazards, imminent danger, safety and health hazards in the workplace, or other job safety and health emergencies, such as toxic waste in the workplace, call OSHA's 24-hour hotline: 1-800-321-OSHA

This information will be made available to sensory impaired individuals upon request. Voice phone (202) 523-8615; TDD message referral phone: 1-800-326-2577

Source: U.S. Department of Labor, Occupational Safety and Health Administration, Washington, D.C.

emergency procedures, and security records. Once developed, these materials and programs should be continually updated and revised by the committee to meet the changing needs of the property.

The committee should meet once a month to monitor the property's security plan and programs. Other committee responsibilities generally include:

- Monitoring, analyzing, and suggesting solutions for recurring security problems

- Maintaining records on such incidents as theft, vandalism, and on-site violence

- Conducting spot security audits and property inspections

- Investigating security incidents

Last but not least, the security committee should maintain open lines of communication with the local police department. The local police can be a good source of security training ideas and ideas on basic property security.

Suspicious Activities

A lodging property, although open to the public, is private property. Hotel operations have a responsibility to monitor and, when appropriate, to control the activities of persons on the premises. Along these lines, a property should establish a policy on how to approach and handle unauthorized persons.

The individuals allowed in guestroom areas are guests, their visitors, and on-duty employees who are performing their jobs in the authorized area. Hotel housekeeping employees can be part of an effective security force—particularly in guestroom areas. The housekeeping staff should be trained to spot suspicious activities and unauthorized or undesirable persons. If an individual is seen loitering, checking doors, knocking on doors, or looking nervous, he/she should be considered suspicious and approached.

Unauthorized or undesirable persons should be approached with caution. If an employee feels threatened or in danger, he/she should not approach the person but rather go to a secure area such as a storage room, lock the door, and call the front desk or security.

If the employee does approach such an individual, he/she should do so politely. The attendant should ask the person if he/she can be of assistance, but avoid getting into a long conversation. If the individual claims to be a guest, the attendant should ask to see a room key. If the person says that he/she is not a guest, or does not have a room key, the attendant should explain the hotel policy and direct the individual to the front desk. The attendant should then watch to see whether that person proceeds to the directed area, and then call the front desk or security.

Employees themselves can present similar security problems. Employees who are not in their designated work area should also be stopped and asked if they need help. Depending on the individual's response and manner, he/she should be reported to security or the housekeeping supervisor.

Friends and visitors of employees should not be allowed in guestroom areas or employee locker rooms. Hotel management should choose a designated area where friends and visitors may wait to meet the employee. This procedure also helps reduce the potential for theft.

Theft

It is impossible to eliminate all employee and guest theft in a hospitality operation. However, management can reduce the volume of furniture, fixtures, equipment, and soft goods stolen from a property by reducing the opportunities to steal. Opportunities present themselves as unlocked doors, lack of inventory control, and plain carelessness.

Guest Theft. Unfortunately, guest theft is all too common in hotels. Some guest theft is considered a form of marketing; other guest theft is not. Most hotels assume that guests will take items which prominently display the hotel logo such as matches, pens, shampoo, ashtrays, and sewing kits. For the most part, these items are provided for the guest's convenience and are actually a form of advertising used by the hotel. However, towels, bathrobes, trash bins, and pictures are not part of the marketing strategy and are not meant to be taken by guests. When these items turn up missing, it can add up to a large expense for a hospitality operation.

To reduce the theft of these items, some properties keep count of the number of towels placed in the room. When the guest requests additional towels, it is noted at the front desk. The room attendant, too, notes how many towels are in the room when cleaning the next day. The room attendant's ability to spot missing items may allow the hotel time to charge the guest for items that have been taken.

As another strategy, some hotels place items such as towels, bathrobes, and leather stationery folders on sale in their gift shops. This may reduce the likelihood of theft since guests have the option of purchasing these items. Also, having these items on sale helps set a standard price that can be levied against guests for a missing item. Other helpful ideas to reduce guest theft are:

- **Use as few monogrammed items as possible.** Most guests are looking for a souvenir when taking towels or bathrobes—and are not really looking to steal. The use of fewer items with logos reduces temptation.

- **Keep storage rooms closed and locked.** Do not allow guests to take any items from storage rooms. Also, amenities stored on carts should be stocked in a secure place or in a locked compartment. Guests walking in the hallway could very easily take home a year's supply of shampoo and soap in a matter of minutes when items are stored on top of the cart.

- **Affix or bolt guestroom items and fixtures to appropriate surfaces.** If it is not nailed, glued, bolted, or anchored to the wall—and it is small enough to fit in a suitcase—it is a prime target for guest theft. The easier an item is to remove, the more likely it is that it will be removed. All pictures, mirrors, and wall decorations should be discreetly affixed to the wall. Lamps should be too large to fit easily into a suitcase or bag. Expensive items such as televisions should be bolted and equipped with an alarm that alerts the front desk or security if an attempt is made to remove the item. It is also a good practice to record serial numbers of items such as TV sets, and to etch the hotel name in a discreet spot on the surface.

Many luxury properties do not secure items such as alarm clocks, television remote controls, or books in order to preserve a certain image. This does not mean they do not have a problem with theft. These hotels, however, usually charge a high enough rate to compensate for stolen items. Despite this fact, room attendants should still inventory items when cleaning, and notify the front desk, security, or the appropriate supervisor or department of any missing items.

- **Secure windows.** The closer the guestroom is to the parking area, the easier it is to remove an item from a room. A classic hotel legend is the 75-pound wood-framed mirror that was removed from the back window of a first floor guestroom. The mirror was too large and too heavy to be walked out the front lobby door but was easily removed without notice through the guestroom window into a car. The moral is to secure all first floor windows—and sliding glass doors—so that they cannot be opened all the way. Whenever possible, limit the number of entrances and exits guests may use to get to their room.

Employee Theft. It is up to management to set the standards for reducing employee theft—and to act as a good example. A manager who takes hotel steaks home to barbecue will not be effective when asking employees not to steal food, linen, and other hotel property. Management should also detail explicit rules and regulations concerning employee theft. The employee handbook should spell out the consequences of stealing hotel property. If the rules state that employees who are caught stealing will be prosecuted and fired then the hotel should follow through. It is important that management does not discriminate against certain employees when enforcing these rules.

Managers should screen applicants before making a job offer. A thorough background check should be conducted, including a check for any criminal convictions. Before asking any questions or making any inquiries, check local, state, and federal laws to ensure that the selected screening techniques are not illegal or prohibited.

Good inventory control procedures can also help control theft. Detailed records that note any unusual or unexplained fluctuations should be kept of all items in stock. At one hotel, for example, it was noted that a consistent amount of toilet paper came up missing each month. Through strict and diligent inventory control, it was discovered that a housekeeping employee took a case of toilet paper each delivery day and sold it by the roll to fellow employees in the hotel parking lot.

It is a good practice to conduct a monthly inventory of all housekeeping supplies including toilet paper, amenities, and linens. If the items in storage do not match the usage rate, or if too little stock is on the shelves, it may be an indication of employee theft. Employees should be aware of the results of monthly inventories—especially when shortages are discovered.

In addition to keeping records of items in stock, records should be kept of stolen or missing items—including those from guestrooms. The record should include the name of the room attendant and any other hotel employees who had access to the room. For example, if a room service employee delivers a meal, that employee's name should be entered into the log.

Keep all storeroom doors locked. Storerooms should be equipped with automatic closing and locking devices. Locks on storerooms should be changed periodically to reduce the opportunity of theft.

If property design permits, management should designate employee entrances and exits. These entrances should be well-lighted, adequately secured, and provided with round-the-clock security. Employee entrances may include a security staff office which monitors arriving and departing employees.

Employees should know what items they may bring onto or remove from the property. Management may establish a claim-checking system for bringing items onto the premises and a parcel-pass system for taking items off the premises. If an employee has permission to remove hotel property, he/she should be issued a signed permit from the supervisor or an appropriate manager before doing so.

Restricting employee parking to a carefully selected area can also help control losses. Keeping the area well-lighted reduces the temptation to steal and also makes the lot safer for employees who leave work after dark. The employee parking area should not be so close to the building that it allows employees to easily and quickly transfer stolen property to their cars.

If the hotel is large or has a very high turnover rate, employees are less likely to know their fellow workers. In such cases, identification badges may be required to prevent strangers who pose as employees to gain admittance to the property.

Bomb Threats

Housekeeping procedures for handling bomb threats should be part of the property's security manual. Housekeeping's role usually consists of helping in the search for any suspicious objects that could be bombs.

Where and how the search is conducted will depend on the way the property received the bomb threat. Information from the caller or letter may give clues on where personnel should search and on what type of bomb or object to look for. Searches often include stairways, closets, ashtrays, trash containers, elevators, exit areas, and window sills. It may be helpful to take a flashlight to inspect areas with little light.

Search team employees look for objects that are normally not found in an area. Housekeeping personnel have an advantage since their daily routines promote familiarity with many hotel areas. If a suspicious looking object is found, it should not be touched or moved; notify the person in charge of the search team or an appropriate supervisor immediately. Notification is best done face-to-face or over the telephone. Avoid using radios, walkie-talkies, or beepers. Some bomb devices are sensitive to these sound waves and may go off.

If nothing is found after completing the search, the teams should regroup in a designated area. An all-clear sign should be given after all search procedures have been performed and management is satisfied that the guests, employees, and property are not under any real threat.

Quite often, guests are not notified when bomb threats are received. This is because many bomb threats are just that—threats. However, bomb threat emergency procedures should still be followed just in case it is a real emergency. Generally, these procedures do not include notifying guests until a search is completed. If a guest does ask an employee what he/she is doing during a search, the employee should respond in a way that does not arouse unnecessary suspicion or fear.

The safety and security manual should include evacuation plans in case a bomb should actually be found or explode on the premises. It should also include provisions for emergency medical services. In these instances, housekeeping employees should follow procedures to assist in rescue efforts. The local police should be notified of all bomb threats. If police respond to such calls, the hotel should follow the directions laid out by police personnel.

Fires

Fires are grouped into four classifications based on the different products of combustion. Class A fires involve wood and paper products. Class B fires involve flammable liquid, grease, and gasoline. Class C fires are electrical in nature. Class D fires are formed by combustible metals and do not usually occur in hospitality operations.

Many hotel fires are fueled by a combination of combustibles. It is very likely that a fire started by Class A combustibles could grow to include Class B and C materials.

Fires start for many reasons. Some fires may be caused by an accident or mechanical malfunction. Others may be the result of arson. In the 1980s, several fires occurred in the lodging industry that involved fatalities. The causes of two of the worst fires were traced to electrical malfunction and arson.

Fire Detection Systems. Because of the tragedy and implications of hotel fires, state and federal legislation has been passed which places more fire safety requirements on hotel operations. These requirements include the installation of smoke detectors, suppression systems, and alarms.

Smoke detectors. Smoke detectors are commonly found in hotel rooms. Some are battery-operated and function independently from the hotel alarm system. Other smoke detectors are hard-wired. When one goes off, an alarm sounds not only in the room or immediate area, but at a control panel located at the front desk or maintenance office.

Smoke detectors are set off by sensing smoke. There are two types: photoelectric and ionization. Photoelectric detectors go off when smoke interferes or blocks off a beam of light located inside of the detector. Ionization smoke detectors also go off when they sense smoke. However, the alarm sounds when the detector senses a shift in electrical conductivity between two plates.[3]

Smoke detectors can go off when a fire is not present. Steam from very long and hot showers or smoke from a guest's cigarette can sometimes set off these alarms.

Fire suppression systems. Fire suppression systems include sprinklers that extinguish a fire with water. Sprinklers are generally triggered by heat—not smoke. Therefore, if a smoke detector sounds an alarm it does not necessarily mean that the sprinkler system will go off. Sprinkler heads are usually located near or on the ceiling in guestrooms, housekeeping storage closets, and laundry rooms, as well as other public areas. Since sprinkler heads extend several inches from the ceiling or wall, employees must be careful when cleaning around them. If a head is broken or knocked while cleaning it could set off the fire alarm and send hundreds of gallons of water into the area.

Pull stations. Alarms can be set off by smoke detectors, heat detectors, sprinkler systems, and by pull stations. Pull stations are located in public areas such as lobbies and corridors, and near elevators and exits. These stations are usually red and require someone to break a glass panel or pull a lever to turn on a fire alarm. Employees should know the location of pull stations and how to sound the fire alarm should they spot a fire or see smoke.

Fire Safety Training. Housekeeping staff may be involved with the evacuation of guests and other employees in the case of fire. In some instances, housekeeping staff may be required to search guestrooms in areas of the hotel where the fire has not reached to ensure that all guests have left the property. Training is important in any fire safety program so employees know how to respond calmly and professionally in an actual emergency.

Fire safety training should also provide instruction on reporting a fire and on what to do if a fire is spotted. The training program should include emergency escape procedures and outline duties for housekeeping staff involved in the process. The training should also describe a method of accounting for all employees after emergency evacuation has taken place. This can be accomplished by instructing all employees to meet at a designated area outside of the hotel.

Since most deaths and injuries from a fire are a result of smoke and toxic fumes—not the flames—employees should be shown how to escape a smoke-filled hallway or room. When trying to leave a smoke-filled room, the employee should stay close to the floor and cover his/her mouth and nose with a wet towel. Employees should be instructed never to use an elevator in a fire and to always leave the property by using the fire escape or stairway.

Employees should be instructed on how to put out a small fire that is confined to a trash can or limited area. Different portable fire extinguishers will do the job for different types of fires. Exhibit 6 lists approved types of fire extinguishers. Most hotel properties have Type ABC extinguishers; these can be used on most types of fires encountered in a lodging operation. Employees should be told where fire extinguishers are located and be trained in how to use them.

OSHA mandates standards for fire safety training with respect to emergency and fire prevention plans. Any additional employee training above and beyond OSHA and local fire and safety laws can benefit the hotel in more than one way. For one, it provides a safer environment for hotel employees. Two, it can be used as a selling point by the hotel since well-trained employees provide a higher level of safety for guests. Three, fire safety training can be used by a hotel that has experienced a fire as an added legal defense. Employee training can demonstrate that hotel management took precautions with respect to fire safety and was not negligent.[4]

Exhibit 6   Types of Fire Extinguishers

Extinguisher Classifications †	A — Water Types (includes antifreeze)		AB — AFFF Foam and FFFP	AB — Carbon Dioxide	BC — Dry Chemical Types (Purple K, Super K, Monnex, Potassium Bicarb, Urea based)		BC — Halogenated Types 1211 1301 1211/1301	ABC — Multipurpose Dry Chemical		ABC — Halogenated Types 1211 1211/1301
Extinguishing Agent	Water Types (includes antifreeze)		AFFF Foam and FFFP	Carbon Dioxide	Dry Chemical Types Purple K Super K Monnex Potassium Bicarb. Urea based		Halogenated Types 1211 1301 1211/1301	Multipurpose Dry Chemical		Halogenated Types 1211 1211/1301
Discharge Method	Stored Pressure	Pump Tank	Stored Pressure	Self Expelling	Stored Pressure	Cartridge Operated	Stored Pressure	Stored Pressure	Cartridge Operated	Stored Pressure
Sizes Available	2½ Gal.	2½–5 Gal.	2½ Gal. (33 Gal.)	5–20 lb. (50–100 lb.)	2½–30 lb. (50–350 lb.)	4–30 lb. (125–350 lb.)	1–5 lb.	2½–20 lb. (50–350 lb.)	5–30 lb. (125–350 lb.)	5½–22 lb. (50–150 lb.)
Horizontal Range (Approx.)	30–40 ft.	30–40 ft.	10–25 ft. (30 ft.)	3–8 ft. (3–10 ft.)	10–15 ft. (15–45 ft.)	10–20 ft. (15–45 ft.)	10–16 ft.	10–15 ft. (15–45 ft.)	10–20 ft. (15–45 ft.)	9–16 ft. (20–35 ft.)
Discharge Time (Approx.)	1 Min.	1–3 Min.	50–65 Sec. (1 Min.)	8–15 Sec. (10–30 Sec.)	8–25 Sec. (25–60 Sec.)	8–25 Sec. (25–60 Sec.)		8–25 Sec. (20–60 Sec.)	8–25 Sec. (25–60 Sec.)	10–18 Sec. (30–45 Sec.)
Operating Precautions and Agent Limitations	Conductor of electricity. Needs protection from freezing. (except antifreeze). Use on flammable liquids and grease will spread fire.		Conductor of electricity. Needs protection from freezing. Not effective on water-soluble flammable liquids such as alcohol, unless otherwise stated on nameplate. AFFF not effective on pressurized flammable liquid/gas fires.	Smothering occurs in high concentrations. Avoid contact with discharge horn. Limited effectiveness under windy conditions. Severely reduced effectiveness at sub-zero (F) temperatures.	Extensive cleanup, particularly on delicate electronic equipment. Obscures visibility in confined spaces.		Avoid high concentrations and unnecessary use.	Extensive cleanup. Damages electronic equipment. Obscures visibility in confined spaces. Limited penetrating ability on deep-seated Class A fires.		Avoid high concentrations and unnecessary use.

NOTE: Protection required below 40°F and above 120°F. *(A — Water Types)*

NOTE: Only dry chemical types are effective on pressurized flammable gases and liquids; for deep fat fryers, multipurpose ABC dry chemicals are not acceptable. *(ABC)*

Notes: Figures in parentheses refer to wheeled units.
†Class D information is not shown.

Source: Adapted from the National Association of Fire Equipment Distributors, "Selection Guide to Portable Fire Extinguishers," Chicago, Illinois, copyright 1988. Used with permission.

Flame-Resistant Materials. Fires need fuel to burn. In a hotel, fuel for a fire can include beds, linens, draperies, carpeting, and cleaning chemicals. As part of a comprehensive fire protection program, properties should purchase and use fire-resistant fabrics and materials whenever possible. Fire-resistant materials are rated with flame spread ratings. A zero flame spread is the lowest. Local, state, and federal codes may dictate the minimum fire-resistant ratings for different material. The local fire department may also provide information regarding flame retardant fabrics and material.

Key Control

Proper key control procedures are important for guest security and privacy. **Key control** also protects the property by reducing the possibility of guest and property theft. Housekeeping is primarily concerned with four categories of keys: emergency, master, storeroom, and guestroom.

Types of Keys. Emergency keys open all doors in the property—even those that guests have double-locked. These keys should be kept in a secure place. Some properties also keep an emergency key off the premises. Distribution and use should occur only in emergency situations such as a fire or when a guest or employee is locked in a room and needs immediate assistance. Most housekeeping personnel do not use emergency keys on a day-to-day basis.

Master keys also open more than one guestroom. Master keys are separated into three levels of access. The highest level is the grand master. This key opens every hotel room and, many times, all housekeeping storage rooms. If the guest has turned the dead bolt, master keys will not open the door. Master keys can be used in emergency situations when it is vital for an employee to enter some or all areas of a hotel. Master keys are kept at the front desk for such emergency purposes.

The next level of master key is the section master. This type of master key opens rooms in one area of a hotel. An inspector may be issued more than one key of this type because he/she may be required to inspect the work of more than one room attendant.

The lowest level of master key is the floor key. Generally, a room attendant is given this key to open the rooms he/she is assigned to clean. If the employee has rooms to clean on more than one floor or area, he/she may need more than one floor key. Floor keys typically open the storeroom for that floor—unless the room is specially keyed or is accessed by another master key.

Guestroom keys are those keys distributed to guests. This type of key opens a single guestroom and, in some cases, other locked areas such as the pool. Guestroom keys are stored at the front desk when not in use.

Key Control Procedures. A log can be used to monitor the distribution of master keys. This log should include the date, time, and the name of the person who signed for a particular key. Every time an employee receives or returns a master key he/she should be required to initial or sign the log. The person issuing the keys should also initial or sign the log for each master key transaction. Exhibit 7 shows a sample key control log. In larger properties, the linen room attendant distributes

Exhibit 7   Sample Key Control Sheet

KEY CONTROL SHEET

Date _____   Page _____ of _____

Key Code	Name	Signature	Time Out	Issued By	Time In	Signature	Received By

Courtesy of Holiday Corporation, Memphis, Tennessee

and secures the keys for the room attendants. At smaller properties, the executive housekeeper or the front desk may assume this function.

Employees issued keys should keep the keys on their person at all times. Key belts, wrist bands, or neck chains are recommended devices for keeping track of master keys. Master keys should never be left on top of a housekeeping cart, in a guestroom, or in an unsecured area. An employee should never loan the key to a guest or to another employee. The room attendant who signed for the master key is the employee who is responsible for it and should never leave the property. Finally, a room attendant should never use a master key to open a room for a guest. If a guest asks an employee to unlock a room, the employee should politely explain the hotel's policy and direct the guest to the front desk.

Room attendants are also responsible for retrieving guestroom keys if the guest leaves the key in the room. Many hotels provide key lock boxes on the room attendant's cart to store guestroom keys. If no lock box is available, room keys should be kept in a secured area—not on top of the cart—until returned to the front desk. If a room attendant finds a room key in the hallway or public area, the front desk should be notified immediately. The key should be returned to the front desk or placed in the lock box.

Lost or Stolen Keys. Lost or stolen keys create security and safety problems. When keys come up missing, the only solution is to rekey the affected rooms. The rekeying of rooms is very expensive and time-consuming—especially when it involves higher levels of master keys.

Card Key Systems. Many hotels use a card key system. This type of room-locking mechanism uses regular door locks and special plastic cards that act as keys to unlock the doors. The plastic cards look like credit cards with holes punched in them; some have a magnetic strip. The system uses a computer which codes the cards to lock and unlock doors.

A card key system is initially expensive to purchase. However, if a card is lost or stolen, the procedure for rekeying is quick and inexpensive. Rather than rekeying the door locks, the computer is used to create new room-lock codes for each room.

Master keys may be created and destroyed through the computerized card system. If a room attendant is responsible for cleaning rooms on more than one floor, the employee only needs one master card key rather than several floor masters.

Lost and Found

Many times, the housekeeping department handles the lost and found function. Lost and found items should be stored in an area that is secure and has limited access. One employee per shift should be assigned to handle the lost and found as part of his/her job.

In large hotels, the linen room clerk may handle the lost and found procedures. In smaller properties, the task may be delegated to the executive housekeeper or front desk personnel. When an employee finds an item left behind by a guest, he/she should immediately turn it over to the lost and found. In no instance should lost and found items be left in an unsecured spot such as on top of a room attendant cart.

Items should be tagged, logged, and secured after they have been turned over to the lost and found. Tags may be numbered or used to identify the item. A log should be used to record the date, time, where the item was found, and by whom. The log should also have space to record if and when the item was recovered by its owner. Exhibit 8 shows a sample lost and found log.

All lost and found property should be kept for at least 90 days. If items are not claimed after 90 days, it is up to management to decide how to dispose of the items properly. Many hotels donate unclaimed lost and found items to local charities. It is important to ensure that the lost and found policy of the hotel complies with local and state laws.

Guestroom Cleaning

Security in guestroom areas is important to maintain for the safety of the guests and employees. Room attendants should respect guest property and should not open guest luggage or packages, or snoop in dresser drawers or closets. Some hotels even have a policy that forbids room attendants to move guest property. In these instances, room attendants are instructed to clean around guest objects.

Exhibit 8   Sample Lost and Found Log

Item No. (optional)	Date and Time Found	Description of Article (include color, size, brand, etc.)	Area/Room No. Where Found	By Whom Found	How Disposed Of (Enter address if mailed)	By Whom	Date

LOST AND FOUND LOG

Courtesy of Holiday Corporation, Memphis, Tennessee

Since guests sometimes hide valuables and belongings in pillowcases or between mattresses, room attendants must be extra careful when removing linens. Other favorite hiding places for guest valuables include the top of closets and under lamps.

If room attendants notice any of the following while cleaning, they should immediately contact their supervisor, security, or the front desk:

- Guns or weapons of any kind
- Controlled substances or drugs
- Unauthorized cooking or unsafe electrical appliances
- Foul odors
- Unauthorized pets
- Ill guests
- Large amounts of cash or valuable jewelry

When cleaning, the room attendant should always keep the door open and the cart rolled in front of the entrance to block access from the outside. If a guest wants to enter the room while the attendant is cleaning, the attendant should politely ask the guest his/her name and ask to see a room key. This ensures that the room being cleaned is that guest's room. If the guest does not have a key, the attendant should tell him/her to contact the front desk. A guest should never be allowed to enter a room just to look around. Again, the attendant should explain that this is the hotel's policy and is enforced for the guest's safety and security.

A room should never be left unattended with the door open. If an employee must leave the room while cleaning, he/she should lock the door on the way out. This procedure should be followed even if the employee is out of the room for only a few minutes.

After cleaning the room, all windows and sliding glass doors should be locked. The guestroom door should also be checked to see that it is locked.

Unfortunately, guests often point the finger at the room attendant if an item comes up missing from the guestroom. This is just one more reason for room attendants to be considerate of guest property and to protect the guest's room from any possible theft. For the most part, an employee who is alert and careful can contribute to the overall safety and security of the operation—and to a guest's safe and trouble-free stay.

Endnotes

1. This discussion on OSHA is adapted from Raymond C. Ellis, Jr., and the Security Committee of AH&MA, *Security and Loss Prevention Management* (East Lansing, Mich.: Educational Institute of the American Hotel & Motel Association, 1986), Chapter 9.

2. This discussion (through the end of the chapter) is adapted from Ellis, *Security and Loss Prevention Management*. The information provided is in no way to be construed as a recommendation by the Educational Institute of the American Hotel & Motel Association or the AH&MA of any industry standard, or as a recommendation of any kind, to be adopted by or binding upon any member of the hospitality industry.

3. Brian Ledeboer and A.H. Petersen Jr., "Detect, Control Fires Before They Become Infernos," *Power*, June 1988, p. 27.

4. Walter Orey, "Full Fire Protection Requires Diligence," *Hotel and Motel Management*, January 12, 1987, p. 32.

Key Terms

emergency key	OSHA (Occupational Safety and Health Act)
guestroom key	safety
job safety analysis	security
key control	security committee
master key	Underwriters Laboratories
MSDS (material safety data sheet)	workers' compensation

Discussion Questions

1. What are the two main reasons that housekeeping and maintenance have more accidents and injuries than other hotel departments?

2. What three simple steps can employees take to prevent accidents and injuries?

3. What are four potentially hazardous conditions that a housekeeping employee may face in any given day?

4. Why is it necessary to establish a continual training program in cleaning chemical safety?

5. What is the purpose of the Occupational Safety and Health Act? Name three areas in a hospitality operation that are covered by OSHA regulations.

6. Compare the primary uses of danger, caution, and safety instruction signs.

7. What are three ways that a property can reduce the incident of guest theft? Of employee theft?

8. What three fire detection systems are required in hotel operations?

9. What part does housekeeping play in a property's key control efforts?

10. What four conditions or items should a trained room attendant report to his/her supervisor or to security if they are spotted during guestroom cleaning?

REVIEW QUIZ

When you feel you have covered all of the material in this chapter, answer these questions. Choose the *best* answer.

True (T) or False (F)

T   F   1.   A job safety analysis supplements job breakdowns with safety tips.

T   F   2.   OSHA regulations are primarily designed to protect employers.

T   F   3.   OSHA requires a hotel to develop a written emergency escape plan.

T   F   4.   According to OSHA regulations, any stairway with four or more steps must have at least one railing.

T   F   5.   Danger signs are used to warn employees of potential hazards in the workplace.

T   F   6.   OSHA officers have the authority to inspect a property without providing advance notice.

T   F   7.   Smoke detectors can go off when a fire is not present.

T   F   8.   When trying to leave a smoke-filled room, you should stay close to the floor and cover your mouth and nose with a wet towel.

T   F   9.   Room attendants are issued section master keys to open the rooms they are assigned to clean.

T   F   10.   A good security practice when cleaning a guestroom is for the room attendant to keep the door open and place the cart so that the entrance to the room is blocked.

Alternate/Multiple Choice

11.   When lifting heavy objects, you should keep your back:

a.   bent.
b.   straight.

12.   The OSHA HazComm standard requires employers to:

a.   provide wholesome meals and clean eating facilities for employees.
b.   inform employees about hazardous chemicals they use on the job.

13.   OSHA standards dictate the occupancy level of individual rooms and buildings on the basis of:

a.   the number of exits.
b.   the number of fires reported at the property.

14. If a ladder is 20 feet tall, the footing should be _____ feet away from the wall.

    a. 2
    b. 3
    c. 4
    d. 5

15. Fires involving wood and paper products are classified as _____ fires.

    a. Class A
    b. Class B
    c. Class C
    d. Class D

Chapter Outline

Planning the OPL
    Laundering Linens
The Flow of Linens Through the OPL
    Collecting Soiled Linens
    Transporting Soiled Linens to the
       Laundry
    Sorting
    Washing
    Extracting
    Finishing
    Folding
    Storing
    Transferring Linens to Use Areas
Machines and Equipment
    Washing Machines
    Drying Machines
    Steam Cabinets and Tunnels
    Flatwork Ironers and Pressing
       Machines
    Folding Machines
    Rolling/Holding Equipment
    Preventive Maintenance
    Staff Training
Valet Service
    Contract Valet Service
    On-Premises Valet Service
Staffing Considerations
    Staff Scheduling
    Job Lists and Performance Standards

Learning Objectives

1. Identify the steps involved in processing linens as they flow through an on-premises laundry operation.

2. Describe two methods of sorting soiled linens.

3. List the steps in a typical wash cycle.

4. Identify the function of three different types of detergents.

5. Distinguish between the functions of alkalies and sours in the wash cycle.

6. Explain the benefits of automatic detergent and solution dispensing controls.

7. Describe the benefits of water reuse washers.

8. Explain the importance of preventive maintenance programs in on-premises laundries.

9. Cite the basic requirements and benefits of providing an on-premises valet service.

10. Explain how linen needs are forecasted.

11. Explain how laundry staff are scheduled.

Managing an On-Premises Laundry

This chapter was written and contributed by Michael T. Floyd,
National Sales Manager—On-Premises Laundry,
Speed Queen Company, Ripon, Wisconsin.

Doing the laundry at home or the laundromat may not be your favorite chore, but it is not a difficult one. Once a week you sort a basket of wash, select a detergent and the proper washer setting, and then dry and fold the items. But imagine doing laundry every day by the truckload, and you begin to have an idea of the scope of a lodging property's laundry. Add to the sheer volume of wash the responsibilities of making it look, smell, and feel good, and getting it to the right place at the right time. Then consider that linen (sheets, towels, tablecloths, and other items) is a housekeeping department's second largest expense, and you will understand why good laundry management is essential to the success of a lodging operation.

Some hotels do not operate on-premises laundries. These properties contract with an outside laundry service which provides them with clean linen on a scheduled basis. The linen supply may be owned by the hotel or rented from the laundry service.

Since the recent trend has been for hotels to operate an on-premises laundry (OPL), this chapter focuses on OPL management. The topics covered include planning the physical layout of the OPL, procedures for laundering different fabrics, the flow of linens through the OPL, typical machines and equipment, and staffing considerations.

Planning the OPL

The best OPL is the one tailored to the needs of the hotel it serves. If possible, representatives from areas of the hotel affected by the laundry operation should be involved in the planning stages of the OPL. Some of the important planning considerations include:

- What is the maximum amount of laundry (output) the OPL will be expected to handle? Output is generally measured in pounds. The number of pounds should be related to the occupancy levels in guestrooms and number of covers

Exhibit 1   OPL: Typical Small Property Layout

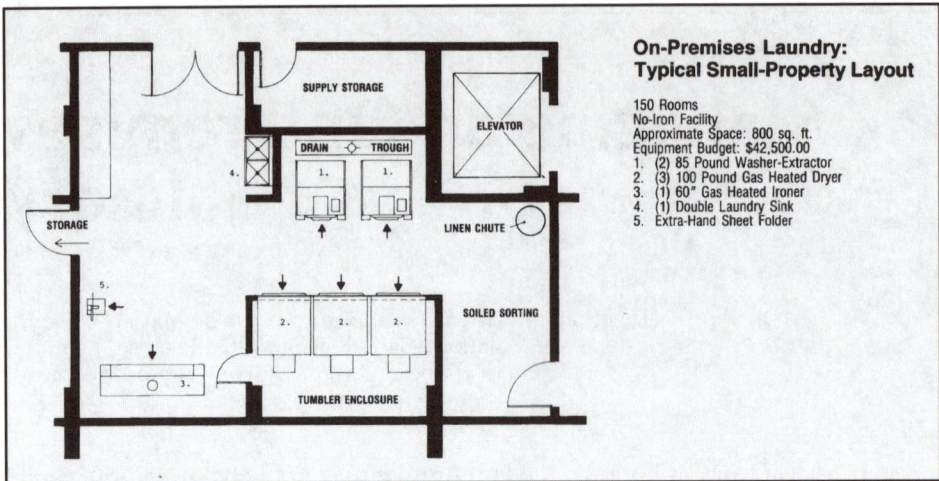

On-Premises Laundry:
Typical Small-Property Layout

150 Rooms
No-Iron Facility
Approximate Space: 800 sq. ft.
Equipment Budget: $42,500.00
1. (2) 85 Pound Washer-Extractor
2. (3) 100 Pound Gas Heated Dryer
3. (1) 60" Gas Heated Ironer
4. (1) Double Laundry Sink
5. Extra-Hand Sheet Folder

Source: Jay D. Chase, "Laundry Service: Consider the Options," *Lodging,* May 1986, p. 38.

in food and beverage outlets. The OPL should be designed to handle maximum output for peak business periods.

- How much space should be devoted to the OPL? Laundry needs, amount of equipment, and the amount of linen kept on-hand in storage, will determine space needs. In addition, many properties allocate extra space in case of growth.

- How much equipment should be purchased? The output levels will determine the amount of equipment necessary to handle the hotel's laundry needs. The type of linens the hotel uses usually determines what kind of equipment is necessary. Energy and water conservation concerns may also affect equipment decisions.

- Will there be valet service? Valet service will require dry cleaning equipment and separate work areas for valet staff.

The size of the property and type of service offered are other important planning considerations. Laundry needs in small properties (under 150 rooms) vary considerably. A very small operation offering economy services may devote between 400 to 800 square feet of space to the OPL. A property offering mid-range service with a food and beverage operation may require between 1,500 and 2,000 square feet for the OPL. On the average, a small property's OPL processes about 400,000 pounds of laundry per year. It contains washer/extractors and drying machines. Small properties frequently rely on no-iron linens to reduce finishing time. However, no-iron linens lose their wrinkle-free characteristics after numerous washings. A small **flatwork ironer** is often needed to keep these linens looking good. A floor plan for an OPL at a small property is shown in Exhibit 1.

Medium-sized properties (150 to 299 rooms) may offer economy to world-class services. Properties with food and beverage outlets generally require more

Exhibit 2   OPL: Typical Medium-Sized Property Layout

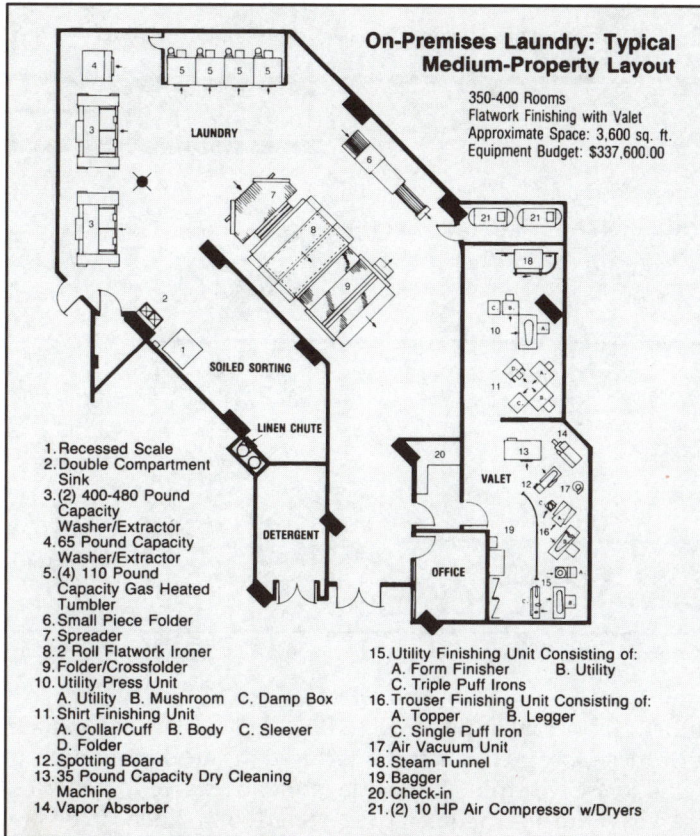

On-Premises Laundry: Typical Medium-Property Layout

350-400 Rooms
Flatwork Finishing with Valet
Approximate Space: 3,600 sq. ft.
Equipment Budget: $337,600.00

LAUNDRY

SOILED SORTING

LINEN CHUTE

DETERGENT

VALET

OFFICE

1. Recessed Scale
2. Double Compartment Sink
3. (2) 400-480 Pound Capacity Washer/Extractor
4. 65 Pound Capacity Washer/Extractor
5. (4) 110 Pound Capacity Gas Heated Tumbler
6. Small Piece Folder
7. Spreader
8. 2 Roll Flatwork Ironer
9. Folder/Crossfolder
10. Utility Press Unit
    A. Utility B. Mushroom C. Damp Box
11. Shirt Finishing Unit
    A. Collar/Cuff B. Body C. Sleever
    D. Folder
12. Spotting Board
13. 35 Pound Capacity Dry Cleaning Machine
14. Vapor Absorber
15. Utility Finishing Unit Consisting of:
    A. Form Finisher   B. Utility
    C. Triple Puff Irons
16. Trouser Finishing Unit Consisting of:
    A. Topper   B. Legger
    C. Single Puff Iron
17. Air Vacuum Unit
18. Steam Tunnel
19. Bagger
20. Check-in
21. (2) 10 HP Air Compressor w/Dryers

Source: Jay D. Chase, "Laundry Service: Consider the Options," *Lodging*, May 1986, p. 34.

linens than hotels offering limited or mid-range services without food and beverage operations. Up to 1.5 million pounds of laundry per year may flow through a medium-sized property's OPL. OPL space varies from 2,000 or 3,000 square feet to as many as 6,500 to 7,000 square feet. That space may accommodate flatwork ironers with folding capabilities, **steam tunnels** or **cabinets**, and valet service equipment. Exhibit 2 shows a floor plan for an OPL at a medium-sized property.

Large properties (300 rooms or more) may devote from 8,000 to 18,000 square feet to the OPL and handle as much as 8.5 million pounds of laundry per year. Large OPLs use more sophisticated equipment than smaller OPLs. A typical floor plan for a large OPL is shown in Exhibit 3.

Laundering Linens

The marketplace offers more fabrics to choose from than ever before. The choice of fabric is also more important than ever before because it directly affects the costs of operating the OPL.

Exhibit 3   OPL: Typical Large Property Layout

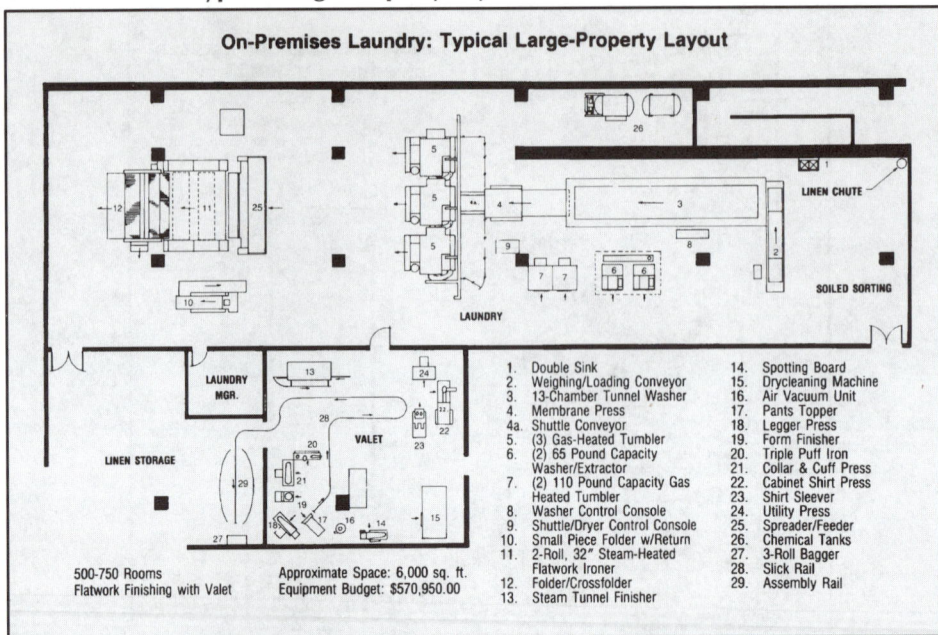

On-Premises Laundry: Typical Large-Property Layout

1. Double Sink	14. Spotting Board
2. Weighing/Loading Conveyor	15. Drycleaning Machine
3. 13-Chamber Tunnel Washer	16. Air Vacuum Unit
4. Membrane Press	17. Pants Topper
4a. Shuttle Conveyor	18. Legger Press
5. (3) Gas-Heated Tumbler	19. Form Finisher
6. (2) 65 Pound Capacity	20. Triple Puff Iron
Washer/Extractor	21. Collar & Cuff Press
7. (2) 110 Pound Capacity Gas	22. Cabinet Shirt Press
Heated Tumbler	23. Shirt Sleever
8. Washer Control Console	24. Utility Press
9. Shuttle/Dryer Control Console	25. Spreader/Feeder
10. Small Piece Folder w/Return	26. Chemical Tanks
11. 2-Roll, 32" Steam-Heated	27. 3-Roll Bagger
Flatwork Ironer	28. Slick Rail
12. Folder/Crossfolder	29. Assembly Rail
13. Steam Tunnel Finisher	

500-750 Rooms
Flatwork Finishing with Valet

Approximate Space: 6,000 sq. ft.
Equipment Budget: $570,950.00

Source: Jay D. Chase, "Laundry Service: Consider the Options," *Lodging,* May 1986, p. 36.

The synthetic fabrics introduced in the 1960s led to the development of no-iron sheets. Because these sheets eliminated or reduced the need for ironing, many properties were able to switch from an outside laundry service to an OPL. Moreover, properties that used the no-iron linens discovered these linens to be more durable than all-cotton linens. This durability cut down the rate at which the linens had to be replaced (and still does).

Today's fabrics range from all-natural fibers (wool and cotton, for example) to a variety of synthetics (such as polyester and nylon). For most properties, the fabric of choice is a polyester/cotton blend (sometimes called **polycotton**) because it requires less care than all-natural fabric yet offers most of its comfort.

However, no-iron linens do not totally fulfill the basic functions guests have learned to expect with 100% cotton linens. For example, polyester napkins are not as absorbent as 100% cotton napkins. Polycottons also pick up and retain stains more easily because the cotton absorbs the stain and the polyester traps it in the fabric. Furthermore, the resins which keep no-iron fabrics from wrinkling tend to break down and are washed out at high temperatures. The resins also retain chlorine from bleach, which weakens the fabric.

No matter what linens the property decides to buy, it is important to make certain that all items needing laundering include thorough instructions for care from the vendor. The following sections discuss some of the most popular fabrics used in hospitality operations. Exhibit 4 contains a summary of general care instructions for the fabrics listed here.

Exhibit 4   General Care of Linen Fabrics

Fiber Group	Cleaning Method	Water Temperature	Chlorine Bleach	Dryer Temperature	Iron Temperature	Special Storage
Acrylic	launder	warm	yes	warm	medium	none
Cotton	launder	hot	yes	hot	high	store dry
Polycotton	launder	hot	yes	warm	medium	none
Nylon	launder	hot	yes	warm	low	none
Polyester	launder	hot	yes	warm	low	none
Wool	dry clean	warm	no	warm	medium with steam	protect from moths; do not store in plastic bags.

Fabric combinations should always be cared for by following the manufacturer's recommendations.

Cotton. Cotton is strong and actually becomes stronger when wet. It is very absorbent and can be starched, which makes it especially good for napkins and tablecloths. It can also be washed and ironed at high temperatures. Some shrinkage (from 5% to 15% in its first washings) does occur. Cotton fabrics have a lower color retention capability than polyester fabrics.

Mineral acids are hard on cotton fibers. These acids form when microscopic mineral particles (called ions) mix with oxygen. Ions are found naturally in many water supplies. Mineral acid damage underscores the need for a source of good water in the OPL.

Wool. Once the fabric of choice for blankets, wool has fallen out of favor in many commercial operations because it is not as durable as some synthetic materials and can be irritating to the touch. Wool is one of the weakest fibers and becomes even weaker when wet. It also shrinks and mats relatively easily. For this reason, many heavy-duty cleanings will break down the fibers of a wool blanket very quickly. Wool does resist soiling better than some other common fabric materials and is very absorbent.

Acrylic. Acrylic is lightweight and does not shrink. Its strength is similar to cotton's, but it decreases when wet. Because it holds moisture on its surface, acrylic is fairly slow-drying.

Polyester. Polyester is one of the strongest common fibers and does not lose its strength when wet. It dries quickly, is wrinkle-resistant, and does not soil easily. Polyester tends to break down at higher drying and ironing temperatures. Polyesters and polyester blends are good choices for uniforms, aprons, and other garments. They are less effective as napery (table linens).

Nylon. Nylon is very strong when wet or dry. It is also easy to wash and quick to dry. Nylon, however, is sensitive to heat.

Blends. Many properties use linens made of cotton/polyester blends. These gain strength with initial washings. Their characteristics depend on the amount and types of fibers blended. Blends can be damaged by high wash temperatures—those greater than 180°F [83°C] or high dryer temperatures—those greater than 165°F [74°C].

The Flow of Linens Through the OPL

Every laundry uses a basic cycle of operation. This cycle includes the following steps:

- Collecting soiled linens
- Transporting soiled linens to the laundry
- Sorting
- Washing
- Extracting
- Finishing
- Folding
- Storing
- Transferring linens to use areas

Exhibit 5 diagrams an abbreviated version of this process. Executive housekeepers or laundry managers should develop procedures for each of these steps to prevent resoiling of clean linens, extend the life of the linens, and keep the OPL efficient and cost-effective.

Collecting Soiled Linens

Room attendants cleaning guestrooms should strip linens from beds and bath areas and put them directly into the soiled linen bags attached to the housekeeping cart. Items should never be piled on the floor where they can be walked on and soiled further or damaged. Putting the linens directly into soiled linen bags prevents room attendants from using towels, sheets, napkins, or other items to blot spills or wipe smudges. *Staff should never use linens for any cleaning purposes.* Misuse of linens can permanently damage items—which can lead to higher replacement costs.

In some properties, room attendants follow procedures for pre-sorting soiled linens. This may simply mean tying a knot in one corner of a heavily soiled item to help laundry workers sort it more easily. Room attendants may also sort linens by soil type and put them into specially marked plastic bags. In food and beverage outlets, buspersons gather soiled linens when tables are cleared. Because tableware can easily get gathered up with linens and thrown into the soiled linen hamper, buspersons should be cautioned to carefully remove all items from tables. Some tableware metals can permanently stain linens. Buspersons should shake napkins and tablecloths over a waste receptacle to remove crumbs and food as soon as possible after the table is cleared. Linens can then be placed in soiled linen hampers for delivery to the laundry.

Exhibit 5   The Flow of Laundry Through the OPL

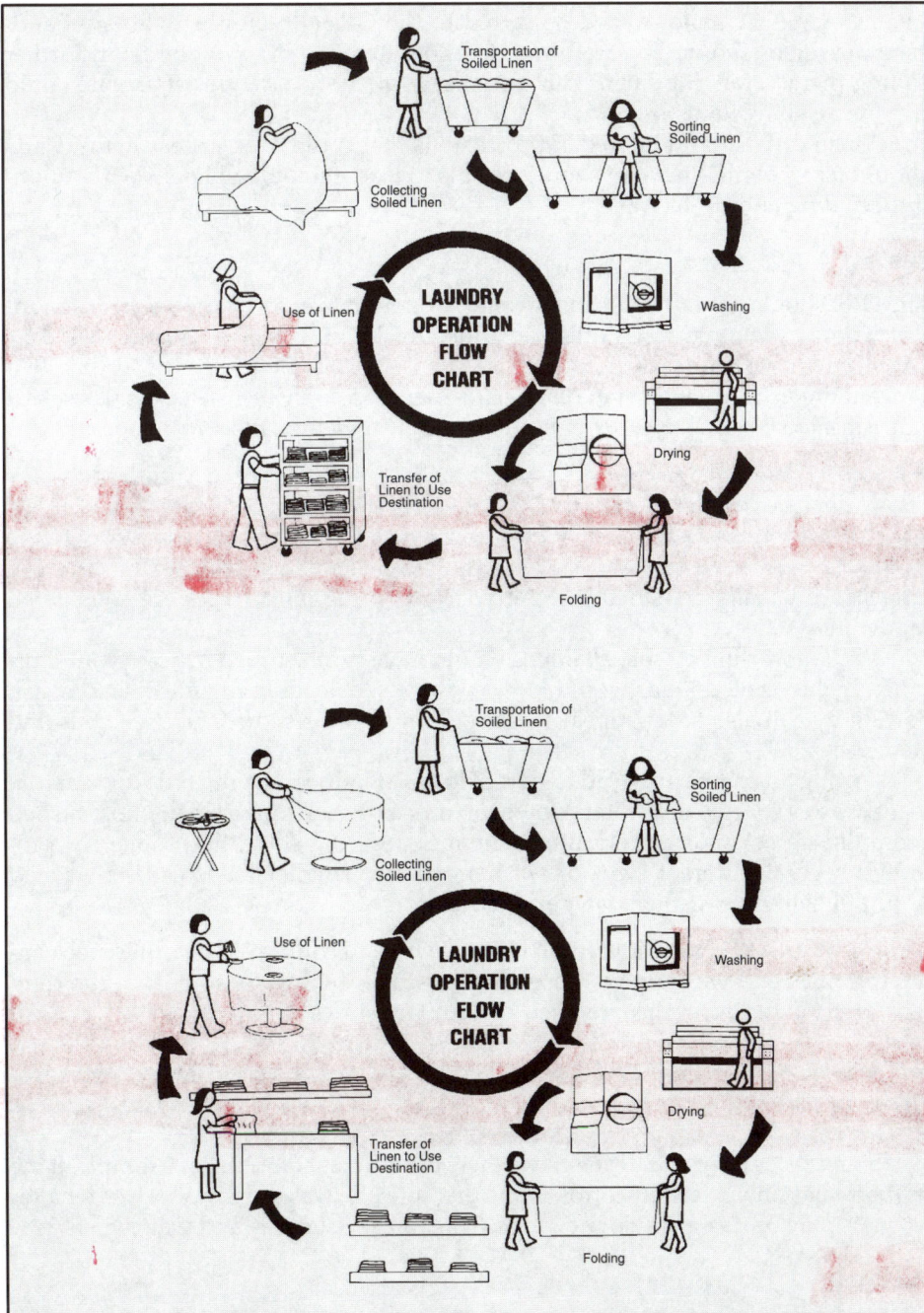

Source: "On Premises Laundry Procedures in Hotels, Motels, Healthcare Facilities, and Restaurants" (pamphlet) (St. Paul, Minn.: Ecolab, Institutional Products Division), undated.

Transporting Soiled Linens to the Laundry

Linens are either hand-carried or carted to the OPL. Employees who are hand-carrying linens should be careful not to allow items to drag on the floor, further soiling them. Dragging linens can also create safety hazards for staff who could trip over trailing items.

Linen carts should be free of protrusions that could snag or tear items. Carts should move easily, and staff should be able to load and unload linens without undue bending and stretching.

Sorting

The OPL should contain a sorting area large enough to store a day's worth of laundry without slowing down other activities in the OPL. Soiled linens should be sorted by the degree of soiling and by the type of fabric. Both types of sorting help prevent unnecessary wear and damage to linens. *Cleaning rags should always be separated and washed by themselves, never with linen that guests will use.*

Sorting by Degree of Soiling. When sorting by degree of soiling, laundry workers divide linens into three categories: lightly, moderately, and heavily soiled. Heavily soiled items require heavy-duty wash formulas and longer wash time. Moderately or lightly soiled linens are washed with gentler formulas and in fewer cycles. (Sheets are usually classified as lightly soiled, while pillowcases are considered moderately soiled.)

Without sorting by soil, all linens would have to be washed in heavy-duty formula. Lightly soiled items would be overprocessed, leading to unnecessary wear. Sorting by soil also saves repeat washing of items to remove stubborn soils and stains.

Sorting, of course, can lead to partial loads of laundry. Doing too many partial loads wastes energy and water. However, if heavily soiled fabrics are not washed promptly, stains could set and ruin the item. Some OPLs solve this problem by providing several different sizes of washers so that smaller loads can be washed promptly without wasting water and energy.

Sorting by Linen Type. Different fibers, weaves, and colors require different cleaning formulas and washing methods. Sorting linens by type ensures that the right temperature and formulas are used on similar fabrics. Wool and loosely woven fabrics, for example, require a mild formula and gentle agitation. Colors should not be washed with chlorine bleach. New colored linens should be washed separately the first few times to avoid dying other fabrics. Some special items such as aprons should be washed in nylon bags to prevent tangling.

Some OPLs purchase different washers for different fabrics. For example, if the property has only a few all-cotton items that must be washed in hot water, smaller washers may be designated for cottons. This can save energy and water costs.

Washing

After linens are sorted, laundry workers collect batches of laundry and deliver them to the washers. Linens should be weighed before they are put into the washer to ensure that washers are not overloaded. Weighing is also important for measuring OPL output.

Laundry workers at some hotels pre-treat soiled linens before washing them. However, pre-treating laundry takes a great deal of time and can increase labor costs dramatically. As a result, most OPLs rely on the chemicals used in the washer to clean linens.

Today's modern washing equipment can overwhelm an inexperienced worker used to doing laundry on a wash-rinse-spin machine at home. OPL washing equipment requires workers to choose from as many as ten cycles and from a range of detergents, soaps, and fabric conditioners. Asking four basic questions makes these choices less confusing and helps determine the proper procedure for doing a particular batch of laundry. The four questions are:

- How much *time* will it take to wash any given item properly? Linens that are heavily soiled will take more time to wash; lightly soiled linens will take less time. Improperly regulating the time it takes to wash linens will result in a poor wash or unnecessary wear on fabric. It can also waste energy and water.

- At what *temperature* should the water be in order to get items clean? In general, laundry workers should "think low," choosing the lowest possible temperature to do the job in order to save energy. However, some detergents and chemicals work properly only in hot water, and some types of soils require higher temperatures. For example, the water temperature for washing oily soils should be 180°F to 190°F [83°C to 88°C]; for moderate to heavy soils, at least 160°F [72°C]. Kitchen rags and linens should be washed at 140°F [60°C].

- How much *agitation* is needed to loosen soils? Agitation is the "scrubbing" action of the machine. Too little agitation—which is frequently caused by overloaded washers—leads to inadequate washing. Overloading also causes unnecessary wear and tear on equipment. Too much agitation can cause fabric damage.

- What *chemicals* will do the best job on particular soils and fabric types? Chemicals may include detergents, bleaches, softeners, and so forth.

The type of soil and fabric will dictate wash time, temperature, degree of agitation, and which chemicals to use. Each of these elements affects the others. For example, using too much detergent for the water level will create too many suds; too many suds will impede the agitator. Exhibit 6 offers an overview of common washing problems, their causes, and some solutions.

Usually, time, temperature, and mechanical action are pre-set by supervisors in the laundry room. Equipment salespeople can also help with pre-setting.

Wash Cycles. The typical wash process consists of as many as nine steps:

1. **Flush (1.5 to 3 minutes)**—Flushes dissolve and dilute water-soluble soils to reduce the soil load for the upcoming suds step. Items are generally flushed at medium temperatures at high water levels.

2. **Break (4 to 10 minutes, optional)**—A high-alkaline **break** (soil loosening) product is added, which may be followed by additional flushes. The break cycle is usually at medium temperature and low water level.

Exhibit 6   Common Laundry Problems

Problem	Cause	Solution
Graying	Too little detergent	Increase amount of detergent; add bleach.
	Wash cycle temperature too low	Increase temperature.
	Poor sorting; transfer of soiling occurs	Rewash with increased detergent at hottest possible temperature. Use bleach suitable for fabric. Implement proper sorting procedures.
	Color "bleeding"	Do not dry. Rewash with detergent and bleach. Sort more carefully by color. Launder new colored fabrics separately the first few times.
Yellowing	Insufficient detergent	Increase the amount of detergent or use an *enzyme* product or bleach.
	Wash cycle temperature too low	Increase wash temperature.
	Use of chlorine bleach on wool, silk, or spandex items	Yellowed items cannot be restored. Avoid chlorine bleach on such items in the future.
Rust stains	Iron and/or manganese in water supply, pipes or water heater	Rewash clothes with a commercial rust-removing product; *do not use chlorine bleach.* To prevent further staining, use a water softener to neutralize iron/manganese in water supply. If iron is in pipes, run the hot water a few minutes to clear the line. Drain water heaters occasionally to remove rust buildup.
Blue stains	Blue coloring in detergent or fabric softener fails to disperse properly	For detergent stains, soak items in a plastic sink or container for an hour in a solution of one part white vinegar per four parts water. For softener stains, rub fabric with bar soap and wash. To prevent stains, add detergents to the washer before clothes, then start washer to ensure better mixing of detergent. Dilute fabric softener before adding to wash.
Poor soil removal	Too little detergent	Increase amount.
	Wash temperature too low	Increase temperature.
	Overloading washer	Wash fewer items per load, sort properly, and use the proper amount of detergent and water temperature.
Greasy or oily stains	Too little detergent	Treat with prewash stain remover or liquid laundry detergent; increase amount of detergent.
	Wash temperature too low	Wash in higher temperature.
	Undiluted fabric softener has come into contact with fabric	Rub fabric with bar soap and wash; dilute fabric softener before adding to cycle.
	Dryer-added softener	Rub fabric with bar soap and wash. Avoid too small a load, improper dryer setting, too hot dryer.
Residue of powder (especially noticeable on dark or bright colors	Undissolved detergent	Add detergent to the washer before clothes, then start washer.
	Non-phosphate granular detergent combines with water minerals and forms residue	Remove stain by mixing 1 cup of white vinegar to 1 gallon of warm water. Soak in a plastic container or sink and rinse. To prevent residue, switch to a liquid detergent.

Exhibit 6 *(continued)*

Problem	Cause	Solution
Stiff, faded, or abraded fabrics	Non-phosphate granular detergent combines with water minerals and forms residue	Remove stain by mixing 1 cup of white vinegar to 1 gallon of warm water. Soak in a plastic container or sink and rinse. To prevent residue, switch to a liquid detergent.
Lint	Improper sorting (mixing *napped* fabrics with others)	Dry items and pat with masking or transparent tape, rewash and use fabric softener in final rinse. Prevent problems by sorting more carefully.
	Tissue in apron or uniform pockets	Check pockets before laundering.
	Overloading washer or dryer	Wash and dry fewer items.
	Insufficient detergent	Increase amount of detergent.
	Clogged washer lint filter or dryer lint screen	Clean filters and screens after use; rewash items.
	Overdrying causes static electricity which attracts lint	Rewash items using fabric softener; remove items from dryers when they are slightly damp.
Holes, tears, or snags	Incorrect use of chlorine bleach	Always use a bleach dispenser and dilute bleach with four parts water; never pour directly on linens.
	Unfastened zippers, hooks, or belt buckles	Fasten zippers, hooks and eyes, and belt buckles before washing.
	Burrs in washer	Inspect washer on a weekly basis and repair as necessary.
	Washer overload	Avoid overloading.
Color fading	Unstable dye	Test items for colorfastness before washing; wash new items separately.
	Water temperature too hot	Use cooler water.
	Improper use of bleach	Test item for colorfastness; use oxygen bleach.
	Undiluted bleach poured on fabric	Dilute bleach.
Wrinkling	Failure to use correct cycle	Use permanent press cycle; cooler temperatures in wash; remove items promptly from dryers and fold immediately.
	Washer/dryer overloading	Do not overload
	Overdrying	Put items back in the dryer on permanent press cycle for 15 to 20 minutes; heat and cool-down time will remove wrinkles. Remove all items promptly.
Shrinking	Overdrying	Reduce drying time and remove items while damp; remove knits (especially cotton) while slightly damp and stretch back into shape; dry flat.
	Residual shrinking	Allow for some shrinking when purchasing items.
	Agitation of wool items	Lower agitation in wash/rinse cycles; regular spinning will not promote shrinkage.
Pilling	Synthetics pill naturally with wear	Prevent unnecessary wear by using fabric softener and spray starch or fabric finish.

Source: Edwin B. Feldman, P.E., ed., *Programmed Cleaning Guide for the Environmental Sanitarian* (New York: The Soap and Detergent Association, 1984), pp. 163–168.

3. **Suds (5 to 8 minutes)**—This is the actual wash cycle to which detergent is added. Items are agitated in hot water at low water levels.

4. **Carryover suds or intermediate rinse (2 to 5 minutes)**—This rinse cycle removes soil and alkalinity to help bleach work more effectively. This cycle rinses linens at the same temperature as the suds cycle.

5. **Bleach (5 to 8 minutes)**—Bleach is added to this hot water, low water level cycle. Bleach kills bacteria, whitens fabrics, and removes stains.

6. **Rinse (1.5 to 3 minutes)**—Two or more rinses at medium temperatures and high water levels are used to remove detergent and soil from the linens.

7. **Intermediate extract (1.5 to 2 minutes, optional)**—This high-speed spin removes detergent and soil from linens, usually after the first rinse step. This cycle should not be used after a suds step because it could drive soils back into the fabric. It should not be used on no-iron linens unless the temperature of the wash is below 120°F [49°C].

8. **Sour/softener or starch/sizing (3 to 5 minutes)**—Softeners and **sours** are added to condition fabric. The cycle runs at medium temperature and at low water levels. Starches are added to stiffen cotton fabrics; sizing is added for polyester blends. Starching/sizing replaces the sour/softener step.

9. **Extract (2 to 12 minutes)**—A high-speed spin removes most of the moisture from the linens. The length of the spin depends on fabric type, extractor capacity, and extractor speed.

Many OPLs are now choosing washers with cold water options. Cold water washes with synthetic bactericidal detergents can:

- Remove stains that hot water will set

- Preserve the wrinkle-free characteristics of no-iron fabric and absorbency of towels

- Save energy costs

Chemicals. Hotel and other commercial OPLs use many more chemicals to wash linens than people use in their washers at home. The hotel laundry "fine-tunes" its chemicals to ensure an effective wash that leaves linens looking as close to new as possible. In general, a laundry's chemical needs depend mainly on the types of linens it uses and the soiling conditions encountered. In addition, hotel OPLs use more **alkali** to enhance the detergent's cleaning power. Alkali, however, is abrasive and must be neutralized with other chemicals.

Generally, it is a good idea to deal with more than one chemical vendor. This ensures that you obtain good advice on chemical usage along with updates on new technology. However, juggling several vendors at the same time may not be the most effective use of the executive housekeeper's time. Some executive housekeepers accept bids from chemical vendors and select one vendor as the sole supplier for the upcoming year. This system works well as long as the vendor selected is keenly aware that if the property's needs are not met, another vendor stands ready to do the job.

The following list provides a brief description of the major categories of chemicals used in laundry operations.

Water. Although not always recognized as such, water is the major chemical used in the laundry process. Two to five gallons of water are used for every pound of dry laundry. Water that is perfectly safe to drink may not be suitable for washing linens. Certain minerals, for example, can stain or wear linens. Other substances can cause odors or "hard" water that hampers sudsing. Many of these substances can also clog pipes and machinery. Fortunately, other chemicals can be added to water to help it clean better. Many OPL operators recommend testing the laundry's water supply to identify potential problems.

Detergents. The term detergent is actually a catchall word for a number of cleaning agents. **Synthetic detergents** are especially effective on oil and grease. Synthetic detergents often contain **surfactants**. These are chemicals that aid soil removal and act as antibacterial agents and fabric softeners. **Builders** or **alkalies** are often added to synthetic detergents to soften water and remove oils and grease. **Soaps** are another kind of detergent. **Neutral** or **pure** soaps contain no alkalies; **built** soaps do. Built soaps are generally used on heavily soiled fabrics; pure soaps are reserved for more lightly soiled items. Hard water reduces a soap's cleaning ability and also leaves a "scum" on fabrics that causes graying, stiffness, and odor. Soaps are destroyed by sours.

Fabric (optical) brighteners. Brighteners keep fabrics looking new and colors close to their original shade. These chemicals are often pre-mixed with detergents and soaps.

Bleaches. Bleaches cause strong chemical reactions that, if not carefully controlled, can damage fabrics. Used properly, bleaches help remove stains, kill bacteria, and whiten fabrics.

There are two kinds of bleaches: **chlorine** and **oxygen.** Chlorine bleach can be used with any washable, natural, colorfast fiber. Chlorine bleach is safe for some synthetics and destroys others. All synthetics should therefore be tested before chlorine bleach is used. Oxygen bleach is milder than chlorine bleach and is generally safe for most washable fabrics. Oxygen bleach works best in hot water and on organic stains. Oxygen bleach should never be used with chlorine bleaches as they will neutralize each other.

A bleach's pH (degree of acidity or alkalinity) and water temperature must be controlled carefully to prevent fabric damage. Dry bleaches contain buffers that control pH, but they are more expensive than the liquid variety.

Alkalies. Alkalies or **alkaline builders** help detergents lather better and keep stains suspended in the wash water after they have been loosened and lifted from the fabric. Alkalies also help neutralize acidic stains (most stains are acidic), making the detergent more effective.

Antichlors. Antichlors are sometimes used in rinsing to ensure that all the chlorine in the bleach has been removed. Polyester fibers retain chlorine, and for this reason are typically treated with antichlors when chlorine bleach is used.

Mildewcides. Mildewcides prevent the growth of bacteria and fungus on linens for up to 30 days. Both of these types of microorganisms can cause permanent stains that ruin linens. Moisture makes a good breeding ground for mildew growth. Therefore, soiled damp linen should be washed promptly and not allowed

Exhibit 7 Common Finishing Problems

Problem	Cause	Solution
Wrinkling (synthetics)	Washing/drying temperatures too high causing breakdown of no-iron characteristics	Reduce heat.
	Insufficient cool-down in dryer	Turn heat down during the last few minutes of drying time; remove them before they are bone dry.
Glazed or fused fibers (synthetics)	Dryer heat is too high	Lower heat (140°F to 145°F; 60°C to 63°C); heat over 160°F (71°C) is too high.
Loss of absorbency	Washing/drying temperatures too high	Reduce heat.
	Too much fabric softener	Use less softener.

to sit in carts for long periods. Clean linens should be dried and/or ironed as they are removed from washers or extractors.

Sours. Sours are basically mild acids used to neutralize any residual alkalinity in fabrics after washing and rinsing. Detergents and bleaches contain alkali, and any residual alkali can damage fibers and cause yellowing and fading. In addition, residual alkalies can cause skin irritation and leave odors.

Fabric softeners. Softeners make fabrics more supple and easier to finish. Softeners are added with sours in the final wash cycle. They can reduce flatwork ironing, speed up extraction, reduce drying time, and reduce static electricity in the fabric. Too much softener can decrease a fabric's absorbency.

Starches. Starches give linens a crisp appearance that stands up during the items' use. If they are used, starches should be added in the final step in the washing process.

Extracting

Extracting removes excess moisture from laundered items through a high-speed spin. This step is important because it reduces the weight of the laundry and makes it easier for workers to lift the laundry and move it to dryers. Extracting also reduces drying time. Most washing machines now have extracting capabilities.

Finishing

Finishing gives the linens a crisp, wrinkle-free appearance. Finishing may require only drying or may include ironing. Linens should be sorted by fabric type before they are dried. Steam cabinets or tunnels are often used to dry blends because they give these fabrics a finished, wrinkle-free look. Some common finishing problems and their solutions are outlined in Exhibit 7.

Drying. Items that are dried generally include towels, washcloths, and some no-iron items. Drying times and temperatures vary considerably for different types of linens. In every case, however, drying should be followed by a cool-down tumbling period to prevent the hot linens from being damaged or wrinkled by rapid cooling and handling. After drying, linens should be removed immediately for folding. If folding is delayed, wrinkles will set in.

Dryers should never be pre-warmed or run when empty. This can lead to "hot spots" which can damage fabric or cause fires. It also wastes energy.

Ironing. Sheets, pillowcases, tablecloths, and slightly damp napkins go directly to flatwork irons. Ironers vary in size and degree of automation. Uniforms are generally pressed in special ironing equipment. Steam tunnels are being used more often for removing wrinkles from polyester blend uniforms.

Folding

Since some properties still do a lot of folding manually, folding usually sets the pace for the linen room. Washing and drying items faster than they can be folded leads to unnecessary wrinkling and resoiling.

Folding personnel must also inspect linens, storing those that are to be reused and rejecting stained, torn, or otherwise unsuitable items. This inspection may increase folding time. Folding and storing should be done well away from the soiled linen area to avoid resoiling clean laundry.

Storing

After folding, the items are post-sorted and stacked. Post-sorting separates any linen types and sizes that were missed in pre-sorting. There should be enough storage room for at least one par. Finished items should be allowed to "rest" on shelves for 24 hours after laundering because many types of linens are more easily damaged right after washing. Once linens are on shelves, yellowing and fading can be spotted quickly.

Transferring Linens to Use Areas

Linens are usually transferred to their use areas via carts. Carts should be cleaned at least once daily, and more often if necessary. Transferring linens just before use and covering carts can help prevent resoiling. It may be a good idea to have separate carts for soiled and clean linens to avoid accidental soiling.

Machines and Equipment

OPL machinery is a major investment in itself and affects the lifespan of another major investment—linens. The choice of OPL machines and equipment could mean the difference between a financially successful and a disastrous OPL. Machines with insufficient capacity, for example, result in damaged linens, unsatisfactory cleaning performance, excessive energy and water costs, or increased maintenance costs. Improperly maintained equipment can also lead to higher linen and equipment costs.

Most laundry equipment manufacturers offer free estimates of the type and amount of equipment needed, based on how many pounds of linen the operation must process in a day. The following discussion offers some basic information about the types of equipment available for OPLs.

Washing Machines

Most washers are made of stainless steel. They are sized by their capacity (that is, the number of pounds of linen they can handle in a single load). Sizes vary from 25- to 1,200-pound capacities. A large capacity washer in a hotel laundry may not resemble a conventional washer designed for home use. Some machines have separate "pockets" which hold several large loads at a time. Some washers, called **tunnel washers**, have several chambers; each chamber is used for a particular wash cycle. As soon as the first cycle is finished on the first load of laundry, the wash moves into the second chamber. The laundry attendant can then load the first chamber with the next batch.

Washers consist of a motor, inside and outside shells, and a casing. The outside shell is stationary and holds the wash water. The inside shell holds the laundry and is perforated to allow water for various cycles to flow in and out.

In the past decade, most washers have been designed so that the perforations are turned away from the articles to be washed. Perforations on older equipment, however, may protrude and cause excess wear and tear on linens. This can reduce the useful life of linens by up to 50%. Even in a medium-sized OPL, older machines can cause enough fabric damage to pay for a new washer in about one year.

The washer's motor rotates either the perforated inner shell (on washwheel washers) or an agitator (agitator washers). The rotating shell or agitator helps the detergent break up soils on fabrics in the wash cycle and remove detergents and other chemicals during the rinse cycles.

Most newer washers have automatic detergent and solution dispensing capabilities. These washers are programmed electronically—either at the factory or in the OPL—to dispense solutions. Other washers require an operator to add detergent and solutions manually. Machines that require manual dispensing usually have fewer **ports** or **hoppers** (openings through which detergents can be poured). Equipment should have at least five ports—two for detergents and one each for bleach, sour, and softener. Chemicals that are simply dumped onto the linens can severely damage them. To ensure proper mixing, many commercial systems are automated.

Whether added automatically or manually, solutions must be added in the right amount and at the right time. As more and more sophisticated chemicals are developed to improve the quality of the wash at a lower cost, measuring solutions and adding them at the right time becomes increasingly important.

While many automated machines are more economical to operate than washers requiring manual dispensing, they can also cause problems if improperly used. Allowing detergent salespeople to tinker with automated machines to "improve" the quality of the wash can, in fact, decrease the quality of the machine's performance. Having the manufacturer's representative check the machine periodically is probably the surest way to get the most out of the machine.

Microprocessors—one of the latest innovations in washers—allow greater control over the washer's functions than more conventional automatic models. For example, water temperature can be regulated more exactly. Microprocessors also allow operators more ease and flexibility in programming combinations of detergents and solutions for specific fabric types and soil levels.

Another new innovation is the reuse washer. This machine can save energy, sewage, water, and chemical costs. A water reuse washer is equipped with insulated storage tanks. Water that can be reused is siphoned into the tanks to maintain the proper temperature and then released into the proper cycle of the next batch of laundry. Control panels allow laundry operators to make adjustments in the water to be reused to account for soil conditions, water hardness, and fabric type. The control also automatically saves reusable water and discharges water that cannot be reused.

Hotels in extremely dry climates are beginning to use OPL waste water to irrigate lawns and gardens. Some treatment is needed to neutralize phosphates and other chemicals before the water can be used on plants. Special care also must be taken to ensure that recycled water is not accidentally used for drinking. Many properties, for example, color used water with a harmless vegetable dye to avoid mix-ups.

Most washers have extraction capabilities. The motor spins the inside shell rapidly to remove most excess water after washing is completed. If the washer cannot remove this water, a separate extractor must be used. Extractors are available in centrifugal, hydraulic, and pressure types.

Many washers offer high-speed extraction capabilities. These machines are often sold and purchased as time savers. In addition, they often offer substantial energy savings because they reduce drying time.

High-speed extraction requires that a machine be able to handle many times its capacity weight during the extraction cycle. For example, machines in hotel OPLs typically handle 70 to 300 pounds of dry laundry. Add to that load the weight of the water and the force of the spin, and the machine may be handling up to half a ton of real weight. Washers with high-speed extractors should be mounted on special soft mount pads and then bolted onto the floor. The soft mount pads act like shock absorbers and ensure that the machine will not loosen from its foundation.

Washers and dryers should also be designed to prevent burns and bruises. Most machines built after 1980 have well-insulated bodies and glass on doors to keep heat from escaping. The insulation also protects employees from burns.

Washing machines do break down; when one is out of operation, the hotel is faced with a series of expensive problems. For one, a broken machine may idle many of the hotel's housekeeping staff—which means wasted labor. Second, a broken machine may hamper the flow of linen and the makeup of guestrooms—which means lost sales. Finally, a breakdown means that the OPL will later be overworked—which translates into overtime expenses.

Three rules help reduce breakdown time. First, buy only strong, industrial equipment from a supplier you trust. Before you make a purchase decision, spend sufficient time reading sales literature to identify the brands and models of machines appropriate for your operation. For example, machine specifications should list the weight of the machines. In general, heavier machines are more durable because an inner frame supports the cylinders and puts less stress on the body of the machine. Second, read, thoroughly understand, and follow the equipment's maintenance requirements. More than 90% of all machine failures can be prevented by following the manufacturer's maintenance recommendations. Finally, consider

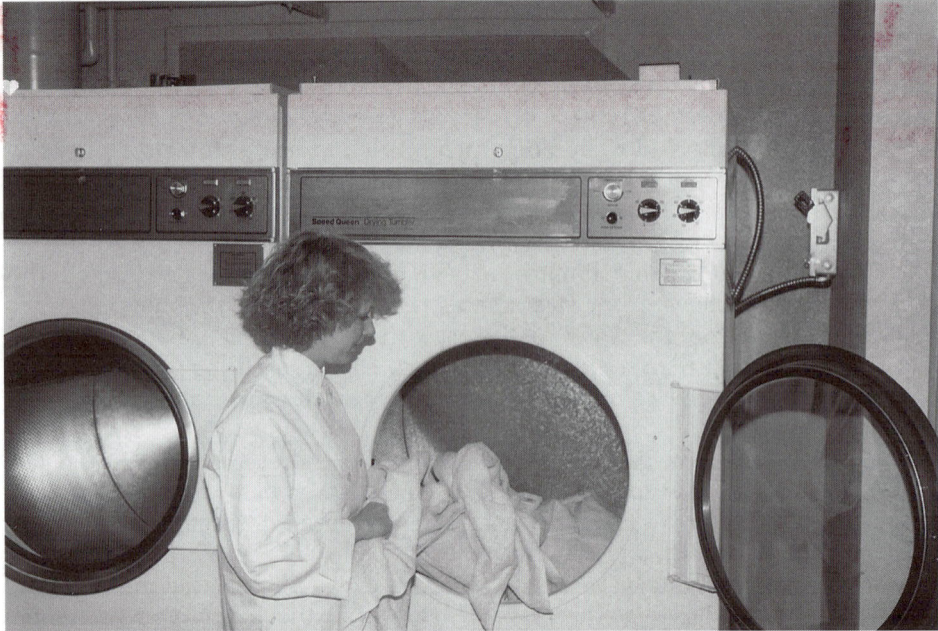

Drying is part of the finishing process. Some linens, like these towels, can be simply dried, folded, and stored. The proper temperature selection is vital to the life of linens. Excessively high temperatures or overdrying, for example, could reduce the absorbency of these towels. (Courtesy of Speed Queen, Ripon, Wisconsin)

buying extended warranties on the equipment. Usually, this cost is less than that incurred by equipment failure. More important, the warranty forces the manufacturer's local representative to maintain a vested interest in keeping equipment operating properly.

Drying Machines

Dryers remove moisture from articles by tumbling them in a rotating cylinder through which heated air passes. Air is heated by gas, electricity, or steam. The air flow must be unrestricted to ensure the dryer's energy efficiency.

Like washers, dryers must be maintained properly. As dryers get older, they frequently receive less maintenance even though they require more. This means they waste more energy. Laundries are usually designed with greater drying capacity than washing capacity because it takes one and a half to two times longer to dry laundry than to wash it. As a result, work can continue relatively smoothly for a short time if a dryer breaks down. However, it is easier to keep dryers properly maintained in the first place than work around broken machinery. Dirt or lint clogging the air supply to the dryer is the most frequent problem. Cleaning air vents twice daily can help eliminate this.

The Occupational Safety and Health Act (OSHA) requires that lint levels in the air be controlled in institutional laundries. Most dryers are equipped with a system of ducts which eject lint into containers and minimize air contamination. Ducts should be checked regularly for leaks, and containers should be emptied regularly.

Steam Cabinets and Tunnels

Steam cabinets or **tunnels** effectively eliminate wrinkles from heavy linens such as blankets, bedspreads, and curtains. A steam cabinet is simply a box in which articles are hung and steamed to remove wrinkles. A steam tunnel actually moves articles on hangers through a tunnel, steaming them and removing the wrinkles as they move through.

Steam cabinets—and tunnels to a lesser extent—can disrupt the flow of laundry through the OPL because they are time-consuming to operate. They also require a worker to load and unload the tunnel or cabinet, which increases the OPL's labor costs. As a result, only very large hotels with valet service or hotels that do frequent loads of curtains, bedspreads, and blankets find steam tunnels cost-effective. Most hotels that use no-iron linens do not require steam cabinets.

Flatwork Ironers and Pressing Machines

Flatwork ironers and pressing machines are similar, except that ironers roll over the material while presses flatten it. Also, items can be fed into ironers but must be placed on the presses manually. Either process is time-consuming and thus used only for items that require ironing. Some ironers also fold the flatwork automatically.

It is important for proper iron operation and maintenance that material arrives in the proper condition from the finishing process. For example, dirt left on linens because of improper rinsing in the wash process can shorten ironer life. Too much sour left in the linens can cause them to roll during ironing, and too much alkali can cause linens to turn brown. Moisture extraction must be controlled, too. Linens should be moist before going into the ironer. Linens that are too dry will cause static electricity to build up on the ironer. On the other hand, linens that are too wet will be difficult to feed into the ironer.

It should be noted that older no-iron linens frequently have to be ironed. In reality, no-iron linens have two distinct lives. Initially, their performance is that of a true, no-iron fabric. However, this condition usually lasts less than half the article's total useful life. Over time, the crispness of no-iron linens is reduced because repeated washings break down the fabric. Because linens are so expensive, many OPLs have discovered that buying an ironer is cheaper than buying new linens.

Folding Machines

The term folding machine is actually a misnomer. It does not actually fold the laundry, but holds one end of the item to be folded so that staff can fold it more easily. The most common folding machine acts as a passive partner, providing the worker with an extra set of "hands" to assist in folding linens.

Folding machines are now available which virtually eliminate tumble drying and hand folding. These space-saving units dry, iron, fold, and often crossfold and stack flatwork. Some have microprocessor controls that determine fold points and trigger other related functions.

Rolling/Holding Equipment

Rolling and holding equipment is used for linen handling. Carts are used in most laundries to move linens and to hold them after they have been sorted for washing, drying, and finishing. Carts must be kept orderly so that staff can move freely through the OPL. They must also be carefully marked so that carts for clean linens are not mixed up with those used for soiled items.

Very large OPLs may have an overhead system of tracks to which laundry bags can be attached to hold linen ready to be sorted, washed, dried, or finished. Overhead systems may be semi- or fully-automated, depending on the size of the OPL.

Automated overhead systems have a number of advantages. First, they allow laundry to move in an orderly manner throughout the OPL. Second, one person can move all the carts simultaneously instead of many people having to move carts to and from washers, dryers, extractors, ironers, and other equipment on the OPL floor. This can represent considerable labor cost savings for large OPLs. Finally, overhead systems also provide extra storage space in case one step in the laundry process gets backed up, preventing a disorderly pile-up of carts on the OPL floor.

Very sophisticated OPLs have automated equipment that will move soiled laundry to the machines. These systems include conveyors, overhead monorails, and pneumatic tubes.

Even though a sophisticated linen handling system may require a large initial cash investment, it could pay for itself very quickly in labor savings. The larger the operation, the more desirable an automated system becomes.

Regardless of the type of system used at your laundry, you should observe a few basic guidelines. First, transport devices should not have sharp corners or other parts that could tear linens. Second, they should be easy to use. They should not require workers to bend excessively or repeatedly to remove items from the bottom of a cart. Carrying should be avoided when items have to be transported. Third, make certain there is adequate headroom and floor space for personnel traffic. Make sure carts can fit comfortably through doorways. Holding space should be situated so that workers do not have to reach high or far back on shelves.

Preventive Maintenance

A detailed and strictly adhered to preventive maintenance program is essential to the efficient operation of an OPL. Lost productivity and expensive repairs easily justify the costs of these programs. The program should include a record of repairs or maintenance procedures and the total cost of each. Manufacturers usually provide literature about their equipment, but they should also offer instructions for making and keeping good maintenance records. This data will identify troublesome units that may have more serious problems. When the total cost of the repairs and maintenance begins to approach the cost of the machine itself, the property should consider replacing it.

Typical examples of daily maintenance procedures include checking safety devices; turning on steam, water, and air valves; checking ironer roll pressure; and cleaning dryer lint screens.

Maintaining water and energy efficiency is an important aspect of preventive maintenance as is reducing repair and downtime costs. Leaking valves, damaged insulation, and constricted gas, air, and water paths can be quite costly. Keep accurate records of utility use to identify such problems. Periodically, check water levels in washers. Too much water results in decreased agitation and poor cleaning. Not enough water causes excessive mechanical action that can damage fabrics.

No matter how good maintenance precautions are, unexpected breakdowns or repair delays can occur. Many properties develop a contingency plan to help cope with unforeseen emergencies. A contingency plan should include an estimate of how long the stock of clean linen will last and at what point an outside laundry will need to be called. Having a number of outside laundry contacts will allow you to have soiled linen cleaned in time to meet the hotel's needs. Some hotels form an emergency laundry network so that they can help each other out in the event of OPL emergencies.

Staff Training

Manufacturers and distributors can often help train employees to use machinery properly. They can also provide safety instructions and updates that can help the executive housekeeper or laundry manager develop good safety procedures. In general, staff should be trained to inspect all equipment daily before start-up and to treat all equipment with care.

Once safety procedures are established, the executive housekeeper or laundry manager must ensure that employees follow them. Such measures as unannounced fire drills, displaying charts outlining safety procedures, quarterly safety meetings, and monthly reviews of and follow-up on all work-related accidents can be good ways to ensure that safety procedures are followed. The executive housekeeper or laundry manager should periodically review procedures with all employees individually. This keeps long-term employees alert, reinforces prior training, and orients new personnel to the proper use of equipment and supplies. Periodic retraining procedures are an important part of any safety program.

More and more workers, especially in service occupations, do not read or speak English. This poses problems for training in general, but is especially critical in the area of safety training. Safety procedures printed in languages besides English should be available if needed. Whenever possible, briefings or safety lectures should be presented by bilingual staff members. Some properties help bridge the communication gap by assigning bilingual "buddies" to workers not proficient in English so that safety procedures can be communicated to those workers.

Valet Service

Valet service means that a hotel will take care of guest laundry needs. Valet service can be handled in two ways. The hotel may contract with an outside laundry or dry cleaning operation to take care of guest needs; or, the hotel may have its own valet service equipment and staff on the premises. Whether contracted or on-premises, valet service can be either same day or overnight. Same day service means that laundry is sent out in the morning and arrives back in the guest's room by evening. Overnight service means that laundry is sent out in the evening and returned by morning.

Contract Valet Service

Hotels that use outside contractors should operate under a formal agreement that specifies exactly what services the outside laundry or dry cleaning operation will provide. Some hotels provide and ask the contractor to use special bags and boxes for laundry stamped with the hotel's name and/or logo. The agreement should also state when laundry will be picked up and returned.

Bell service staff or room attendants may deliver clean laundry to guestrooms. In small properties, the front desk may activate the message light on the guest's telephone. When the guest reports for the message, the laundry is delivered.

On-Premises Valet Service

Hotels that provide on-premises valet service cite four main advantages to their operation. They say that it is often quicker and promotes more goodwill with guests than contracting with an outside operation. Further, the dry cleaning equipment required by a valet service allows the OPL to handle employee uniforms as well as special linen items. Most important, however, is the revenue the valet service generates. An efficient valet service helps defray the overall OPL costs. In fact, the decision to operate on-premises valet service often rests solely on whether it will turn a profit.

Offering valet service requires the housekeeping department virtually to set up its own laundry business. It must:

- Set times for laundry pickup and delivery
- Determine how laundry will be delivered to guestrooms
- Figure bills to be attached to clean laundry (though the hotel's controller usually sets the price rates)
- Determine the final hotel liability policy in accordance with state and local laws
- Handle lost and damaged items
- Field guest comments and complaints

Whether or not the hotel can provide on-premises valet service often depends upon the amount of space in the OPL. Valet staff will need their own work space for sorting, tagging, pre-treating spots, washing, drying, and finishing. Extra space will be needed for equipment. Valet staff also need to be specially trained. Often a valet supervisor is responsible for training and overseeing valet attendants.

Some hotels have a valet extension number that guests can call to get laundry pickup. Soiled laundry is collected by a valet staff runner, who returns it to the valet service area of the laundry room. There, valet staff tag, sort, and pre-treat (if necessary) each item. Some valet services include minor mending jobs such as sewing on a button. When laundry is finished, properly packaged, and ticketed with a bill, the runner returns items to guestrooms.

Staffing Considerations

Proper staffing is critical to the efficiency of the OPL. Labor costs—the number one expense in any hotel operation today—must be carefully controlled in order for the OPL to remain cost-effective. Scheduling too many employees can severely cut into the hotel's profits. Having too few employees can also eat up profits in overtime pay and inefficiency that could ultimately affect guests and result in lost business.

Staff Scheduling

To efficiently schedule the laundry staff, executive housekeepers or laundry managers must be able to forecast the hotel's daily linen needs for three or four weeks in advance.

Forecasting Linen Needs. The first step in forecasting the hotel's daily linen needs is to review past records and determine the average number of pounds of linen used per occupied room and per dining room cover. The second step is to obtain occupancy forecasts from the rooms division and cover forecasts from the food and beverage division. These forecasts should include special events that will affect the hotel's linen needs. These events might be an unusual number of banquets and parties, economic circumstances that keep occupancy high or low, construction projects around the hotel, conventions, and so on.

Multiplying the number of expected occupants (or covers) by the average number of pounds of linen used per occupied room (or cover) yields the total number of pounds of linen that the laundry will have to process the next day.

Scheduling Staff. Once you know the daily needs for the schedule period, you need to know one more thing: how many workers it will take to handle the load. By keeping productivity records over a period of time, you should be able to develop some ratios that will help you determine the number of staff needed to process various amounts of linen.

Along with these ratios, you will need to set minimum and maximum staff levels for the OPL. For example, two people in the OPL may not be able to keep the laundry moving through the sorting, washing, drying, finishing, and folding stages smoothly—no matter how light the linen demand. By the same token, the OPL may be too small to allow more than a certain number of workers to work efficiently together at the same time.

When you need more people than your OPL can handle comfortably in one shift, you can schedule two or three equally staffed shifts—or you can schedule one or two shifts with the maximum number of workers and another shift with just enough workers to meet the remainder of the linen demand. If you find yourself unable to meet the hotel's demands with three fully staffed shifts, you need a bigger OPL.

Exhibit 8   Sample Job List: Washer

Reports to: Head washer
Tasks: Employee must be able to:
1. Inform the head washer about all matters pertaining to the wash room.
2. Prepare soiled linen for wash.
3. Load and set cycles on washers.
4. Check wash cycles regularly for proper performance.
5. Unload washers.
6. Keep the wash room neat and clean.
7. Check and stock supplies.

Many managers prefer to schedule two or three equally staffed shifts instead of one or two full shifts and another partially staffed shift. They say that the full shifts tend to overload machines and partial shifts underload them, causing unnecessary wear and tear on machines and inefficient energy use. Some managers reason, too, that it is more efficient to run a full staff some days and close down on others—rather than to operate with a partial staff. Union rules may, however, dictate the way the OPL is staffed.

Other Staffing Considerations. Besides determining laundry needs and the number of workers needed to fulfill those needs, there are other staffing considerations.

Many properties cross-train laundry personnel so that each worker can do every job in the OPL. Cross-training allows for some job variation and lets workers cover for one another during vacations, illnesses, or other leaves. While the laundry manager or executive housekeeper is in charge of training employees, each area of the OPL—sorting, washing, finishing, etc.—may have a team leader who supervises the workers in that area. Workers are often rotated to different areas on a regular basis and work with different staff members as well.

Another important item to consider is when to schedule shifts. If the laundry is not located in the basement of the hotel or in a separate building, for example, it probably should not operate at night when guests could be disturbed.

Whether to stagger schedules is another consideration. There are some advantages to having one or two workers begin their shift early and then bringing in other workers at intervals of two or three hours. Shift staggering can provide full staffing in the middle of the day when the laundry load is heaviest.

Job Lists and Performance Standards

Besides determining the proper number of workers for the laundry, the executive housekeeper or laundry manager develops the job lists and performance standards for various positions in the OPL. Exhibits 8 through 12 present sample job lists for various positions in the OPL.

As soon as the equipment is installed in any OPL, performance standards should be developed for all activities, and employees should be thoroughly trained. The equipment supplier can often provide information that can be used to develop performance standards and help train employees. Proper training and

Exhibit 9   Sample Job List: Laundry Attendant

Reports to: Laundry manager
Tasks: Employee must be able to:
1. Inform the laundry manager about any malfunctions on finishing equipment or problems with safety mechanisms.
2. Check equipment for proper functioning.
3. Finish linens according to property quality standards.
4. Keep work areas clean and neat.
5. Keep maintenance records on machinery and turn these over to the laundry manager.

Exhibit 10   Sample Job List: Linen Distribution Attendant

Reports to: Laundry manager
Tasks: Employee must be able to:
1. Fold and stack linens on shelves.
2. Maintain pars on shelves as needed.
3. Inspect the quality of the linens on shelves.
4. Make reports on linen quality to the laundry manager.
5. Report unusual linen shortages to the laundry manager.
6. Keep the work area neat and clean.
7. Return laundered uniforms to the uniform distribution area.
8. Fill requests from food and beverage outlets as needed.
9. Keep records of food and beverage requests and turn over to laundry manager.

Exhibit 11   Sample Job List: Head Washer

Reports to: Laundry manager
Tasks: Employee must be able to:
1. Supervise all personnel in the washing and sorting areas of the OPL.
2. Make reports about all washing and sorting activities to the laundry manager.
3. Oversee:
   - Sorting and washing procedures
   - Filling linen needs for guestrooms and food and beverage outlets.
   - Maintaining adequate supplies of clean uniforms
   - Setting formulas and cycles for types of linen and types of soils
4. Make sure all workers assigned to a shift are present.
5. Make sure staff keep areas and equipment clean and neat.
6. Maintain employee performance and machine performance records.
7. Check supply levels.

Exhibit 12   Sample Job List: Laundry Manager

Reports to: Executive Housekeeper
Tasks: Employee must be able to:
1. Record laundry costs.
2. Make reports and recommendations when requested.
3. Oversee the preventive maintenance program.
4. Approve distribution of linens to guestrooms and food and beverage areas.
5. Direct all OPL staff.
6. Prepare the OPL budget.
7. Hire and train new OPL employees.
8. Develop methods for increasing OPL efficiency.
9. Coordinate all maintenance and repairs of machinery.
10. Supervise the OPL safety program.
11. Evaluate OPL staff performance.

performance standards become more vital to efficient OPL operation as machinery becomes more complicated.

Typical laundry room performance standards might cover, for example, the steps an employee should take in loading a particular machine. This might include a chart showing the number of sheets, pillowcases, towels, or other items that constitute a load; an explanation of the use of a linen scale for weighing loads; the steps for checking the safety of a load before washing; and the proper way to close and secure the washer door.

Other performance standards particularly suited to laundry operations are preventive maintenance procedures (see the section on preventive maintenance in this chapter), linen handling procedures, inventory control procedures, time card control, chemical handling procedures, and linen sorting procedures.

Key Terms

alkalies
antichlors
bleach
break
builders
fabric (or optical) brighteners
flatwork ironer
flushes
hoppers
mildewcides

polycotton
ports
soaps
sizing
sours
steam cabinet
steam tunnel
surfactants
synthetic detergents
tunnel washer

Discussion Questions

1. What factors must be taken into account when laundering different types of linen fabrics?

2. What are the steps in the flow of linens through an OPL?

3. What are two ways in which soiled linens can be sorted before washing? Why are both sorting procedures important?

4. What are the nine steps of the typical wash cycle?

5. What is the function of the various chemicals used in the wash process?

6. What are the basic types of laundry equipment used in an OPL?

7. Why is a preventive maintenance program important to the operation of an OPL?

8. What kinds of equipment and personnel are necessary for an OPL to provide on-premises valet service?

9. How are linen needs forecasted?

10. What factors must be considered when scheduling OPL staff?

REVIEW QUIZ

When you feel you have covered all of the material in this chapter, answer these questions. Choose the *best* answer.

True (T) or False (F)

T   F   1. No-iron linen blends are more durable than all-cotton linens.

T   F   2. All-cotton linens pick up and retain stains more easily than polycotton linens.

T   F   3. Polyester fabrics do not retain color as well as all-cotton fabrics.

T   F   4. All-cotton linens can be washed and ironed at high temperatures.

T   F   5. Many synthetic fabrics are more durable than wool.

T   F   6. An effective laundry practice is to sort and wash cleaning rags with guestroom linens.

T   F   7. In the laundry wash cycle, pure soaps are used for lightly soiled items; built soaps are used for heavily soiled items.

T   F   8. Alkalies are sometimes added to the laundry wash cycle to help detergents lather better.

T   F   9. The speed by which linens can be folded determines the pace by which linens flow through the laundry process.

T   F   10. Extractors are used in laundry operations to remove wrinkles from heavy linen items, such as blankets.

Alternate Choice

11.   Linens should be sorted by degree of soiling _____ by type of fabric.

     a. and
     b. or

12.   The bleach step in the wash cycle comes:

     a. before the suds step.
     b. after the suds step.

13.   A mild bleach that is safe to use with most washable fabrics is:

     a. chlorine bleach.
     b. oxygen bleach.

14.   Laundry operations use sours to neutralize:

     a. acids.
     b. alkalies.

15. Loss of absorbency is a common laundry problem with some linen fabrics that may be caused by:

   a. washing and drying temperatures that are too low.
   b. washing and drying temperatures that are too high.

Part IV

Technical/Reference Guide for Executive Housekeepers

Chapter Outline

Common Housekeeping Chemicals
  Water
  Bathroom Cleaners
  All-Purpose Cleaners
Safety Equipment
OSHA's Hazard Communication Standard
  Listing Hazardous Chemicals
  Obtaining MSDSs From Chemical
    Suppliers
  Labeling All Chemical Containers
  Developing a Written Hazard
    Communication Program

Learning Objectives

1. Describe three categories of common housekeeping chemicals.

2. Identify the common additives to all-purpose cleaners and their applications.

3. Recognize when personal protective gear may be required and what type is appropriate to wear.

4. Describe the purpose of the OSHA Hazard Communication Standard.

5. List and briefly describe the five steps which hotels must take to comply with the Hazard Communication Standard.

6. Identify typical information that must be included on a manufacturer's material safety data sheet (MSDS).

7. Explain required labeling practices for chemical containers.

8. Recognize what may be involved in developing a hazard communication program for a hospitality operation.

Housekeeping Chemicals and Hazard Communication Responsibilities

Joe is a new room attendant at a high-priced hotel with luxurious furnishings in its guestrooms—carved moldings, thick wool carpets, and marble vanity tops in the bathrooms. While Joe is cleaning his rooms one day, he runs out of cleaner for the marble vanity. He has some powder cleanser on his cart that he uses for scrubbing the toilet bowl, so instead of going back to the housekeeping supply area for more marble cleaner, he sprinkles some of the powder on the vanity and scrubs it with a clean rag.

Rita is cleaning a badly stained toilet bowl with an ammonia-based cleaner. She scrubs vigorously at the stains, but they will not come out. She pours more of the ammonia cleaner into the bowl, but the stains are still there. Finally, she reaches for a container of chlorine bleach, dumps some into the bowl, and leans over to scrub again.

Both these scenarios underscore a vital element of chemical use in the housekeeping department: proper training in the safe and effective use of chemicals. Joe obviously does not know that using the abrasive cleanser will mean the expensive marble vanity top will have to be replaced—and that guestroom revenue may be lost while repairs are being made. Rita's case is far more serious. She does not know that mixing ammonia and bleach will produce a deadly gas that will very probably result in her death.

The federal government's Occupational Health and Safety Act (OSHA) and various state laws require hotels to inform and train employees about the chemicals they use and how to use them properly. This chapter presents a brief description of chemicals commonly used in housekeeping departments and focuses on what properties must do to comply with the federal government's OSHA Hazard Communication Standard. Since some states have laws that are more stringent than the federal OSHA regulations, managers should check with the proper state officials to determine the specific requirements that apply to their properties.

Common Housekeeping Chemicals

Water

Water is an essential chemical in most cleaning solutions. Ironically, it can also be the trickiest to use because water is not the same from town to town, from state to

Housekeeping Success Tip

Bill Foster
Director of Security
Sheraton Scottsdale Resort
Scottsdale, Arizona

❝When the Safety Committee of the Sheraton Scottsdale Resort was first assigned to bring our resort into compliance with the OSHA Hazard Communication (HazComm) Standard, Title 29, Code of Federal Regulations, it seemed as prohibitive as the title itself.

Reading the standard further confused the committee. Like most government regulations, it is very wordy and uses terminology that would only be familiar to a chemist.

We found commercial companies willing to help businesses establish a HazComm program, but we felt their fees were excessive. At last we discovered that the U.S. Government Printing Office in Washington, D.C., had a Hazard Communication Standard Compliance Kit at a very reasonable price. The kit, which comes unbound on three-hole punched paper with divider tabs, is a step-by-step guide to setting up a HazComm program.

The first task was to identify all the hazardous chemicals in the hotel. This was very time-consuming because the total numbers were staggering. However, a side benefit of the inventory was that we discovered a number of unnecessary chemicals that we could eliminate from our supplies.

When all chemicals were finally listed by name, location, use, and manufacturer, we requested a material safety data sheet for each chemical from the manufacturer.

When we received the MSDSs, we realized that to be effective, we would have to simplify the form so the average employee would understand the information. Some of the technical information—specific gravity and vapor density, for example—would really confuse employees more than help them. So I developed a simplified form which listed the hazards, first aid measures, and protective measures. When we add new chemicals to our inventory, our safety committee, which meets once a week, fills out the simplified forms for the new chemicals. In order to maintain compliance, we keep both the original MSDSs and the simplified forms in a central library.

Upon completion of the central library, we created departmental libraries. To do this, we duplicated everything in the central library and then sorted both the manufacturer's MSDSs and the simplified versions by the various departments that used each chemical. We felt that having a departmental library would satisfy the obligation to offer ready access to the program and the MSDSs to employees.

Once all the information was complete and containers had been labeled, we developed training plans to disseminate the information to our employees.* As the actual training progressed, it became evident that most employees did not understand the hazards of using various chemicals. We decided to slow the training process and go into more depth than originally planned.

Housekeeping Success Tip (continued)

One of our biggest problems was the housekeeping department, where the sheer number of personnel and a great number of non-English speakers made training difficult and time-consuming.

The kit we purchased gave us an organized approach to creating our own HazComm program and helped us complete our assignment systematically. Without the kit, we would not be near completion. I'm sure it would help others as much as it helped us.**"**

---

*Editor's note: A complete outline of the training plan used by the Sheraton Scottsdale Resort in the housekeeping department is offered in the appendix to this chapter.

---

state, or from country to country. Water absorbs minerals that affect its ability to clean or to react with cleaning agents.

Some common water minerals include calcium, iron, sulfur, and phosphates. Calcium can inhibit a detergent's cleaning ability, requiring more detergents to be added for cleaning jobs. Iron and sulfur can cause discoloration; sulfur also causes a "rotten egg" odor. Phosphates can actually enhance the cleaning power of some detergents so that less detergent is necessary.

Bathroom Cleaners

Room attendants sometimes use ammonia-based or chlorine-based cleaning compounds to clean guestroom bathrooms. It is important to know that these common cleaning chemicals should not be used together. *Ammonia should never be mixed with chlorine-, fluoride-, or bromine-based chemical cleaners.* When these chemicals are combined, they form a highly toxic gas. For example, when ammonia and chlorine are mixed in water, deadly phosgene gas is formed.

If possible, the housekeeping departments should purchase and use *either* ammonia-based cleaners *or* chlorine-, fluoride-, or bromine-based cleaners. By supplying room attendants with only one or the other, the risk of mixing the chemicals is almost eliminated. However, sometimes this is impossible due to the variety of surfaces that room attendants need to clean. Therefore, training and employee awareness is the best defense against potential chemical hazards.

All-Purpose Cleaners

There are a number of all-purpose cleaners on the market today. As their name suggests, they can be used to wash walls, scrub floors, clean tubs and showers, and even wash windows and mirrors. All-purpose cleaners are generally concentrated and can be diluted with water to adapt to different cleaning needs. Some all-purpose cleaners contain additives which limit their uses. A few of the more common additives are discussed in the following sections.

Abrasives. Abrasives are gritty substances used to remove heavy soils and polishes. Abrasives can be used safely on stainless steel, ceramic tile, and on some china. However, it can damage softer surfaces, such as marble or fiberglass. Most experts warn properties to avoid abrasives that could mar porcelain and synthetic surfaces.

Acids. Weak citric acids and vinegar can be used to clean glass, bronze, and stainless steel.

Alkalies. Like those used in the laundry, alkalies in cleaning agents boost the cleaning ability of detergents. They also have disinfecting powers. Alkalies in all-purpose cleaners typically have a pH between 8 and 9.5. The **pH scale** measures the acidity or alkalinity of a substance compared to water. A pH of 7 is neutral. Acids have pHs of less than 7 to 0, which is the most acidic. Alkalies have values of more than 7 to 14, which is the most alkaline.

Degreasers. Degreasers (also called emulsifiers or stabilizers) refer to a number of different products that act on a variety of greases and soils. Solvents have many degreasing capabilities.

Delimers. Delimers remove mineral deposits that can dull, scale, and/or discolor surfaces.

Deodorizers. Deodorizers or room fresheners are designed to conceal the smell of cleaners in the room. Some fresheners may leave a film on surfaces in the guestroom, and should be avoided. Fresheners can become overpowering if used every day. Also, many powder fresheners contain cornstarch which attracts insects.

Disinfectants. Disinfectants kill bacteria, molds, and mildew. Cleaners with disinfectants are usually expensive. The added expense is often unnecessary, since not every surface needs to be disinfected. A good all-purpose cleaner used with proper cleaning, rinsing, and drying procedures is usually sufficient for lodging operations.

Fiberglass Cleaners. Many newer tub/shower units are made of fiberglass. Special cleaners are available to clean fiberglass without scratching the surface area.

Metal Cleaners. Some oil-based metal cleaners remove soils but leave a thin, protective coating on the surface of the metal. This protective coating often picks up fingerprints; when transferred to clothing, this coating can damage many fabrics. Water-based metal cleaners avoid these problems and do a good job.

Wetting Agents. Wetting agents break down the surface tension of the water and allow water to get behind the dirt to lift it off the surface.

Safety Equipment

Housekeeping employees may use chemicals that require wearing protective gear. Personal protective gear may be used for covering eyes, face, head, hands, and, in some cases, the entire body. Protective gear should be worn when using hazardous or toxic chemicals. It should also be worn when performing duties and using equipment that may result in injury from flying objects.

Gloves, goggles, or face shields may be required when diluting chemicals for cleaning purposes or when mixing chemicals for treating swimming pools. The chemical manufacturer must specify what type of equipment is needed when using the product.

When cleaning overhead areas such as ceiling vents, goggles and dust/mist respirators may be needed. Dust/mist respirators may also be used when cleaning very dusty areas. These respirators fit over the employee's mouth and nose and prevent dust and other small airborne particles from being inhaled.

OSHA's Hazard Communication Standard

The U.S. **Occupational Safety and Health Act (OSHA)** is a broad set of rules that protects workers in all trades and professions from a variety of unsafe working conditions. The federal government often revises or expands these regulations and notifies employers when they must comply with new rules.

OSHA regulations require hotel employers to inform workers about the possible risks posed by chemicals they may use to do their jobs. The rule also requires that employers provide training in the safe use of these chemicals. This regulation is called the **Hazard Communication** (often shortened to **HazComm**) **Standard**. Compliance with the regulations offers some real benefits for the property. One survey revealed that 9.7% of all employees in the lodging industry sustain some on-the-job injury. It is estimated that if the HazComm program prevents just one of these accidents, the cost of the program is recovered.[1]

Moreover, failure to comply with OSHA's Hazard Communication Standard can be expensive. Fines may be $1,000 for each violation, and every uninformed employee is a potential violation. OSHA now conducts a review of a property's HazComm compliance with every inspection. In addition, states may have laws that are as stringent as or more stringent than OSHA's.

OSHA requires that hotels take five steps to comply with the HazComm Standard. Hotels must:

1. Read the standard.

2. List the hazardous chemicals at the property. This includes doing a physical inventory of chemicals used, checking with the purchasing department to make sure the list is complete, setting up a file on hazardous chemicals, and developing procedures for keeping the file current.

3. Obtain **material safety data sheets (MSDSs)** from the suppliers of chemicals used at the property and make them available to employees.

4. Make sure all chemical containers are properly labeled.

5. Develop and implement a Hazard Communication Program which explains MSDS information and labeling procedures to employees and informs them about hazards and protective measures.

Exhibit 1   Form for Listing Hazardous Chemicals and Indexing MSDSs

List of Hazardous Chemicals
and
Index of MSDSs

Hazardous Chemicals	Operation/Area Used (Optional)	MSDSs on File

Listing Hazardous Chemicals

Properties may assign one person to inventory all the chemicals used by the property, or department heads may be responsible for inventorying chemicals used in their specific areas. Exhibit 1 shows the inventory sheet provided by OSHA which can be used to compile a list of hazardous chemicals.

The name of each chemical that appears on the MSDS and on the label should be included on the inventory form. The common or trade name for the chemical can also be included. Some substances on a list of hazardous chemicals used in the housekeeping department might include:

- Substances in aerosol containers

- Caustics such as laundry alkali

- Cleansers and polishes

- Degreasing agents (emulsifiers) used in the laundry
- Detergents
- Flammable materials such as cleaners and polishes
- Fungicides and mildewcides used on carpets or in the laundry
- Floor sealers, strippers, and polishes
- Pesticides

Solutions and chemicals that come from manufacturers are not the only substances that should be listed. Any hazardous substance produced by certain tasks should also be included.

The MSDS will indicate whether a chemical is hazardous. To determine whether a chemical is hazardous, look for words such as "caution," "warning," "danger," "combustible," flammable," or "corrosive" on the label.

If an MSDS for a particular chemical is not available, or if the MSDS does not indicate whether the chemical belongs on the list, OSHA recommends treating the substance as if it were a hazardous chemical.

Some chemicals are exempt from the HazComm Standard. For example, rubbing alcohol in a first-aid station would be exempt from the standard because it is intended for personal use by employees. Also, common consumer products, such as general lubricating oil, would be exempt from the HazComm Standard if intended for personal and infrequent use.

The list of hazardous chemicals will be part of the property's Hazard Communication Program. The list must be available to employees whenever they wish to see it, and a procedure for keeping the list up-to-date must be in place.

Obtaining MSDSs from Chemical Suppliers

Once the inventory of hazardous materials is completed, the person in charge of the inventory should check to see if MSDSs are on file for each of the materials listed. If the property has an MSDS for a chemical, the person in charge of the inventory can put a check in the last column of the inventory sheet presented in Exhibit 1. If there is no MSDS for the chemical, the property must contact the chemical manufacturer to obtain one. A sample request letter is presented in Exhibit 2.

All MSDSs should be examined carefully to make sure they are complete and clearly written. If data on MSDSs provide an insufficient basis for training employees on the chemical's hazards, the person in charge of the inventory should contact the manufacturer for additional information or clarification.

MSDSs can be filed by hazard, ingredients, or work areas—whatever way will make it easy for employees to find MSDSs in case of emergencies. MSDSs must be available during *all* workshifts. They should not be stored in areas that will be locked during the afternoon/evening shifts.

Many chemical manufacturers use the MSDS form developed by OSHA. A copy of the form is presented in Exhibit 3. Some manufacturers develop their own forms for MSDSs. A sample manufacturer's MSDS is shown in Exhibit 4. When appropriate, manufacturer's MSDSs must include the following information:

Exhibit 2   Sample Letter Requesting an MSDS

Orange Grove Motel
1234 Leisure Avenue
Vacationland, FL 12345

Acme Hotel Products
5678 Industrial Park Way
Chemical City, NJ 54321

Dear Sir/Madam:

As you are aware, OSHA requires employers to provide training to their employees concerning the hazards of chemicals or other hazardous materials.

To properly train our employees, we need a material safety data sheet (MSDS) for one of your products, _____.

Your prompt attention is necessary to maintain a proper level of safety for our employees. Please send the MSDS for _____ no later than _____.

Sincerely,

Catherine Smith
Executive Housekeeper

- Chemical identity
- Hazardous ingredients
- Physical and chemical characteristics
- Fire and explosion hazard data
- Reactivity data
- Health hazards
- Precautions for safe handling and use
- Control measures

Chemical Identity. OSHA requires that manufacturers list the chemical and common name(s) for the substance. The form must also include the manufacturer's name and address and a telephone number so information can be obtained in case of an emergency.

Exhibit 3   Sample Material Safety Data Sheet—OSHA

Material Safety Data Sheet	U.S. Department of Labor
May be used to comply with OSHA's Hazard Communication Standard, 29 CFR 1910.1200. Standard must be consulted for specific requirements.	Occupational Safety and Health Administration (Non-Mandatory Form) Form Approved OMB No. 1218-0072

IDENTITY *(As Used on Label and List)*	Note: *Blank spaces are not permitted. If any item is not applicable, or no information is available, the space must be marked to indicate that.*

Section I

Manufacturer's Name	Emergency Telephone Number
Address *(Number, Street, City, State, and ZIP Code)*	Telephone Number for Information
	Date Prepared
	Signature of Preparer *(optional)*

Section II — Hazardous Ingredients/Identity Information

Hazardous Components (Specific Chemical Identity; Common Name(s))	OSHA PEL	ACGIH TLV	Other Limits Recommended	% *(optional)*

Section III — Physical/Chemical Characteristics

Boiling Point		Specific Gravity (H$_2$O = 1)	
Vapor Pressure (mm Hg.)		Melting Point	
Vapor Density (AIR = 1)		Evaporation Rate (Butyl Acetate = 1)	
Solubility in Water			
Appearance and Odor			

Section IV — Fire and Explosion Hazard Data

Flash Point (Method Used)	Flammable Limits	LEL	UEL
Extinguishing Media			
Special Fire Fighting Procedures			
Unusual Fire and Explosion Hazards			

(Reproduce locally)	OSHA 174, Sept. 1985

(continued)

Exhibit 3 *(continued)*

Section V — Reactivity Data

Stability	Unstable		Conditions to Avoid
	Stable		

Incompatibility (*Materials to Avoid*)

Hazardous Decomposition or Byproducts

Hazardous Polymerization	May Occur		Conditions to Avoid
	Will Not Occur		

Section VI — Health Hazard Data

Route(s) of Entry:	Inhalation?	Skin?	Ingestion?

Health Hazards (*Acute and Chronic*)

Carcinogenicity:	NTP?	IARC Monographs?	OSHA Regulated?

Signs and Symptoms of Exposure

Medical Conditions
Generally Aggravated by Exposure

Emergency and First Aid Procedures

Section VII — Precautions for Safe Handling and Use

Steps to Be Taken in Case Material Is Released or Spilled

Waste Disposal Method

Precautions to Be Taken in Handling and Storing

Other Precautions

Section VIII — Control Measures

Respiratory Protection (*Specify Type*)

Ventilation	Local Exhaust		Special	
	Mechanical (*General*)		Other	

Protective Gloves	Eye Protection

Other Protective Clothing or Equipment

Work/Hygienic Practices

Exhibit 4   Sample Manufacturer's Material Safety Data Sheet

BUTCHERS®

Material Safety Data Sheet

00611

Manufacturer's Name	:	THE BUTCHER COMPANY	**HMIS Rating**
Address	:	120 Bartlett Street	Health 1
		Marlborough, MA 01752-3013	Flammability 0
			Reactivity 0
Telephone Number	:	(508) 481-5700	

Person Responsible
for Preparation   : Bonita C. Patterson

Date Prepared   : December 1, 1988

IDENTITY

Common Name HOT SPRINGS Cleaner

Proper Shipping Name; Hazard Class; Hazard ID No.: NA

INGREDIENT INFORMATION

Principal Hazardous Component(s)	CAS No.	%	Threshold Limit
Tetrasodium ethylenediaminetetraacetic acid	64-02-8	1-3	NE
Sodium metasilicate	6834-92-0	1-3	NE

SARA Title III Section 313 and 40 CFR Part 372 Notification:
Ingredients in this product are not currently subject to notification

PHYSICAL & CHEMICAL CHARACTERISTICS

Boiling Point   212°F   **Specific Gravity** 1.05   **Vapor Pressure** ND

Percent Volatile by Volume 92   **Vapor Density**   ND   **Evaporation Rate** ND

Solubility in Water   Complete

Appearance and Odor   Green fluorescent liquid; herbal fragrance

Flash Point   >200°F   (T.C.C.)   **Extinguisher Media** NA

Special Fire Fighting Procedures   This product is not flammable.

Unusual Fire and Explosion Hazards   None known to The Butcher Company.

ND=Not Determined, NE=Not Established, NA=Not Applicable

(continued)

Exhibit 4 *(continued)*

00611
HOT SPRINGS Cleaner   Page 2

REACTIVITY DATA

Stability Stable **Conditions to Avoid** None known to The Butcher Company

Incompatibility None known to The Butcher Company

Hazardous Decomposition Products Normal products of combustion.

Hazardous Polymerization Will Not Occur
    Conditions to Avoid None known to The Butcher Company

HAZARD DATA

Signs and Symptoms of Exposure
Direct contact of product with eyes may cause irritation. Prolonged or repeated
contact of product with skin may cause irritation. Preexisting skin disorders
may be aggravated.

Chemicals Listed as Carcinogens or Potential Carcinogens   None

Emergency and First Aid Procedures

1. **Inhalation** Remove to fresh air.
2. **Eyes** Flush with water for at least 15 minutes. Call a physician.
3. **Skin** Flush with water. Call a physician if irritation develops.
4. **Ingestion** Drink large quantities of water. Do not induce vomiting. Call a
    physician.

SPECIAL PROTECTION INFORMATION

Respiratory Protection None required if good ventilation is maintained.
Protective Gloves If prolonged or repeated contact is possible, wear rubber or other
impervious gloves.
Eye Protection Where eye contact may occur, wear chemical splash goggles.

SPECIAL PRECAUTIONS AND SPILL/LEAK PROCEDURES

Handling and Storage Use good personal hygiene practice. Wash contaminated clothing
and equipment before reuse.

Release or Spill Before attempting clean-up, refer to Hazard Data above. Use mop or
wet vacuum to collect material for proper disposal. Rinse area with water.

Waste Disposal Dispose of this material in accordance with federal, state and local
regulations.

Courtesy of the Butcher Company, Marlborough, Massachusetts

Hazardous Ingredients. The manufacturer must list certain hazardous substances that are found in the chemical.

Physical and Chemical Characteristics. The physical and chemical properties of the chemical listed in this section can help workers identify the chemical by sight or smell. This increases workers' understanding of the chemical's behavior and alerts them to take necessary precautions.

Fire and Explosion Hazard Data. Knowing if and under what circumstances the chemical could catch fire and/or explode is extremely important when training employees to handle chemical emergencies. OSHA requires manufacturers to recommend fire fighting procedures and the substance (extinguishing media) to be used to extinguish the fire. The manufacturer must also note any unusual fire or explosion hazards.

Reactivity Data. The chemical manufacturer must provide information about the chemical's stability. A stable chemical is one that will not burn, vaporize, explode, or react in some other way under normal conditions. The MSDS should describe under what circumstances the chemical could become unstable. Temperature extremes or vibrations, for example, could cause some chemicals to ignite.

Health Hazards. How the chemical could enter the body (routes of entry) and its **acute** and **chronic hazards** must be listed on the MSDS. Acute hazards are those which could affect the user immediately; chronic hazards are those that could affect the user over repeated, long-term use of the chemical. In addition, the manufacturer must list whether the chemical is a carcinogen according to the National Toxicity Program (NTP), the International Agency for Research on Cancer (IARC), or OSHA; whether it will aggravate any other medical conditions; and what emergency and first aid procedures should be taken if dangerous exposure to the chemical occurs.

Precautions for Safe Handling and Use. Chemical manufacturers must provide advice on handling, storing, and disposing of chemicals. Information on the proper handling of accidental spills must also be included.

Control Measures. How to use the chemical safely must be outlined. Manufacturers may recommend that users wear protective gloves, goggles, clothing, or use protective equipment. A section on good work/hygiene practices must also be included.

Labeling All Chemical Containers

Chemical manufacturers, in addition to providing MSDSs, must provide proper labeling for chemicals. OSHA requires employers to check these labels for completeness and accuracy. The label must contain the name of the chemical, hazard warnings, and the manufacturer's name and address. If no label is provided, the employer must prepare a label from the MSDS or ask the manufacturer to provide one.

The point of the labeling requirement in the HazComm regulations is to provide an "early warning system" for users of the chemical. Labels must note the chemical's physical and health hazards *during normal use* in order to comply with OSHA regulations. (It is nearly impossible to list all the hazards that could occur during accidents or improper use.) For example, the label may not simply state, "avoid inhalation." It must explain what effects inhaling the chemical could have. However, if the label is too detailed, employees may not see important cautions.

To make the label quick to read and easy to comprehend, some properties use a labeling system with color, letter, and number codes. For example, a red label might indicate that the chemical poses a physical hazard; the letter F might indicate that the chemical is flammable; and the number four (on a scale of one to four) might indicate that the hazard is relatively severe. Employees must be trained to read and understand whatever labeling system is used at the property.

The OSHA labeling provision also requires employers to ensure that employees properly label containers into which they pour chemicals. In some states, the portable containers into which chemicals are poured for immediate use *by the person pouring them* do not need to be labeled. Other states require labeling of all portable containers. Labeling requirements should be ascertained on a state-by-state basis.

Developing a Written Hazard Communication Program

Only employees who must use hazardous chemicals in the course of their jobs must be trained to use them properly. Front desk agents, bellpersons, accounting clerks, and others who do not routinely use chemicals every day do not have to be trained.

Lectures, films, and videotapes can all be part of employee training. However, training must provide time for questions and discussion and help employees apply what they learn. For a small fee, the federal government will provide any property with a Hazard Communication Standard Compliance Kit. This kit contains an extensive glossary of terms frequently found on MSDSs. The kit's reference number is 929-022-00000-9, and the publication can be obtained by writing to:

> Superintendent of Documents
> Government Printing Office
> Washington, D.C. 20402-9325
> Telephone (202) 783-3238

There are many successful Hazard Communication Programs at hotels around the country. The appendix at the end of this chapter examines what one property—the Sheraton Scottsdale Resort in Scottsdale, Arizona—did to develop its HazComm program.

Endnotes ───────────────────────────────

1. "Is Your Hotel a Hazard?" *Lodging Hospitality*, October 1988, p. 228.

Key Terms ───────────────────────────────

acute hazard	MSDS
chronic hazard	OSHA
HazComm Standard	pH scale

Discussion Questions

1. What are some common chemicals used by housekeeping staff? What functions do each of these chemicals perform?

2. Why should ammonia never be mixed with chlorine-, fluoride-, or bromine-based chemical cleaners?

3. What types of safety equipment are available for employees to use when handling chemicals?

4. What is the function of the U.S. Occupational Safety and Health Act (OSHA)?

5. How does OSHA's HazComm Standard benefit employers and employees alike?

6. What are the five steps required to comply with OSHA's HazComm Standard?

7. Where do material safety data sheets (MSDSs) come from and what kinds of information do they contain?

8. How is an acute hazard different from a chronic hazard?

9. What are OSHA's requirements for chemical container labeling? Why is it important to label chemical containers?

10. What steps would a property have to take to develop an effective HazComm Training Program?

Appendix

The Sheraton Scottsdale Resort's HazComm Program

T HE SHERATON SCOTTSDALE RESORT has established a company policy that will en-sure our compliance with OSHA's Hazard Communication Standard, Title 29, Code of Federal Regulations 1910.1200. The purpose of this program is to meet the company's goal of providing all employees with a safe and healthy work environ-ment. This goal will be accomplished by:

- Establishing a hazardous substance library

- Training employees to understand MSDSs

- Training employees to understand and read labeling information

- Providing a comprehensive hazardous substance training program

- Listing resort areas where exposure to hazardous substances may occur

The chairperson of the resort's Safety Committee, acting on behalf of the general manager, has overall responsibility for coordinating this program.

Hazardous Substance Library/MSDSs. Under this program, all toxic and hazard-ous substances listed in 29 CFR 1910, Subpart 2, will be inventoried. A material safety data sheet will be obtained from the manufacturer for each substance inven-toried. A complete library consisting of the hazardous substance inventory, origi-nal and simplified versions of the MSDS, and a listing of hazardous substance storage areas will be maintained in the security department.

A departmental library consisting of all substances inventoried and all MSDSs pertaining to each work area will be maintained by that department manager and available for employee use. It will be the responsibility of the department manag-ers to train employees how to read and interpret MSDSs.

It shall be the responsibility of the Safety Committee chairperson to revise the libraries as required.

Labeling. All substances, when received from the manufacturer, will be inspected by each department manager to ensure proper labeling. It shall also be the respon-sibility of each department manager to ensure proper labeling on any container to which a hazardous substance is transferred. All areas used for storage of hazard-ous substances shall display a warning sign.

Training. Every new employee required to work with a hazardous substance will receive training by his/her department manager in the use and potential hazards of each substance. As new products are added to the chemical inventory, addi-tional employee training will be provided. Each employee will be provided with all protective measures suggested by the manufacturer and trained in their use.

Each training program shall include, but not be limited to, the following:

- Physical hazards of substances

- Health hazards of substances

- Protective procedures
- Proper use of substances
- Proper disposal of containers
- The location of the central library and how to read the MSDSs
- The Hazard Communication Program of the Sheraton Scottsdale Resort and what it means to the employee

At the completion of the training segment, each employee will be tested to ensure an understanding of this program.

Hazard Communication Training: Housekeeping Department

The purpose of this course is to raise awareness of safety and proper chemical handling in the housekeeping department and to reduce the number of accidents that occur at the Sheraton Scottsdale Resort. The Safety Committee feels that the implementation of this course will create a better and safer environment for our employees and show that we care about our employees and our property.

Housekeeping Training Program Outline

Training Time: 1 to 10 hours

A. Introduction to the training program
   1. What you will learn
   2. What you will be able to accomplish
   3. How the evaluation system will work

B. Outline of the training
   1. Review provisions of the Hazard Communication Standard
   2. Review material safety data sheets (MSDSs)
   3. Discuss the location and availability of the written Hazard Communication Program chemical list (MSDS library)
   4. Review labels and how to read them
   5. Define and inform employees of the location of hazardous substance storage areas
   6. Review procedures for proper handling, application, and disposal of chemicals
   7. Discuss safety equipment: Safety Committee Chairman
      a. Fire extinguishers
      b Rubber gloves
      c. Closed-toe shoes
      d. Goggles

8. Define physical and health hazard

9. Discuss areas and circumstances that create hazardous operations

10. Identify the presence of hazardous chemicals in the work area

11. Discuss proper emergency procedures

12. Test employees on training

13. Explain the evaluation system

14. Closing comments

15. Question and answer session

Housekeeping Lesson Plans

Learning Outcomes:

1. Meet the company's goal of providing all employees with a safe and healthy work environment.

2. Satisfy the employees' "Right to Know" regarding chemical hazards in the workplace.

3. Provide techniques to work with chemicals in a safe manner.

4. Identify chemical hazards.

5. Provide proper accident responses.

What to Do	How to Do It	Remarks
1. Review the provisions of the Hazard Communication Standard.	Hand out a copy of the property's Hazard Communication Program. Call on an employee to read it aloud. Discuss the Hazard Communication Program.	Lead a brief discussion on the Hazard Communication Program and what it means to employees and their jobs.
2. Review material safety data sheets (MSDSs).	Hand out a copy of a manufacturer's MSDS and the corresponding simplified MSDS. Ask an employee to read the simplified version aloud. Discuss both the simplified and the manufacturer's MSDS.	Provide a full explanation of the manufacturer's and simplified MSDS. Explain to employees that copies of both versions are on file in the central library and in the departmental library. Note that the departmental library is available to all employees on all shifts at all times.

What to Do	How to Do It	Remarks
3. Discuss the location and availability of the written Hazard Communication Program chemical list (MSDS library).	Identify the location of the central library in the security office and the departmental library in the executive housekeeper's office.	Explain that it is important to know the location of the library to assist in research of chemical content, danger signs, and precautions for safe handling and use. Instruct employees to notify their supervisors immediately in case of an accident so that appropriate action may be taken.
4. Review labels and how to read them.	Pass out sample labels. a. Read the label instructions. b. Call on an employee to interpret the label. Discuss the proper labeling of containers to which chemicals have been transferred.	Stress the importance of knowing the name of the chemicals used. It is important that labeling instructions are followed. Discuss where to locate further information not on labels.
5. Define and inform employees about hazardous substance storage areas.	Have employees list all areas where chemicals are stored. Provide a map with storage areas highlighted. Compare the map to employees' list of storage areas.	Stress the importance of knowing where chemicals are stored. Discuss proper storing requirements and their importance.
6. Review proper handling, application, and disposal of chemicals.	Have employees categorize all chemicals by areas of use. Refer to MSDS for proper disposal, safety equipment, and safe handling. Review application of chemicals and safety precautions.	Reinforce employees' knowledge of chemicals already being used.

What to Do	How to Do It	Remarks

7. Discuss safety equipment:
  a. Fire extinguisher
  b. Rubber gloves
  c. Closed-toe shoes
  d. Goggles
	Use visual aids on proper use of fire extinguishers and their locations. Discuss proper protective clothing.	Stress the importance of knowing how to use a fire extinguisher. Emphasize that wearing protective clothing helps guard against injury.

8. Define physical and health hazard.	Hand out definitions of physical and health hazards. Read aloud and discuss their relation to the employees' jobs.	Ensure that employees understand the difference between the two types of hazards.
9. Discuss areas and circumstances that create hazardous operations.	Identify and brainstorm a list of possible areas that could be dangerous. Discuss why these areas could be dangerous. Examples might include wet floors or fumes in enclosed areas. Discuss how to avoid hazards; for example, mop up wet spills, work in a well-ventilated area.	Stress that safety is everyone's responsibility. Taking proper precautions can safeguard employees or possibly guests from accidents.
10. Identify the presence of hazardous chemicals in the work area.	Discuss chemical warning signs to look for, such as smell, or physical reactions such as nausea and headaches.	Refer to the MSDS. Explain to employees what section to look under.
11. Discuss proper emergency procedures.	Have employees refer to the MSDS. Point out the section regarding accidents. Discuss what to do when an accident occurs. Discuss the supervisor's role in reporting accidents.	Review the location of the security office. Stress the importance of informing the supervisor and processing the accident report. This is to protect employees as well as the hotel.

What to Do	How to Do It	Remarks
12. Test employees on training.	Hand out written tests to the employees. Explain the test and instruct employees on the time limit and passing grade requirements. Explain the procedures in case an employee does not pass.	Employees must have a perfect score to pass the test. If an employee does not pass, he/she must go through training again. The time limit on completing the test is 30 minutes.
13. Explain the evaluation system.	Inform employees on how the evaluation system works: • That it is part of their written job breakdown/evaluation form. • That they will be evaluated from day-to-day observation of their work. • That they will be evaluated on the number of accidents that occur in their work areas.	Stress the importance of proper follow-up and evaluation. Employees can evaluate themselves on using the proper techniques for working with chemicals.
14. Closing comments.	Review the training and the importance of safety awareness. Stress that awareness on the employees' part will help create a better work environment for everyone.	Review briefly to cover all items discussed and shown. This will help enforce training.
15. Question and answer session.	Answer questions and address concerns of employees. Encourage employees to ask questions and seek answers regarding proper safety measures.	Allow enough time to cover all concerns or questions an employee might raise. Be prepared to offer assistance.

Housekeeping Hazard Communication Program Test

1. What type of reaction would inform you of the presence of hazardous chemicals?

   a. nausea
   b. laughter

2. Where is the MSDS central library located?

   a. housekeeping office
   b. security office

3. Where should you go to fill out an accident report?

   a. human resources
   b. security office

4. Who should be trained on the handling of chemicals?

   a. supervisors
   b. employees
   c. both a and b

5. Why is it important to transfer labeling information?

   a. to ensure proper handling
   b. to know what the bottle contains
   c. both a and b

6. What kinds of accidents should be reported?

   a. all
   b. only the serious ones
   c. none

7. Who should mix chemicals?

   a. employees
   b. supervisors
   c. only people who have completed the training program

8. Why is it important to read chemical labels?

   a. to prevent accidents
   b. to know the proper use of chemicals
   c. both a and b

9. List three department chemical storage areas.

   a. _____

   b. _____

   c. _____

REVIEW QUIZ

When you feel you have covered all of the material in this chapter, answer these questions. Choose the *best* answer.

True (T) or False (F)

T F 1. The presence of phosphates in the local water supply may inhibit the cleaning power of detergents.

T F 2. Tough bathroom stains can be safely removed by using a mixture of ammonia-based and chlorine-based cleaners.

T F 3. Alkalies register values greater than seven on the pH scale.

T F 4. Ceramic tile can withstand abrasive cleansers.

T F 5. A weak alkali solution is effective when cleaning glass, bronze, or stainless steel surfaces.

T F 6. Degreasers are also called emulsifiers.

T F 7. Delimers are used to kill bacteria, mold, and mildew.

T F 8. Items intended for personal use by employees are exempt from the HazComm Standard.

T F 9. If an MSDS is not available for a chemical, the chemical should be considered hazardous.

T F 10. Acute hazards are those associated with repeated, long-term use of a chemical.

Alternate/Multiple Choice

11. Calcium, iron, and sulfur are common:

a. water minerals.
b. wetting agents.

12. The pH scale measures the acidity or alkalinity of a substance compared to:

a. water.
b. phosphorus.

13. The HazComm Standard requires employers to:

a. keep their employees away from hazardous chemicals.
b. inform their employees about hazardous chemicals.
c. compensate employees who must use hazardous chemicals.
d. train every hotel employee in the safe use of hazardous chemicals.

14. MSDS stands for:

    a. major safety detail sheet.
    b. mandatory standard for danger and safety.
    c. material safety data sheet.
    d. material safety danger standard.

15. Reactivity data describe a chemical's:

    a. color and odor.
    b. health hazards.
    c. proper container.
    d. stability.

Chapter Outline

Preparing to Clean
    Assembling Supplies
    Room Assignments
Cleaning the Guestroom
    Entering the Guestroom
    Beginning Tasks
    Making the Bed
    Dusting
    Cleaning the Bathroom
    Vacuuming
    Final Check
Inspection
    Inspection Program Technology
Deep Cleaning
Turndown Service and Special Requests

Learning Objectives

1. Explain the importance of a well-organized and properly stocked room attendant's cart.

2. Identify the function and contents of a room attendant's hand caddy.

3. Explain how room attendants are assigned rooms to clean.

4. Explain how room attendants determine the order in which to clean their assigned rooms.

5. Describe procedures that are followed when a guest refuses cleaning services.

6. Identify the steps involved in guestroom cleaning.

7. Describe the sequence typically followed by room attendants when cleaning guestroom bathrooms.

8. Explain the function of a guestroom inspection program.

9. Describe how guestrooms can be scheduled for deep cleaning.

10. Identify typical turndown service procedures.

Guestroom Cleaning

N<small>O OTHER FEATURE OR SERVICE</small> provided by a property will impress the guest more than a spotlessly clean and comfortable guestroom. The condition of the guestroom conveys a critical message to guests. It shows the care that the property puts into creating a clean, safe, and pleasant environment for its guests. This places a big responsibility on the housekeeping department. After all, the guestroom is the main product that a property sells. Housekeeping plays a greater role than any other department in ensuring that this product meets the conditions that guests need and expect.

To maintain the standards that keep guests coming back, room attendants must follow a series of detailed procedures for guestroom cleaning. A systematic approach can save time and energy—and reduce frustration. In this respect, room cleaning procedures not only ensure quality for the guest, but ensure efficiency and satisfaction for the employee performing the task.

The sequence of room cleaning consists of preparatory steps, actual cleaning tasks, and a final check. Room inspections are also an integral part of the overall process of guestroom cleaning. In some properties, the responsibilities of room attendants extend to providing special services and amenities. Regardless of the range of services, a room attendant should recognize the value and logic behind the organization of cleaning activities. Adhering to a careful routine can save time and ensure a professional job.

Preparing to Clean

In most properties, the room attendant's workday begins in the **linen room**. The linen room is often considered the headquarters of the housekeeping department. It is here that the employee reports to work; receives room assignments, room status reports, and keys; and checks out at the end of his/her shift. Here, too, the room attendant prepares for the workday by assembling and organizing the supplies that are necessary for cleaning.

Assembling Supplies

Like most craftsmen, a room attendant requires a special set of tools to do his/her job. For the professional room attendant, these tools come in the form of the various cleaning supplies and equipment, linens, room accessories, and **amenities** that are necessary for preparing a guest's room.

In a sense, the **room attendant's cart** could be regarded as a giant tool box stocked with everything necessary to do an effective job. Just as a carpenter would avoid going on-site with an inadequate supply of wood and nails, so would a room

---

Housekeeping Success Tip

Elizabeth Struc, CHHE
Director of Services
Berkeley Marina Marriott
Berkeley, California

    ❝I believe that employees develop pride in their work when they have a clear sense of responsibility. This sense can be encouraged when managers clearly delineate what they expect from employees—and listen to what employees say. Employees who are proud of their work will often contribute ideas and insights that can make getting the job done just that much easier.

    Recently, we had a problem with soap scum building up on tile grout in guest bathrooms. We tried every product available on the market with little success. Finally, at the suggestion of an employee, I called a salesperson who had helped me with other cleaning problems. He called his chemical supplier, and together, we created a product that works wonders on built-up soap scum. We call it the Marriott Tile Cleaner.

    Another time, we tried everything on the market to clean the ceramic floor tile in public areas. At the suggestion of a public area attendant, we tried the in-house laundry pre-spotter. Lo and behold, the tiles came out sparkling clean! Now, all we do is scrub the floor tiles once a month and feel proud of their appearance through the next scheduled cleaning.

    Listen to your employees. Be willing to try new and different ways of doing things—even though they seem to stray from the norm. Many employees have creative suggestions for solving problems that can actually get the job done more efficiently and effectively.**❞**

---

attendant avoid going to an assigned room with an inadequate supply of cleaning items.

A well-organized and well-stocked cart is a key to efficiency. It enables the room attendant to avoid wasting time looking for a cleaning item or making trips back to the linen room for more supplies. The specific amounts of items loaded onto a cart will vary according to the types of rooms being cleaned, the amenities offered by the property, and, of course, the size of the cart itself. A room attendant's cart is generally spacious enough to carry all the supplies needed for a half-day's room assignments.

Stocking the Cart. Carts are typically stored in the linen room along with the housekeeping supplies. In large properties, supplies are often centralized in a particular area and issued to room attendants each morning. Most carts have three shelves—the lower two for linen and the top for supplies. It is just as important not to overstock a cart as it is not to understock. Overstocking increases the risk that some items will be damaged, soiled, or stolen in the course of cleaning. Items typically found on a room attendant's cart include:

- Clean sheets, pillowcases, and mattress pads
- Clean towels and washcloths
- Clean bath mats
- Toilet and facial tissue
- Fresh drinking glasses
- Soap bars
- Clean ashtrays and matches

In most cases, all the cleaning supplies for the guestroom and bathroom are positioned in a **hand caddy** on top of the cart. This way, the room attendant does not have to bring the entire cart into the room in order to have easy access to supplies. Items conveniently stocked in the caddy may include:

- All-purpose cleaner
- Spray window and glass cleaner
- Bowl brush
- Dusting solution
- Cloths and sponges
- Rubber gloves

A laundry bag for dirty linens is usually found at one end of the cart and a trash bag at the other. A broom and vacuum are also positioned on either end of the cart for easy access. For safety and security reasons, personal items and room keys should not be stored on the cart.

Exhibit 1 illustrates one efficient stocking arrangement for a room attendant's cart. In all cases, carts should be stocked according to a property's specifications. Room attendants must also be sure to stock the proper eye, hand, and face protection. Each property should inform room attendants of the proper policies regarding the use of such protective gear and for handling cleaning chemicals.

Alternative Carts. Some hotels now use an integrated transporting and storing system as an alternative to the traditional room attendant cart. The equipment is modular in design and consists of various containers, caddies, and shelves that can be easily removed and arranged within a larger service cart. These components are loaded to convenient levels to allow for the efficient movement of linens and supplies when servicing guestrooms. Like a boxcar, a separate, detachable component accompanies the main unit and is used to catch trash and soiled linen.

These carts are furniture-grade in appearance and can be secured with a locking tambour door. Given these unobtrusive features, some properties pre-load carts and deliver them directly to guestroom floors for pickup by the room attendants. Among other advantages, these carts are lightweight and easy to clean.[1]

Exhibit 1   Sample Stocking Arrangement for Room Attendant's Cart

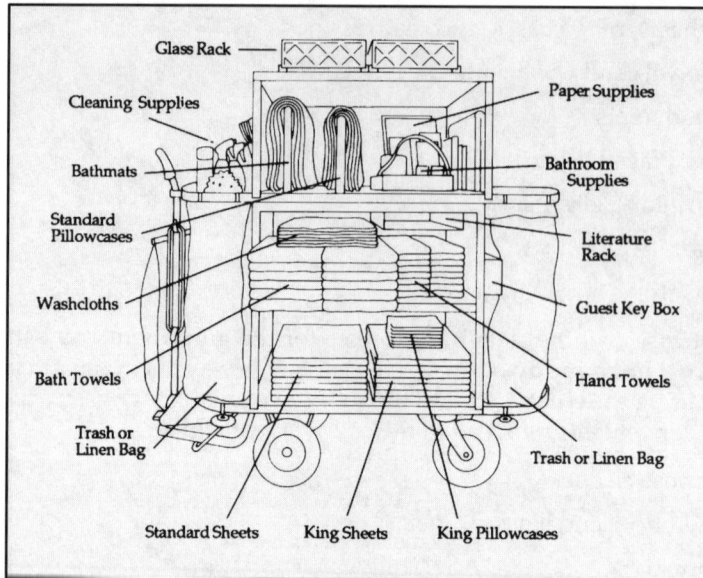

Courtesy of Holiday Corporation, Memphis, Tennessee

Room Assignments

After assembling supplies, the room attendant is ready to begin cleaning guestrooms. The order in which he/she cleans rooms will be determined by the room status report.

The **room status report** (sometimes called the housekeeping report) provides information on the occupancy or condition of the property's rooms on a daily basis. It is generated through two-way communications between the front office and the housekeeping department. For example, when a guest checks out, the front desk notifies housekeeping by phone or through a computer system. In turn, once a room is clean and back in order, the flow of information is reversed so the front office will know the room is again ready for sale.

The room status report is generally easy to read and uses simple codes to indicate room status. There are several categories of room status, but, for the most part, a room attendant's cleaning schedule will be determined by these three:

Check-out: A room from which the guest has already checked out
Stayover: A room in which the guest is scheduled to stay again
Due out: A room from which a guest is due to check out that day

Another designation commonly used is early makeup. This refers to rooms for which a guest has reserved an early check-in time or to a request for a room to be cleaned as soon as possible. Abbreviations used to indicate these categories on the room status report will vary from property to property.

A floor or shift supervisor uses information from the room status report to draw up room assignments for housekeeping personnel. Room assignments are generally listed according to room number and room status on a standardized form. The number of rooms assigned a room attendant is based upon the property's work standards for specific types of rooms and cleaning tasks. The room attendant uses the assignment sheet to prioritize the workday and to report the condition of each assigned room at the end of the shift. In the sample form illustrated in Exhibit 2, room attendants are provided space to make written comments on each room and to indicate room items needing repair.

After reviewing the assignment sheet, a room attendant will have a sense of where he/she should begin cleaning. For the most part, the order in which rooms are cleaned is the order which best serves guests. Check-outs are usually done first so the front office can sell the rooms as guests arrive. The exceptions to this rule are rooms needing early makeup. In most properties, early-makeup rooms are cleaned before check-outs. After early makeups and check-outs, a room attendant will generally clean stayovers. Due outs are usually the last rooms cleaned. Sometimes, room attendants may be able to wait until the guest has actually checked out to avoid duplication of efforts.

In all cases, the room attendant should avoid disturbing the guest. A "Do Not Disturb" sign clearly indicates that the room attendant should check back on the room later in the shift. Other rooms which room attendants must delay servicing include rooms which the guest has double-locked from the inside. Many properties have room attendants leave a card on the door which indicates that attempts at service have been made. These cards may also offer fresh towels or service later in the evening. Usually, a room attendant will report such rooms to department headquarters if he/she is unable to service the room by 2:00 or 3:00 P.M.

When a guest refuses service, a floor supervisor or other management person should call to arrange a convenient time for cleaning. Such calls are also made to check that the guest is not experiencing a situation that requires intervention, such as a serious illness or accident. Upon contacting the guest, the floor supervisor or manager should also ask the guest if he/she would like fresh towels and soap. In many hotels, the guest is asked to initial the supervisor's report to indicate that he/she refused service. Under no circumstances should a room remain unserviced for more than two days without the approval of the general manager.

Cleaning the Guestroom

Room attendants must follow a system to consistently produce spotlessly clean guestrooms. A systematic plan saves time and can prevent the room attendant from overlooking a cleaning task—or even from cleaning an area twice.

To be most effective, guestroom cleaning should follow a logical progression from actually entering the guestroom to the final check and departure. Exhibit 3 shows a general sequence followed in some properties for guestroom cleaning. The easiest and most direct manner to explain these guestroom cleaning tasks is from the perspective of the room attendant.

Exhibit 2   Sample Room Assignment Sheet

(FRONT)	(BACK)

HOUSEKEEPING ASSIGNMENT SHEET

NAME_____

ROOM #	ROOM STATUS	COMMENTS

ROOM STATUS   V-VACANT
                        O-OCCUPIED

Back column labels (top to bottom): Tub, Bed, Dresser, HVAC, Carpet, Door, Commode, Art, Table, Drapes, Lights, Faucet, Television, Room No.

Courtesy of Holiday Corporation, Memphis, Tennessee

Entering the Guestroom

Guestroom cleaning begins the moment the room attendant approaches the guest-room door. It is important to follow certain procedures when entering the guest-room that show respect for the guest's privacy.

When approaching a guestroom, first observe whether the guest has placed a "Do Not Disturb" sign on the knob. Also, be sure to check that the door is not double-locked from the inside. If either condition exists, respect the guest's wishes

Exhibit 3   General Sequence for Guestroom Cleaning

Step 1:	Enter the guestroom.
Step 2:	Begin cleaning. Tidy and air out the room.
Step 3:	Strip bed.
Step 4:	Make the bed.
Step 5:	Dust the guestroom.
Step 6:	Clean the bathroom.
Step 7:	Vacuum.
Step 8:	Make the final check.
Step 9:	Close the door and make sure it is locked.
Step 10:	Note room status on assignment sheet and proceed to next room.

and return later to clean the room. If this is not the case, knock on the door and announce "Housekeeping." Never use a key to knock since it can damage the surface of the door. If a guest answers, introduce yourself and ask what time would be convenient to clean the room. Note that time on your status sheet or schedule. If no answer is heard, wait a moment, knock again, and repeat "Housekeeping." If there is still no answer, open the door slightly and repeat "Housekeeping." If the guest does not respond after this third announcement, you can be fairly certain that the room is empty and can begin to enter.

However, just because a guest doesn't answer doesn't always guarantee that a guest is not in the room. Sometimes, the guest may be sleeping or in the bathroom. If this is the case, you should leave quietly and close the door. Should the guest be awake, excuse yourself, explain that you can come back later, discreetly close the door, and proceed to the next room.

When you do finally enter, position your cart in front of the open door with the open section facing the room. Doing so serves a triple purpose: it gives you easy access to your supplies, blocks the entrance to intruders, and, in the case of stayovers, alerts returning guests of your presence. If the guest does return while you are cleaning, offer to finish your work later. Also, make sure that it is, in fact, the guest's room, by checking his/her room key. This is done for security purposes to prevent unauthorized persons from entering the room.

Beginning Tasks

Most room attendants begin their system of cleaning by airing out and tidying up the guestroom. After entering the room, turn on all the lights. This makes the room more cheerful, helps you see what you are doing, and allows you to check for light bulbs which need to be replaced. Draw back the draperies and check the cords and hooks for any damage. Open the windows so the room can air out while you are cleaning. Check the air conditioning and heater to make sure they are working properly and are set according to property standards.

Next, take a good look at the condition of the room. Make note of any damaged or missing items such as linens or wastebaskets. If anything of value is gone or if something needs repair, notify your supervisor.

Remove or replace dirty ashtrays and glasses. Always make sure that cigarettes are fully extinguished before dumping them in the appropriate container. As you replace the ashtrays, be sure to replenish matches. Collect any service trays, dishes, bottles, or cans that might be scattered around the room. Follow your property's procedures for taking care of these items properly. Some properties have room attendants set these items neatly in the hallway and call room service for pickup. Empty the trash and replace any wastebasket liners. In occupied rooms, straighten any newspapers and magazines. Never throw out anything in an occupied room unless it is in the wastebasket. In rooms where the guest has checked out, visually scan the room and check the dresser drawers for personal items which may have been left behind. Report these items to your supervisor, or hand them in to the lost and found depending on the hotel's policy.

Making the Bed

Making the bed is the next task you do in guestroom cleaning. It is important to start cleaning here—especially in stayover rooms. If the guest returns while you are elsewhere in the room, the freshly made bed will give the room a neat appearance—even if other areas have not been touched. In check-out rooms, some properties recommend that you strip the bed shortly after entering and remake it near the end of your cleaning. This way the bed has a chance to air out.

Before you begin, remove any personal items from the bed and place them aside. Remove the bedspread and blanket and place them on a chair to keep them clean and free from dust and dirt. If the blanket or bedspread is dirty—or if you notice any holes or tears—be sure to replace it. Strip the bed of dirty linen and place the pillows on the chair with the bedspread and blanket.

Once the bed is stripped, you should check the mattress pad and the mattress. Make a note to inform your supervisor if the mattress shows any stains, burns, or damage. If the mattress pad needs changing, remove the old pad and lay a fresh one on the bed so it unfolds right-side up. Spread the pad evenly over the center of the bed and smooth out any wrinkles.

The most efficient way of making a bed is to completely finish one side before beginning on the next. This system saves time walking back and forth around the bed. Begin by placing the bottom sheet on the mattress and mitering that sheet in the upper left-hand corner of the bed. **Mitering** is a simple way to make a smooth, neat, professional corner. A step-by-step method for mitering a corner is illustrated in Exhibit 4. Next, move to the foot of the bed—still on the left-hand side—and miter that corner of the sheet.

Place a fresh top sheet on the bed, wrong-side up. Then, place the blanket on top of the sheet. At the head of the bed, turn the top sheet over the blanket about six inches. Smooth your hand over the bed so the surface appears even and without wrinkles. Miter the top sheet and blanket in the bottom left-hand corner of the bed and tuck them in along the side of the bed. Now, working clockwise, walk to the other side of the bed. Miter the bottom sheet at the right foot of the bed, followed

Exhibit 4   Step-by-Step Approach to Mitering

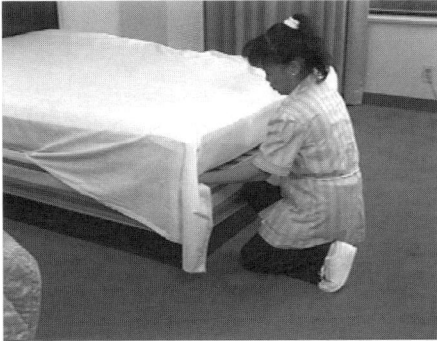

Step 1 Begin with the sheet hanging loosely over the corner. Tuck in the sheet along the foot of the bed, right up to the corner.	**Step 2** Take the loose end of the sheet, about one foot from the corner, and pull it straight out, forming a flap. **Step 3** Pull up the flap so it is flat and wrinkle-free.
Step 4 Tuck in the free part at the corner, making sure it is snug. **Step 5** Pull the flap out toward you and down over the side of the bed.	**Step 6** Tuck in the flap and make sure the corner is smooth and snug.

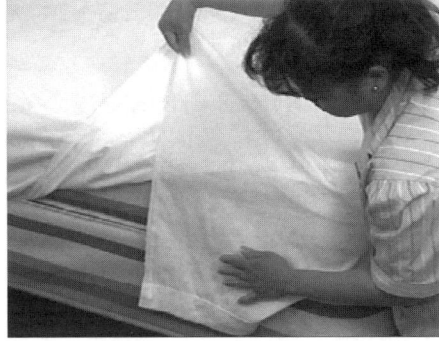

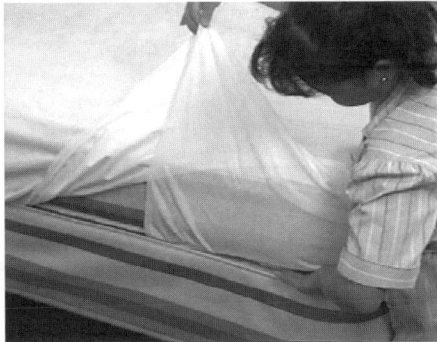

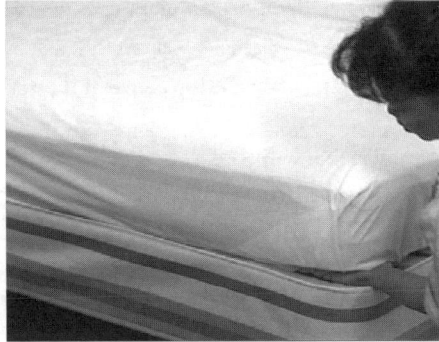

by the top sheet and blanket. Move down the right-hand side of the bed and miter the bottom sheet in the top right corner. Fold the top sheet over the blanket so it is even with the left-hand side. Finally, make sure the blanket and top sheet are neatly tucked in along the sides and at the foot of the bed.

Now, center the bedspread evenly over the bed. Fold the bedspread down from the head of the bed, leaving enough room to cover the pillows. Fluff the pillows and put on the pillowcases. Work the pillow down into the case so that no part of the pillow ticking is showing. For sanitation reasons, never hold the pillow under your chin or with your teeth. Position the pillows at the head of the bed with

Housekeeping Success Tip

Patricia L. Harper, CHHE
Rooms Division Manager
Holiday Inn Denver Downtown
Denver, Colorado

❝When I was executive housekeeper for Holiday Inn Denver Downtown (a property managed by Mariner Hotel Corporation based in Dallas, Texas), our company offered a workshop on 'Training the Trainer.' I left the workshop very excited about instituting the program in the housekeeping department because we had been experiencing a very high rate of turnover.

After writing out room cleaning procedures, I passed them to the trainers. I told them to conduct the training according to the following plan:

1. *Demonstrate* the task.

2. While demonstrating, *explain why* we do what we do.

3. Have the trainees *perform* the task and *explain why* they are doing it that way.

4. *Critique* their performance and offer *feedback*.

These steps can be repeated any number of times that the trainee and the trainer feel comfortable with.

I feel that 'training the trainer' establishes rapport between the trainer and trainee because of the free flow of communication. The new employee feels 'in the know' about the property's procedures, and 'training the trainer' decreases employees' insecure, new-on-the-job feelings.

We enjoyed a turnover rate of less than half of our previously high rate and the satisfaction that we did our best to make new hires feel at home.❞

the open ends facing the sides of the bed. Pull the bedspread over the pillows and tuck in the bedspread. Notice that this method of finishing the bed avoids any hand contact with the cases after they are put on the pillows.

The thorough room attendant will take a few moments to check the bed for smoothness. Step back and look carefully at the surface of the bedspread and the line of the pillows. Smooth out any last-minute wrinkles. Finally, if there are two beds in the room, check the second bed and change it if necessary.

Dusting

Like bed-making, the task of dusting requires a systematic and orderly approach for efficiency and ease. Some room attendants start dusting items beside the bed they just finished making and work clockwise around the room. This reduces the chance of overlooking a spot. In all cases, begin with the highest surfaces so that dust doesn't fall on the items you have already cleaned. If your property uses a dusting solution, spray a light amount into the dust cloth. Never spray dusting solution directly onto an object since it can stain or cause stickiness.

The items needing dusting and their location will vary from property to property. As a general rule, the following should be dusted and/or polished:

- Picture frames

- Mirrors

- Headboards

- Lamps, shades, and light bulbs

- Bedside tables

- Telephone

- Windowsills

- Window and sliding glass door tracks (if applicable)

- Dresser—including inside the drawers

- Television and stand

- Chairs

- Closet shelves, hooks, and clothes rod

- Top of doors, knobs, and sides

- Air conditioning and heating units, fans, or vents

You should also clean all mirrors and glass surfaces in the room using glass cleaner—including the front of the television set. When you clean the set, turn it on for a moment to make sure it works properly. Some properties also use a special cleaner or disinfectant for telephone surfaces. As you dust your way around the room, note any bedroom supplies and amenities which may be needed and replenish per your property's specifications. Finally, check the walls for spots and marks and remove any smudges with a damp cloth and all-purpose cleaning solution.

Cleaning the Bathroom

A clean bathroom is important for more than simply appearance. Health and safety considerations on the local, state, and federal level necessitate that the room attendant take extra care when scrubbing, rinsing, and drying bathroom surfaces.

Bathrooms are usually cleaned in the following sequence: shower area, vanity and sink, toilet, walls and fixtures, and floor. Like most cleaning tasks, it is important to work from top to bottom to avoid spotting or dirtying areas already cleaned. The necessary cleaning equipment should be conveniently stocked in the hand caddy. Cleaning items usually consist of an approved all-purpose cleaner for bathroom surfaces; cloths and sponges; glass and mirror cleaner; rubber gloves; and protective eye covering. Some properties also use an odorless disinfectant. Do not use a guest towel for cleaning.

For personal safety, never stand on the edge of the tub when cleaning. When cleaning the inside of the tub, some properties recommend placing a used cloth bath mat in the tub to stand on. As you wash and wipe the tub or shower walls, continually check their condition so you can report any needed repairs to your supervisor. If the tub has a drain trap, be sure to check it for hair. After cleaning the tub itself, clean the shower head and tub fixtures. Make sure to leave the shower head aimed in the correct position. To prevent spotting, and to add sparkle, immediately wipe and polish the fixtures with a dry cloth. Also clean the shower curtain or shower door. Pay special attention to the bottom where mildew may accumulate. Always reposition the door or curtain when you are finished cleaning.

You should exercise the same exacting care when cleaning the vanity and mirror as you do when cleaning the shower area. Clean the countertop and basin, making sure that you remove any hair from the sink stopper and drain. Wipe up any spillage or spots from toothpaste or soap. Rinse and polish the chrome fixtures so they shine. Finish the vanity area by cleaning the mirror with glass cleaner.

Next, clean the toilet bowl and exterior surfaces. Some cleaning procedures recommend applying an all-purpose cleaner before any other cleaning task. This way, the cleaner has time to stand while you clean other bathroom areas. All-purpose cleaners are preferable over acid bowl cleaners for use on a daily basis. When used consistently, acid bowl cleaners can destroy bathroom surfaces. These cleaners also present hazards to employees who use them, most noticeably in terms of causing skin irritation. Most properties use bowl cleaners once or twice a year in deep cleaning programs—and then only under strict supervision.

Regardless of the method followed, flush the toilet to remove any residue and apply the cleaner around and under the lip of the bowl. Scrub the toilet with the brush around the insides and under the lip—then flush again. Use a cloth damp with cleaning solution to clean the top of the seat, the lid, and the sides of the tank. Finally, clean the exterior of the bowl, working down the sides to the base.

Towels, washcloths, bath mats, toilet and facial tissue, and guest amenities should be replenished according to property standards. Spot-clean for fingerprints and other obvious smudges on the walls, especially around light fixtures and electrical outlets. Wipe down the walls and clean both sides of the bathroom door. Starting with the far corner of the bathroom and working back toward the door, mop or wipe down the floor—including the baseboards. Then, gather your things and make your final check of the bathroom. Stop for a moment and visually scan all surfaces from the ceiling to the fixtures to the floor. Check that you've left the bathroom in the best possible condition before turning out the lights.

Vacuuming

Before vacuuming, loosen dirt around baseboards with a broom or rag so it is easier to pick up. Run the vacuum over all exposed areas of the carpet that you can reach, including under the tables and chairs and in the closet. Don't worry about inaccessible areas such as under the bed or dresser. Since cleaning these areas requires moving or lifting heavy furniture, most properties vacuum these areas on a special project basis. However, it is your responsibility to check under beds and furniture for guest belongings or for any debris which must be removed.

You should start at the farthest end of the room and vacuum your way back— just as you did when you wiped down the bathroom floor. As you vacuum, be careful not to bump the furniture. Some properties recommend closing windows and draperies and turning off lights as you work your way back to the door. Working in this fashion saves steps. It also eliminates the need to walk back across the floor after it has been vacuumed, thus preventing the footprints and tracks that can appear in certain types of carpet.

Final Check

The final check is a critical step in guestroom cleaning. It makes the difference between just cleaning the room and doing a professional job.

After reloading your vacuum and cleaning supplies on your cart, take a few moments to give the room a careful look from the guest's perspective. Start at one point in the room and trail your eyes in a circular fashion from one corner to the next until you have visually inspected each item. By doing so, you may discover something you overlooked or that was difficult to spot on the first cleaning.

Make sure that all the furnishings are back in their proper places. Look for little things like making sure the lampshades are straight and their seams are turned toward the back. Smell the air for any unusual odors. If you detect any unpleasant smells, report them to your supervisor. Spray air freshener if needed. Remember that your last look is the guest's first impression. When you are satisfied that the guestroom is neat and thoroughly cleaned, turn off the lights, close the door, and check to see that it is locked. Note the condition and status of the room on your assignment sheet, and proceed to the next room on your schedule.

Inspection

Guestroom inspection ensures that the desired results of an established cleaning system are consistently achieved. The purpose of a room inspection is to catch any problems that may have been overlooked during cleaning before the guest does. A well-conducted and diplomatic inspection program can also motivate employees. Most room attendants take pride in their work and enjoy having the opportunity to show it off to others. Quality cleaning jobs should be noted during inspections and the appropriate personnel recognized.

Inspection programs can take many forms. In some properties, rooms are spot-checked randomly; in others, every room is checked daily. Inspections should be conducted by personnel on the supervisory level such as a floor or shift supervisor, section supervisor, executive housekeeper, or even a manager from outside the housekeeping department. Each inspector is usually responsible for a certain number of rooms and should be aware of the current status of each room he/she is assigned. As a general rule, check-out rooms are inspected soon after room attendants report that they have been cleaned. Rooms that are occupied or have refused service are inspected on varying schedules. For these rooms, the executive housekeeper or inspector will contact the guest to arrange a convenient time for guestroom cleaning and/or inspection. Vacant rooms should also be inspected on a varying schedule based on the number of days the room remains empty between sales.

Exhibit 5 Sample Room Inspection Report

Room Inspection Report

ROOM NO._____

TYPE._____ DATE INSPECTED._____

CONDITION: ☐ EXCELLENT ☐ ACCEPTABLE ☐ UNACCEPTABLE

	BEDROOM	CONDITION			BATHROOM	CONDITION
1	Doors, locks, chains, stops			21	Doors	
2	Lights, switches, plates			22	Lights, switches, plates	
3	Ceiling			23	Walls	
4	Walls			24	Tile	
5	Woodwork			25	Ceiling	
6	Drapes and hardware			26	Mirror	
7	Windows			27	Tub, caulking, grab bars	
8	Heating/air conditioning setting			28	Shower head and curtain	
9	Phone			29	Bath mat	
10	TV and radio			30	Vanity	
11	Headboards			31	Fixtures/faucets/drains	
12	Spreads, bedding, mattress			32	Toilet: flush/seat	
13	Dressers, nightstand			33	Towels: facial/hand/bath	
14	Promotional material			34	Tissue: toilet/facial	
15	Lamps, shades, bulbs			35	Soap	
16	Chairs, sofa			36	Amenities	
17	Carpet			37	Exhaust vent	
18	Pictures and mirrors					
19	Dusting					
20	Closet					

OTHER._____

_____

_____

_____

INSPECTED BY:_____
(Signature)

Room inspections not only help to identify ordinary problems with cleaning but also help to identify areas in the room needing deep cleaning or maintenance. A room inspection report should be completed which notes such items as the condition and proper operation of furniture, fixtures, and equipment; the appearance of ceilings and walls; the condition of carpet and floor coverings; and the cleanliness of window interiors and exteriors. Exhibit 5 shows a sample inspection form. Depending on the property's policies and procedures, the inspector may also be responsible for filling out any work orders or maintenance requests that are needed.

An inspection program is never any better than the follow-up that is given to an identified problem. Each situation noted on the inspection report or maintenance request should be initialed by the manager who is directly responsible for that area. As a general rule, this should occur no later than 24 hours after the inspection.

Inspection Program Technology

A technology which has significantly affected the retail trade promises to lend the same ease and efficiency to the hospitality industry over the next several years. Just as bar code technology saves time and ensures accuracy in countless check-out lanes, so can it save time and ensure accuracy in a hospitality inspection program. A **bar code** is the group of printed and variously patterned bars, spaces, and numerals

Exhibit 6 Sample Inspection Forms Using Bar Code Technology

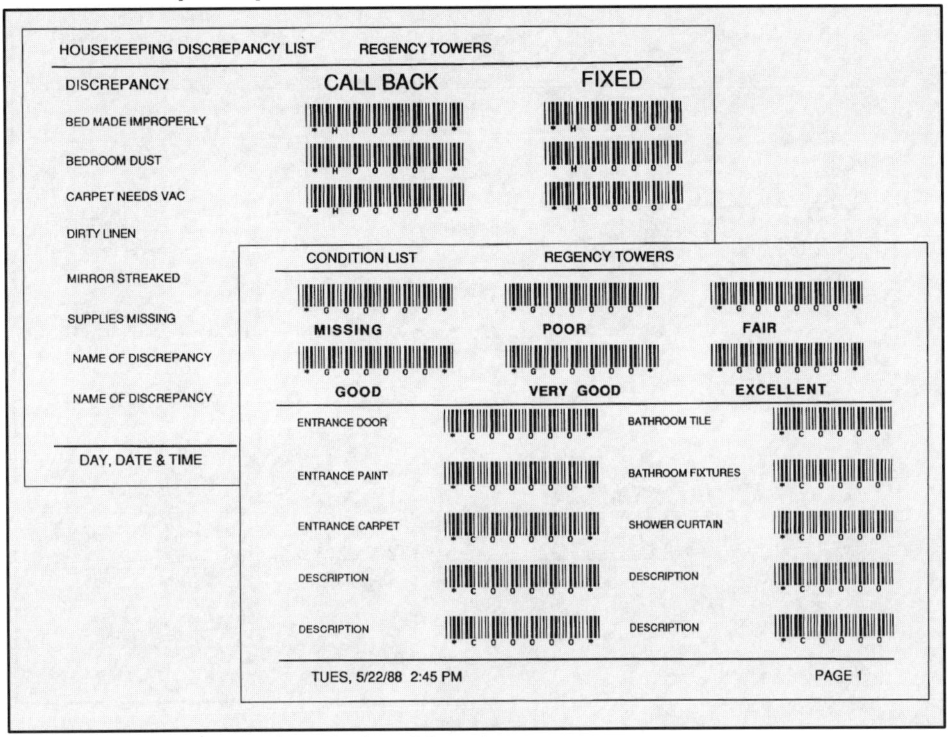

HOUSEKEEPING DISCREPANCY LIST REGENCY TOWERS

DISCREPANCY	CALL BACK	FIXED
BED MADE IMPROPERLY		
BEDROOM DUST		
CARPET NEEDS VAC		
DIRTY LINEN		
MIRROR STREAKED		
SUPPLIES MISSING		
NAME OF DISCREPANCY		
NAME OF DISCREPANCY		

DAY, DATE & TIME

CONDITION LIST REGENCY TOWERS

MISSING POOR FAIR

GOOD VERY GOOD EXCELLENT

ENTRANCE DOOR		BATHROOM TILE	
ENTRANCE PAINT		BATHROOM FIXTURES	
ENTRANCE CARPET		SHOWER CURTAIN	
DESCRIPTION		DESCRIPTION	
DESCRIPTION		DESCRIPTION	

TUES, 5/22/88 2:45 PM PAGE 1

Courtesy of Bar Code Technology, Eastham, Massachusetts

that appear on the packaging of almost every retail item. These codes are designed to be scanned and read into a computer system as identification for the objects they label.

For a hotel, instead of storing price and inventory information, bar codes can be used to store room inspection data. Inspectors or maintenance personnel would gather and record room status information by scanning bar codes with a special device about the size of a credit card—rather than recording information onto forms. This information would later be read by a computer and compiled into various reports to track housekeeping and maintenance activities.

In a property using a bar code inspection system, each guestroom is identified with a small, permanently mounted bar code tag. The tag is placed in a discreet spot, such as on the door frame. The inspector or maintenance person is equipped with a bar code reader and a set of cards that list items or conditions that need to be inspected, attended to, or repaired. Like the guestroom itself, each of these items would have a corresponding bar code. Exhibit 6 shows sample discrepancy and condition lists that can be customized to fit the needs of individual properties.

Upon entering the room, the inspector scans the room's bar code tag. This automatically records the room number, time, and date in the bar code scanner. The condition of each inspected item is noted by scanning the appropriate bar code or

Exhibit 7   Sample Inspection Report Generated through a Bar Code System

Location	Time In	Time Out	Time Spent	Cleaner	Housekeeping	Fixed	Call Back	Maintenance	Ticket #
	Inspector Report: Jones, Ann				The Regency Towers				4 / 6 / XX
101	9:00 AM	9:15 AM	0:15	Smith, Nancy					
103	9:20 AM	9:30 AM	0:10	Smith, Nancy	Heat On	•			
102	9:40 AM	10:00 AM	0:20	Hall, Judy				Faucet Washer	300025
105	10:05 AM	10:30 AM	0:25	Smith, Nancy					
Break	10:40 AM	10:55 AM	0:15						
106	11:00 AM	11:17 AM	0:17	Hall, Judy	Dust	•			
					No Comment Card	•			
107	11:25 AM	11:45 AM	0:20	Smith, Nancy					
Lunch	12:00 PM	1:00 PM	1:00						
108	1:10 PM	1:16 PM	0:06	Hall, Judy					
110	1:25 PM	1:45 PM	0:20	Hall, Judy					
112	1:50 PM	2:05 PM	0:15	Hall, Judy					
111	2:10 PM	2:28 PM	0:18	Smith, Nancy	Heat On	•			
113	2:32 PM	2:55 PM	0:23	Smith, Nancy					
115	3:00 PM	3:15 PM	0:15	Smith, Nancy					
114	3:22 PM	3:35 PM	0:13	Hall, Judy	Guest Supplies		10001	Bathroom Tile Loose	300026
116	3:40 PM	4:05 PM	0:25	Hall, Judy					
118	4:10 PM	4:22 PM	0:12	Hall, Judy					
120	4:25 PM	4:30 PM	0:05	Hall, Judy	Bathroom Tile		10002		
119	4:35 PM	4:50 PM	0:15	Smith, Nancy					

SUMMARY

Number Rooms	Total Time	Inspect Time	Percent Inspect	Maintenance Problems
17	6:45	5:49	87%	2

# Rooms	Cleaner	# Problems	# Fixed	# Callbacks
8	Smith, Nancy	2	2	
9	Hall, Judy	4	2	2

Monday, 4/11/XX 4:45 PM   PAGE 1

Courtesy of Bar Code Technology, Eastham, Massachusetts

combination of bar codes on the inspection cards. For example, if the bed is improperly made, the inspector would scan the "call back" bar code next to that item—indicating that the room attendant needs to redo the bed. At the end of the visit, the inspector "scans out" by scanning the room bar code a second time.

The information stored on the scanner can be retrieved by inserting the card into a special reader attached to a computer system. Depending on the program and property needs, the information can be presented in a summary or report format that provides management with an overview of the condition of each inspected room. Exhibit 7 shows one type of inspector report generated through a bar code system.

Bar code technology lends itself to a great deal of flexibility and can be customized to meet the specific needs and procedures of any property. Some properties coordinate bar code inspection programs with maintenance and engineering activities; others adapt the technology for such purposes as equipment tracking and security inspections. The information gathered and compiled can be as detailed or simple as required.

Deep Cleaning

Routine cleaning can maintain a guestroom's fresh and spotless appearance for a period of time. But after a while, a room will need **deep cleaning**. In theory, deep cleaning resembles the spring cleaning conducted in private homes or the cleaning conducted by apartment complexes between leases.

Deep cleaning removes the dust and dirt that accumulates from everyday wear and tear—and attends to cleaning needs identified during a guestroom inspection. It includes activities such as dusting in high and hard-to-reach areas, cleaning vent fans and filters, vacuuming under beds and heavy furniture, shampooing carpets, turning mattresses, wiping down walls and baseboards, cleaning and vacuuming drapes, cleaning carpet edges, and washing windows and casements. In some properties, deep cleaning is done by room attendants on a special project basis; others use teams in which each employee does a particular deep cleaning task.

Because of the thoroughness involved, deep cleaning requires special scheduling. Deep cleaning also requires more time. It may take twice as long for deep cleaning as it does for routine cleaning. Frequency schedules should be maintained which indicate when and how often a particular deep cleaning task needs to be performed. The schedule for cleaning depends on the quality of the routine cleaning, occupancy, age of furniture and fixtures, and general wear and tear on the room.

Some properties schedule deep cleaning by giving room attendants one extra cleaning task per room per day. For example, a room attendant might be asked to move the nightstand and sweep that area of the carpet. Another way to schedule deep cleaning is to give each room attendant one room to deep clean as part of their daily assignment. Here, a team approach may be necessary since some tasks need the "muscle" and coordination of more than one individual.

Many properties schedule rooms for deep cleaning during low occupancy periods. In this way, alternating blocks of rooms can be closed off and given the "works." The number of rooms taken out of circulation varies depending on occupancy and budget factors. However, throughout the course of a year, all guestrooms should receive attention. If the hotel cannot afford to deep clean all rooms each year, it is possible to implement deep cleaning on a two- or three-year cycle. If this approach is used, special care must be taken to ensure that typical room cleaning tasks meet or exceed the property's standards.

Deep cleaning by block presents a perfect time for housekeeping and maintenance to work closely together. The executive housekeeper and chief engineer can inspect rooms scheduled for deep cleaning and identify special maintenance and cleaning needs. Deep cleaning is also a good time to wash sheers and dry clean drapes. When doing so, label window coverings with the room number and date. Since deep cleaning can be very expensive, both housekeeping and the engineering and maintenance department should schedule the expense into their budgets.

Turndown Service and Special Requests

As the name implies, **turndown service** involves turning down the guest bed and freshening the guestroom for the evening. Some properties—particularly luxury hotels and resorts—have a shift of room attendants whose primary duty is to provide turndown service. This second shift is generally smaller than the day shift and services more rooms per hour. In some hotels, an employee on the turndown shift might service close to 20 rooms per hour, depending on the tasks involved.

Procedures for turndown service include:

- Cleaning the bathroom and restocking it with fresh towels
- Rotating or restocking amenities
- Tidying the guestroom
- Emptying wastebaskets
- Folding back the bedspread, blanket, and top sheet
- Fluffing the pillow
- Drawing the drapes

As an added touch, some properties have the room attendant leave a fresh blossom or a chocolate mint on the pillow to wish the guest "sweet dreams."

In addition to turndown service, housekeeping may be called upon to provide other types of special amenities. These items vary from property to property depending on the markets the operation attempts to reach and satisfy. There are several categories of amenities ranging from conveniences and services to luxuries. Among the items stocked and distributed by some housekeeping departments are hair dryers, irons and ironing boards, sewing kits, spot removers, playing cards, chess sets, backgammon tables, and other conveniences to make the guest's stay more pleasant.

To a great extent, the success of a property depends on the cleanliness, appearance, and ambience of its rooms. Maintaining standards for guestroom cleanliness is accomplished through the meticulous cleaning systems of housekeeping personnel. The procedures which follow provide a framework for the various tasks involved in guestroom cleaning.

Endnotes

1. Robert Propst, *The New Back-of-the-House, Running the Smart Hotel* (Redmond, Wash.: The Propst Company, 1988).

Key Terms

amenity	mitering
bar code	room attendant's cart
deep cleaning	room inspection
hand caddy	room status report
linen room	turndown service

Discussion Questions

1. What are the advantages of taking a systematic approach to guestroom cleaning?

2. When stocking a cart, what items typically go on the bottom two shelves? On the top shelf? In the hand caddy?

3. After reviewing a room status report, what types of rooms should the room attendant clean first? Second?

4. How are rooms that refuse service (including those with a "Do Not Disturb" sign) typically handled?

5. What purposes underlie positioning the cart in front of the guestroom door when cleaning?

6. What is the first *major* task a room attendant will do after entering a guestroom? Why?

7. Why is it important to work from the top down when cleaning items and surfaces?

8. Why is it important to make a final check after cleaning the guestroom? What are some of the conditions that a room attendant should look for?

9. What is the purpose of guestroom inspection? What benefits can employees derive from the process?

10. What are three ways that deep cleaning can be scheduled at a property?

Procedures: Guestroom Cleaning

The procedures presented in this section are for illustrative purposes only and should not be construed as recommendations or standards. While these procedures are typical, readers should keep in mind that each property has its own procedures, equipment specifications, and policies regarding protective gear which are designed to fit individual needs.

Stocking the Room Attendant's Cart

Equipment

- room attendant's cart
- list of assigned rooms
- specifications for supplies
- supplies

Procedures

Step 1
Check list of assigned rooms.

Step 2
Refer to the list of room cleaning supplies specified by the property.

Step 3
Begin loading cart from the bottom up. Stock according to recommended quantities.

Step 4
Place mattress pads, sheets, and pillowcases on bottom shelf.

Step 5
Place bath mats, towels, face cloths, and washcloths on middle shelf.

Step 6
Place room supplies and amenities on top shelf.

Step 7
Stock hand caddy with cleaning supplies such as all-purpose cleaner, cloths and sponges, bowl brush, glass cleaner, and dusting solution. Position on top shelf.

Step 8
Position vacuum, broom, and other sweeping supplies on side of cart.

Entering the Guestroom

Equipment

- properly stocked room attendant's cart

Procedures

Step 1
Check the room status.

Step 2
Check for a "Do Not Disturb" sign. Do not knock if a sign is on the door.

Step 3
Announce presence. Knock firmly and say "Housekeeping." Do not use a key to knock on the door.

Step 4
Wait for a response. If you don't hear an answer, knock again and repeat "Housekeeping."

Step 5
Wait a second time for a response. If you still do not receive an answer, open the door slightly and repeat "Housekeeping."

Step 6
If the guest is asleep or in the bathroom, leave quietly and close the door.

Step 7
If the guest is awake but dressing, excuse yourself, leave, and close the door.

Step 8
If the guest answers your knock, ask when you may clean the room.

Step 9
If the room is unoccupied, position your cart in front of the door and leave the door open. Begin cleaning.

Step 10
If the guest returns while you are cleaning, offer to finish later. Ask to see the guest's room key to verify that the key and room numbers match.

Cleaning the Guestroom: Beginning Tasks

Equipment

- hand caddy stocked with cleaning supplies
- light bulbs
- clean ashtrays and matches
- clean water glasses
- wastebasket liners

Procedures

Step 1
Remove hand caddy from cart and carry into the room.

Step 2
Turn on all the lights. Replace any burned-out light bulbs.

Step 3
Open the draperies. Check drapery cords and hooks.

Step 4
Open windows if appropriate. Check for breakage and dirt.

Step 5
Check the air conditioning and heating unit for proper operation. Set according to property standards.

Step 6
Check the general room condition.

- Note any damaged or missing items.
- Notify your supervisor if anything of value is gone or if something needs repair.

Step 7
Remove and replace dirty ashtrays. Replenish matches.

Step 8
Remove and replace dirty glasses.

Step 9
Collect any food service trays and dishes.

- Set items neatly outside the door.
- Call room service for pickup.

Step 10
Empty the trash and replace wastebasket liners.

Step 11
Straighten newspapers and magazines in stayover rooms.

Step 12
For check-out rooms, report any personal belongings left behind by the guest to your supervisor.

Stripping the Bed

Equipment

- dirty clothes hamper located on room attendant's cart

Procedures

Step 1

Remove any clothing or personal items from the bed.

Step 2

Remove the bedspread and blanket and place them on a chair.

Step 3

Check the bedspread and blanket for stains, tears, or holes. Replace if necessary.

Step 4

Remove cases from pillows. Place the pillows aside with the bedspread and blanket.

Step 5

Remove the sheets.

Step 6

Put soiled linen into the dirty linen bag on the cart.

Step 7

Check under the bed for trash or guest items. Remove them to a convenient place.

Making the Bed

Equipment

- clean bed linen

Procedures

Step 1
Strip the bed.

Step 2
Check the mattress pad for stains and damage.

Step 3
Change the mattress pad if necessary.

- Lay a fresh pad on the bed.
- Unfold pad right-side up and spread it evenly over the center of the bed.
- Smooth out any wrinkles.

Step 4
Notify your supervisor if you note stains or damage to the mattress.

Step 5
Center the bottom sheet on the mattress so equal amounts of sheet hang over each side of the bed.

Step 6
Miter the bottom sheet at the upper left corner of the bed.

Step 7
Miter the bottom sheet at the lower left corner of the bed.

Step 8
Tuck in bottom sheet along the left side of the bed.

Step 9
Place the top sheet on the bed, wrong-side up. Place the blanket on top of the sheet.

Step 10
At the head of the bed, turn the top sheet over the blanket about 6 to 8 inches.

Step 11
Miter the top sheet and blanket at the lower left corner of the bed.

Step 12
Tuck in the top sheet and blanket along the left side of the bed.

Step 13
Walk to the other side of the bed.

Step 14
Miter the bottom sheet at the lower right corner of the bed.

Step 15
Miter the top sheet and blanket at the lower right corner of the bed.

Step 16
Miter the bottom sheet at the upper right corner of the bed.

Step 17
Tuck in bottom sheet along the right side of the bed.

Step 18
Turn the top sheet over the blanket as in Step 10.

Step 19
Make sure the blanket and sheet are tucked in neatly along the sides and foot of the bed.

Step 20
Center the bedspread. Make sure the seams and pattern of the spread are straight.

Step 21
Fold the bedspread down from the head, leaving enough room to cover the pillows.

Step 22
Slip the cases over the pillows.

Step 23
Place the pillows at the head of the bed and bring the bedspread over them. Tuck in the bedspread beneath the pillows.

Step 24
Take a moment to check the bed for smoothness both up close and from a distance. Smooth out any wrinkles.

Dusting

Equipment

- clean dust cloths
- dusting solution
- glass cleaner
- spray disinfectant (optional)

Procedures

Step 1

Using a cloth sprayed with dusting solution, dust items located on walls or high off the floor. Work clockwise around the room. Items include:

- Headboard
- Picture frames
- Lamps, shades, and light bulbs
- Any air conditioning or heating units you can reach

Step 2

Dust and polish mirrors. Spray glass cleaner on a clean cloth and wipe down the mirror.

Step 3

Check the windows carefully. Clean with glass cleaner if necessary.

Step 4

Dust the windowsill.

Step 5

Dust and polish the dresser. Open the drawers and dust the inside surfaces.

Step 6

Dust the nightstand. Start with the top surface and work your way down the sides to the legs or base.

Step 7

Clean and dust the telephone. Check proper operation by picking up the receiver and listening for the dial tone. Use spray disinfectant on the mouthpiece and earphone (optional).

Step 8

Dust the top and sides of the television set and the stand it rests on.

Step 9

Clean the front of the television set with glass cleaner. Turn on the set to make sure it works properly, then turn it off.

Step 10

Dust any tables, beginning with top surface and working your way down to the base and legs.

Step 11

Dust wood or chrome surfaces on chairs, beginning at the top and working your way down the legs.

Step 12

Clean both sides of the connecting door to an adjoining guestroom, if applicable.

- Wipe from top down.
- Polish the knobs and remove any smudges around the knob area.
- When finished, make sure the door is closed and locked.

Step 13

Clean the closet

- Dust both the top and underside of the closet shelf. Remove any smudges on the surfaces.
- Wipe down the closet rod.
- Dust hangers and hooks.
- Clean and dust both sides of the closet door.

Step 14

Wipe down light switches and clean any smudges on surrounding wall area.

Step 15

Clean both sides of the guestroom door.

Step 16

Restock the room with guest supplies.

Cleaning the Bathroom

Equipment

- all-purpose cleaner
- clean cloths and sponges
- glass cleaner
- bowl brush
- clean bath towels, hand towels, washcloths, and bath mat

Procedures

Step 1

Turn on lights and fan. Replace any burned-out light bulbs. Check fan for proper operation.

Step 2

Remove used towels, washcloths, and bath mat.

Step 3

Empty trash and wipe container.

Step 4

Flush the toilet. Apply all-purpose cleaner around and under the lip of the bowl. Let it stand while you attend to other cleaning tasks.

Step 5

Clean the shower area.

- Check the shower head to make sure it is positioned correctly.
- Wash the tub or shower walls and soap dishes using a damp cloth and all-purpose cleaner. Check condition of walls as you clean.
- Rinse the tub or shower walls and soap dishes with sponge.
- Clean both sides of the shower curtain or shower door. Pay special attention to the bottom where mildew may accumulate. Wipe dry.
- Clean shower curtain rod or clean the tracks and frame of the shower door.
- Scrub the bathtub with all-purpose cleaner. Remove and clean the drain trap.

- Clean bathtub fixtures. Polish dry to remove water spots.
- Hang clean bath mat over edge of the tub.
- Reposition shower curtain or shower door to the center of the tub.

Step 6

Clean the vanity and sink area.

- Run some warm water into the sink. Add the correct amount of all-purpose cleaner.
- Clean the countertop area of the vanity.
- Clean the sink. Remove drain trap and clean.
- Clean sink fixtures. Polish dry to remove water spots.
- Wipe dry the countertop area of the vanity.
- Clean mirror with glass cleaner.

Step 7

Clean the toilet.

- Scrub the insides of the toilet and under the lip with the bowl brush. Flush.
- Using cleaning solution and a cloth, clean the top of the seat, the lid, the tank, and the outside of the bowl.
- Wipe dry all the outside surfaces.
- Close the lid.

Step 8

Clean bathroom walls and fixtures.

- Dust light fixtures.
- Using a clean damp cloth, spot-clean fingerprints and smudges.
- Wipe down electrical outlets and light switches, paying close attention to the surrounding wall area.

(continued)

Cleaning the Bathroom *(continued)*

- Wipe and polish towel bars.
- Dust all exposed piping.
- Clean both sides of the bathroom door.

Step 9

Restock bathroom supplies.

- Replenish the towels.
- Replenish guest amenities.
- Replenish toilet and facial tissue supplies.

Step 10

Clean the floor.

- Spray bathroom floor and baseboards with all-purpose cleaning solution.
- Starting with the farthest corner and working your way toward the door, scrub the floor and wipe baseboards.

Step 11

Make one final check. Visually scan all areas of the bathroom for areas you may have overlooked. Turn off the lights and the fan.

Vacuuming and Cleaning Baseboards

Equipment

- damp rag or cloth
- vacuum cleaner

Procedures

Step 1

Clean the baseboards. Begin in the closet area and work your way around the room. Wipe all exposed areas of the baseboard to remove surface dust and dirt.

Step 2

Take vacuum to the farthest corner in the guestroom. Begin vacuuming.

Step 3

Vacuum your way back to the door. Cover all exposed areas of the carpet you can reach including under tables and chairs, behind the door, and in the closet.

Step 4

Close windows and turn off lights along the way.

Checking the Guestroom

Equipment

- none

Procedures

Step 1
Check that all supplies and equipment are properly loaded back on the room attendant's cart.

Step 2
Stand by outside door. Visually scan the guestroom, beginning at one point in the room and working your way back to the beginning point.

Step 3
Attend to any cleaning task or item you may have overlooked.

Step 4
Smell the air. Spray air freshener if necessary.

Step 5
Make sure all lights are turned off.

Step 6
Leave the room and close the door. Check to be sure it is locked.

Step 7
Indicate status of room on room status sheet.

Guestroom Inspection: Bedroom

Equipment

- room inspection report
- pen or pencil

Procedures

Step 1

Check guestroom entrance door.

- Note any scratches, marks, smudges, or dust on surface.
- Check for "Do Not Disturb" sign on inside knob of door.
- Check proper operation of locks, chains, and door stops.

Step 2

Check condition and cleanliness of light switches, plates, and surrounding wall area.

Step 3

Scan ceiling, walls, woodwork, and trim for any damage, dirt, or dust.

Step 4

Check drapes for tears or stains. Check that hooks are in place and that traverse rods work correctly.

Step 5

Check window sills and windows for cleanliness and freedom from streaks. Make sure windows are locked and that locks work correctly.

Step 6

Make sure heating and air conditioning unit is free from dust and dirt, operates correctly, and that it is set according to property standards.

Step 7

Make sure telephone is clean and works correctly.

Step 8

Check the bed.

- Make sure that the bed has fresh linen.

- Check condition and appearance of bedspread; check edges for frays or tears.
- Check headboard for dust.
- Look under the bed for trash or guest items.

Step 9

Check room furniture for scratches, damage, and dust. Check tapestry for stains or tears.

Step 10

Check lamps for scratches, damage, and dust. Make sure light bulbs work and are the proper wattage.

Step 11

Turn on television set and radio to check for proper operation and reception. Turn off and check for scratches, damage, and dust.

Step 12

Check carpets and baseboards for dirt, stains, dust, and streaks.

Step 13

Check that closets are clean and have the proper amount of hangers.

Step 14

Check pictures and mirrors for dust and streaks.

Step 15

Check that bedroom amenities such as stationery and matches are properly stocked.

Step 16

Make a final check around the room to make sure that all items are well-positioned, and that all areas from ceiling to floor are clean and well-maintained.

Step 17

Complete a work order request and/or notify the appropriate department for any item needing attention or repair.

Guestroom Inspection: Bathroom

Equipment

- room inspection report
- pen or pencil

Procedures

Step 1
Check bathroom door for scratches, marks, smudges, or dust on surface.

Step 2
Check condition and cleanliness of light switches, plates, and surrounding wall area. Check vent fan for dirt and dust.

Step 3
Scan ceiling, walls, and tile for any damage, dirt, or dust.

Step 4
Check shower area.

- Check tub and fixtures for water marks, soap film, and hair.
- Check fixtures for correct position and operation. Make sure they do not leak.
- Check shower curtain for mildew and for proper position.
- Make sure a clean bath mat is in place.
- Check caulking between tub and tile for cracks or dirt.

Step 5
Inspect vanity and sink area.

- Check sink and counter area for water marks, soap film, and hair.
- Check mirror for streaks and spots.
- Make sure fixtures operate correctly and do not leak.

Step 6
Check toilet for cleanliness. Flush to check proper operation.

Step 7
Check floor and baseboards for dirt and dust.

Step 8
Make sure towels, face cloths, and washcloths are clean and neatly arranged on towel racks.

Step 9
Check toilet and facial tissue supply.

Step 10
Check that bathroom amenities such as soap, shampoo, and mouthwash are properly stocked.

Step 11
Make a final check of the bathroom to make sure all items are well-positioned, and that all areas from ceiling to floor are clean and well-maintained.

Step 12
Complete a work order request and/or notify the appropriate department for any item needing attention or repair.

Turndown Service

Equipment

- properly stocked room attendant cart
- special turndown amenities

Procedures

Step 1

See procedure for entering the guestroom. When announcing your presence, substitute "Turndown Service" for "Housekeeping."

Step 2

Remove any guest items from the bed. Set neatly aside on the dresser or a chair.

Step 3

Pull back the bedspread so 15 to 18 inches hang over the foot of the bed. Bring this slack part of the spread back over the fold so the fabric faces right-side up.

Step 4

Pull back the sheets.

- For a bed sleeping one guest, turn down the sheets on one side only, usually the side near the night stand or phone.
- For a bed sleeping two, turn down the sheets on both sides.

Step 5

Place the amenity on the pillow. For beds sleeping two, be sure to leave amenity on both pillows.

Step 6

Remove and replace dirty ashtrays. Replenish matches.

Step 7

Remove and replace dirty glasses.

Step 8

Collect any food service trays and dishes.

- Set items neatly outside the door.
- Call room service for pickup.

Step 9

Empty the trash and replace wastebasket liners.

Step 10

Straighten newspapers and magazines.

Step 11

Remove dirty linen in bathroom. Restock with fresh linen.

Step 12

Straighten and wipe down vanity area. Dry and polish fixtures.

Step 13

Straighten and wipe down tub area if necessary. Dry and polish fixtures.

Step 14

Check toilet and facial tissue supply. Replenish if necessary.

Step 15

Close the drapes.

Step 16

Turn on bedside lamp.

Step 17

Turn radio to recommended easy listening station. Adjust to a low volume.

Step 18

Visually scan the guestroom, beginning at one point in the room and working your way back to the beginning point. Attend to any turndown task you may have overlooked.

Step 19

Leave the room and close the door. Check to be sure it is locked.

REVIEW QUIZ

When you feel you have covered all of the material in this chapter, answer these questions. Choose the *best* answer.

True (T) or False (F)

T   F   1.   A room attendant's cart is generally spacious enough to carry all the supplies needed for a full day's room assignments.

T   F   2.   Drinking glasses, fresh soap bars, and facial tissue are stocked in the room attendant's hand caddy.

T   F   3.   A stayover is a room from which a guest will check out later in the day.

T   F   4.   The quantity of rooms assigned to room attendants for cleaning is based on the room status report.

T   F   5.   When a guest refuses service, a floor supervisor should call the guest and arrange a convenient time for cleaning the room.

Alternate/Multiple Choice

6.   When stocking a room attendant's cart, linens typically go on the:

   a.   top shelf.
   b.   bottom shelves.

7.   Assignments for room attendants are based on the:

   a.   room inspection report.
   b.   room status report.

8.   Guestrooms that room attendants usually clean first are:

   a.   check-outs.
   b.   due-outs.
   c.   layovers.
   d.   stayovers.

9.   Guestrooms that room attendants usually clean last are:

   a.   check-outs.
   b.   due-outs.
   c.   layovers.
   d.   stayovers.

10.   Drawing drapes, fluffing pillows, and folding back bedspreads are procedures for:

   a.   guestroom final check.
   b.   guestroom inspection.
   c.   routine guestroom cleaning.
   d.   turndown service.

Chapter Outline

Front-of-the-House Areas
    Entrances
    Lobbies
    Front Desk
    Corridors
    Elevators
    Public Restrooms
    Swimming Pool Areas
    Exercise Rooms
Other Functional Areas
    Dining Rooms
    Banquet and Meeting Rooms
    Administration and Sales Offices
    Employee Areas
    Housekeeping Areas
Special Projects

Learning Objectives

1. Explain the importance of maintaining spotless and well-kept public areas.

2. Identify factors that affect how frequently entrances and lobbies are cleaned.

3. Describe housekeeping's typical cleaning responsibilities at the front desk area.

4. Identify the tasks, in addition to maintaining carpets and floors, that housekeeping staff perform when cleaning corridors.

5. Identify the tasks involved in cleaning ice/vending machine areas.

6. Explain problem cleaning areas in elevators.

7. State how frequently public restrooms should be cleaned.

8. Describe typical housekeeping duties in pool areas and exercise rooms.

9. Identify housekeeping's typical cleaning responsibilities in relation to dining rooms, banquet and meeting rooms, and administration and sales offices.

10. Explain the importance of maintaining clean employee areas and housekeeping areas.

Public Area and Other Types of Cleaning

Most people—including guests—trust first impressions. In a hotel, a guest's first impression often revolves around what he/she sees and experiences in the property's public areas.

Public areas consist of a property's entrances, lobbies, corridors, elevators, restrooms, and health facilities. Other areas which the guest sees include dining areas, banquet and meeting rooms, and sometimes, administration and sales offices. In some properties, public areas are engineered to convey a particular mood through such dramatic features as high ceilings, plant-laden balconies, mezzanines, decorative fabric panels, textured walls and floors, and ornate furniture and fixtures. But all levels of architecture and design aside, nothing can make or break an impression more than the cleanliness and condition of a property's public areas.

The condition of **public**—or **front-of-the-house**—**areas** makes a strong statement for the rest of the property. Spotless and well-kept public areas signal guests to expect the same level of care and attention in their guestrooms. It shows, too, that the property most likely maintains the same standards of cleanliness for its "employees only" areas—or for its rooms and corridors in the back of the house. To a large extent, the responsibility for cleaning public and other functional areas rests with the housekeeping department.

Front-of-the-House Areas

Establishing and maintaining housekeeping procedures in public areas is just as important as it is in guestrooms, but much less standardized. The housekeeping needs of public areas vary considerably from property to property because of architectural differences, lobby space allocations, activities, and guest traffic. These and other factors also affect scheduling routines, requiring many of the cleaning tasks to be performed at night or on a special project basis. Among the typical front-of-the-house areas that need daily—if not hourly—housekeeping attention are entrances, lobbies (including the front desk), corridors, elevators, public restrooms, swimming pools, and exercise rooms.

Entrances

Hotel entrances demand stringent attention since they are among the most heavily trafficked areas in a property. Entrances must be kept clean both for aesthetic and for safety reasons.

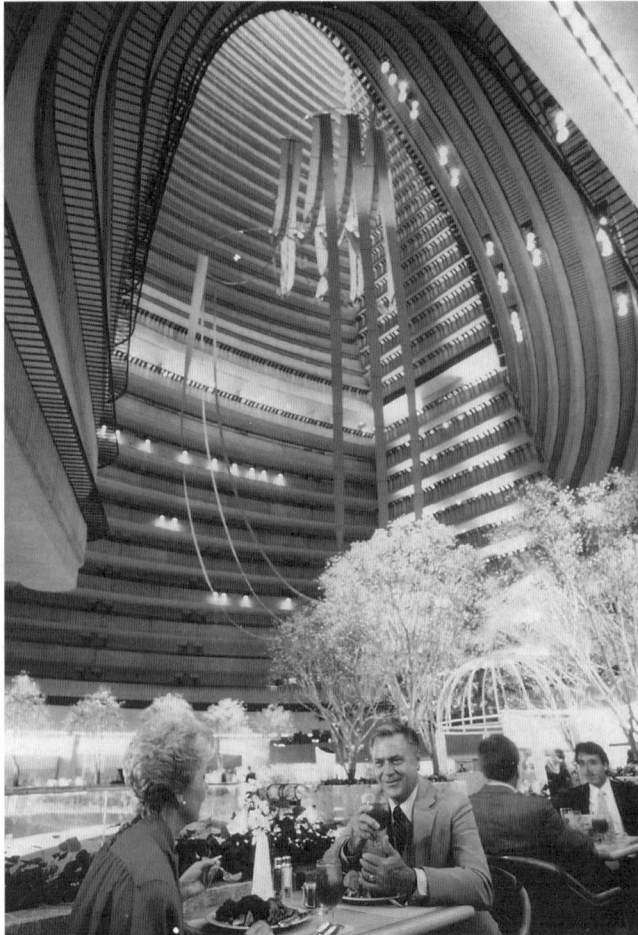

Unique architectural design presents special cleaning considerations for housekeeping. (Courtesy of Marriott Marquis, Atlanta, Georgia)

The frequency of cleaning hotel entrances is largely contingent on the weather. Rain and snow will require that these areas receive attention more often than days when the sun is shining. Salt and mud tracked in during the winter and spring also contribute to the deterioration of floor coverings and surfaces—particularly carpets.

Mattings or runners placed near entrances can alleviate some of the headaches involved in keeping these areas free of puddles, footprints, and outside dirt—and protect guests from slips and falls. If properly positioned, these floor coverings can also protect inside carpet. Inclement weather will demand that runners and mattings be frequently mopped or changed. During all types of weather, an attendant should be assigned to see that the runners and mattings are laying flat, and that entrances are mopped and tidied throughout the day. Attendants should also frequently clean fingerprints and smudges from door surfaces—particularly glass areas. Even one set of fingerprints can spoil the appearance of an otherwise clean entranceway. Thorough cleaning of door surfaces, including door tracks, is conducted very early in the morning to avoid inconveniencing guests.

Lobbies

Lobbies require continual cleaning both because they are heavy traffic areas and because they are the "gateway to the hotel." Many lobbies are a hub of activity where guests check in, socialize, relax, or in the case of some properties, window-shop at special-interest or novelty stores.

Since lobbies are such dynamic areas, properties generally schedule cleaning for the late night and early morning hours—meaning from 10:30 P.M. to 7:00 A.M. In this way, guests incur minimal inconvenience—and the staff can clean with little or no interference. Some cleaning tasks, however, must be performed during daylight hours to maintain lobby aesthetics. These tasks include emptying ashtrays and sand urns, attending to heavily trafficked floor areas, and straightening furniture.

Generally speaking, cleaning duties in the lobby can be performed once every hour, once every 24 hours, or once a week. In most properties, there are no hard or fast rules for how frequently a certain task should be performed. Some properties assign a lobby attendant to patrol the area and attend to cleaning activities as necessary. Activities usually assigned on an hourly or daily basis include:

- Emptying and cleaning ashtrays
- Emptying and wiping down wastebaskets
- Cleaning glass and window areas
- Wiping and dusting lobby telephones
- Polishing drinking fountains
- Polishing railings
- Removing fingerprints or spots from walls
- Dusting furniture and table fixtures
- Polishing doorknobs and wiping surrounding areas
- Dusting and cleaning doorjambs and tracks
- Vacuuming carpet
- Sweeping tile or hardwood floor areas
- Straightening furniture

Among the tasks typically assigned to public area attendants on a weekly basis are:

- Polishing wooden furniture
- Vacuuming upholstered furniture

- Vacuuming or cleaning drapes or window coverings

- Cleaning window sills

- Dusting ceiling vents

- Dusting in high or hard-to-reach areas

- Cleaning carpet edges and baseboards

Some public area cleaning involves using a stepladder to reach high spots on walls, lighting fixtures, or decorative wall pieces. In all instances, employees should follow prescribed safety procedures when performing these cleaning tasks. Properties with unique architecture and interior design may hire outside contractors to attend to special cleaning needs on a periodic basis.

Front Desk

Like lobby cleaning, front desk cleaning must be scheduled during non-peak hours to avoid interrupting the flow of business. The front desk area should be cleaned with the same degree of attention as the lobby since it is such a pivotal area for shaping the guest experience. Although this area is technically part of the lobby, the front desk has its own set of cleaning needs and peculiarities.

Front desks vary in design. Some properties incorporate simple, straightforward designs into their decor while others opt for more elaborate fluted, curved, or grooved surfaces. The latter will generally take longer to clean than a more basic design since they may require special cleaning techniques and equipment. Whether rectangular or round, smooth or textured, every front desk needs the careful attention of the public area attendant to maintain a spotless appearance.

Vacuuming and emptying wastebaskets and ashtrays should be done both in front of and behind the desk. In some properties, housekeeping employees will also be responsible for wiping down and polishing front desk surfaces. Special care should be taken to remove fingerprints, smudges, and shoe and scuff marks, especially near the base of the desk. Any nicks, scratches, or other surface damage should be reported to the appropriate housekeeping supervisor. In all cases, public area attendants must never move any papers or other business-related items, or touch or unplug equipment in the front desk area.

Corridors

Other sections of the hotel which most guests see before stepping foot into the guestroom are the public corridors or halls. In some properties, corridors are considered "guest space" and therefore, become an extension of guestroom cleaning tasks.

A big part of corridor cleaning involves attending to the floor. In most cases, floors are covered with durable, attractive carpet designed for easy care and maintenance. Floors should be vacuumed at least once a day based on guest traffic and occupancy. Carpet shampooing is generally scheduled on a special project basis—most likely during an off-season or a low-occupancy period.

Many properties recommend that, when cleaning baseboards, the attendant begin at one point in the corridor, work his/her way completely down one side of

A circular front desk is among the variations in front desk design that a public area attendant may encounter. (Courtesy of Ramada Inn, Ithaca, New York)

the hall, and then work back down the other side to his/her starting point. The attendant should also wipe down any trim around guestroom doors, paying special attention to fingerprints and smudges. Light fixtures should be dusted and any burned-out light bulbs replaced. Air supply vents and sprinklers should also be dusted and checked for proper functioning. Attendants should note the condition of emergency exit lights and report any damages to his/her supervisor.

Spot-cleaning the walls for smudges and fingerprints can be done much like baseboard cleaning, with the attendant working his/her way down alternate sides of the corridor. As a final step, the attendant should clean the front and back of the exit door, wipe any dirt and dust from the tracks, and check to see that it opens and closes properly.

Ice/Vending Machine Areas. Some properties locate ice and vending machines in easy-to-find or low traffic areas of guest corridors. These devices—and their surrounding areas—will require the attention of a public area attendant to ensure cleanliness and proper functioning.

Ice machines should be checked and cleaned daily. Before cleaning, the attendant should make sure the unit functions correctly by checking the dispenser mechanism. Any clogs or other malfunction which cannot be readily corrected should be noted and reported. The attendant should also check for any water on the floor. Sometimes when ice machines malfunction, melted ice leaks out and can cause floor damage. When cleaning, the attendant should remove any unmelted ice from the dispenser area and wipe the area dry. Exterior surfaces should be cleaned with the recommended cleaning solution and wiped dry. Particular attention should be paid to handles where fingerprints and smudges occur. Sometimes, housekeeping staff clean and sanitize actual mechanisms. If so, procedures should be designed according to the manufacturer's recommendations.

Vending machines can be either leased or owned by the hospitality operation. In either case, the public area attendant will usually be responsible for cleaning and dusting the exterior surfaces and checking mechanical functions. If the machine is property-owned, the attendant should make a note of any items which need to be restocked and report them to his/her supervisor. Attendants may also be required to clean areas underneath and behind the machine. Such tasks are usually scheduled on a special project basis since they require heavy lifting and moving.

Elevators

Elevators require frequent cleaning because of their volume of use. Like the lobby and front desk area, the best time to clean elevators is late at night or very early in the morning to avoid high traffic periods.

Depending on interior design features, elevator surfaces may consist of carpet, vinyl, wallpaper, glass, mirrors, or a combination of materials. Each surface will require the public area attendant to follow a specific set of cleaning procedures to achieve the most effective results. In most instances, the attendant should clean from the top down to avoid resoiling areas already cleaned.

Among the problem cleaning areas in elevators are the sides near the floor. These areas are prime targets for scuff marks, scratches, or tears. The latter conditions should be noted and reported to the appropriate supervisor. Hand railings should be polished to remove fingerprints, as should elevator controls and surrounding wall areas. When cleaning glass and mirrors, the attendant should step back and check the surface for streaks. Both the inside and outside of the elevator door should be wiped down, including the door tracks where dirt and dust collect.

Floor carpets are perhaps the most difficult to keep clean in elevators because of concentrated wear and tear. Some properties use standard vacuums to sweep elevator carpets while others supply attendants with high-powered portable canister vacuums. In all instances, vacuuming should be completed quickly to lessen the time the elevator is out of service. Some properties carpet elevator floors with removable floor coverings. This way, the carpet can be removed for washing instead of being shampooed in place. A matching piece of carpet, of course, is carefully laid and secured while the removable piece is being serviced.

Public Restrooms

Describing housekeeping procedures for public restrooms is easier than for some public areas simply because public restrooms are relatively constant from property to property. Some properties evoke a special atmosphere by decorating with ornate fixtures and mirrors, allocating lounge space filled with upholstered furniture and plants, and by providing conveniences such as hand blow-dryers and changing tables.

Any special feature or convenience will require the property to devise an efficient cleaning procedure. To a large degree, the structure and size of a public restroom is determined by the service level of the property. Despite the features, the goal in cleaning any public washroom is to maintain a sanitary, safe, and attractive atmosphere for the visiting guest. This section will cover the most basic elements in public restroom cleaning.

At the very minimum, public restrooms should be cleaned twice daily: once in the morning and once in the evening. In some properties, more frequent cleanings are required to maintain a pleasant environment and to ensure proper sanitation and safety levels. Sometimes, these additional cleanings consist of "touch-up" cleanings every one or two hours—depending on the traffic.

The equipment needed for cleaning public restrooms is basically the same as that required for cleaning guest bathrooms: an approved all-purpose cleaner for bathroom surfaces; bowl brush; cloths and sponges; glass and mirror cleaner; rubber gloves; and protective eye covering. Some properties also use an odorless disinfectant. Attendants will also need a bucket, mop, and floor cleaning chemicals.

Before entering the restroom, the attendant should check to see that the facility is vacant. If the attendant is cleaning a bathroom for members of the opposite sex, he/she should knock on the door, announce "housekeeping," and wait for a response. Generally, it is assumed safe to enter if a response is not heard after three announcements. When entering, the attendant should prop the door open and place an approved floor sign at the entrance which indicates the restroom is being cleaned.

Many properties recommend applying the appropriate cleaner to toilets and urinals before attending to any other task. This way, the cleaner has time to work while other areas are being cleaned. After applying the cleaner, trash containers should be emptied, wiped, and relined with new bags. Any ashtrays—both floor urns and individual stall units—should be emptied and cleaned.

Basins and surrounding countertop areas should be cleaned next. Most attendants begin by dispensing the recommended amount of cleaner into the sink, adding warm water, and wiping the area clean with a sponge or cloth. In some properties, room attendants dispense a pre-mixed solution onto surfaces using a trigger sprayer. Drain traps should be removed and checked for hair and other debris; faucets should be checked for drips. If there is any exposed piping, attendants should ensure the pipes are free of dust, dirt, and leaks. When cleaning the countertop area, the attendant should check for stains and damage as he/she wipes down the area. Fixtures should be wiped with cleaning solution and a damp rag and then polished. All sinks and surrounding countertops should be wiped dry with a clean cloth. Finally, mirrors should be wiped down with the recommended cleaning solution. Careful attention should be paid to leaving the mirror free of streaks and water spots.

Toilets and urinals should be cleaned using the bowl brush and a clean rag for each fixture. After swabbing the entire bowl or urinal, the attendant should flush the unit several times while rinsing the bowl brush in fresh water. Outer areas of the fixture should be wiped with a damp cloth or sponge working from the top to the base. As a final step, handles should be wiped and polished.

Partitions are most efficiently cleaned using a spray bottle to dispense the cleaning solution and a damp cloth or sponge. Most attendants clean each stall panel separately, beginning in the right top corner and working across and down with wide sweeping movements. Special attention should be paid to areas around door locks and paper dispensers to remove any smudges or fingerprints. An attendant should *never* stand on a toilet to clean high or hard-to-reach areas of partitions. Special cleaning tools with long or extendable handles are available for this purpose. After wiping surfaces clean, the attendant should look for scratches or other surface marks and make a note to report them to his/her supervisor.

Once partitions are clean, the attendant can begin to wipe down walls. Cleaning solutions will vary depending on whether surfaces are tiled, painted, or paneled. In each case, the attendant should work systematically, beginning at one point and cleaning his/her way back to the start. Any scuffs, nicks, or other damage to wall surfaces should be noted.

At this point, dispensers for toilet seat covers, toilet paper, tissue, paper towels, and soap should be restocked. A wet cleaning of the dispenser is usually performed when it is empty to avoid damaging or soiling the paper products. In most instances, a light dusting or simple polishing is appropriate to remove any surface marks or smudges.

Among the last steps in cleaning public washrooms is sweeping and mopping the floor. Beginning in the farthest corner, the attendant should sweep all exposed floor areas—which includes running the broom along the baseboards. Doing so prepares the floor more fully for wet cleaning and removes loose dust and dirt that can muddy mop water. When mopping, it is important to use clean warm water and the appropriate amount of cleaning solution. Solution can be mixed in the water or applied directly onto the floor. As with sweeping, the attendant should begin mopping in the farthest corner and work his/her way back toward the entrance. Mopping around fixtures should be approached slowly and carefully to avoid splashing dirty water onto clean surfaces.

A second bucket of clear hot water should be prepared for rinsing. The attendant should rinse the mop, wring out the excess water, and mop the surfaces a second time. During this stage, it is important to rinse and wring the mop out frequently to ensure that the floor is thoroughly cleansed. Frequently wringing out the mop also prevents excess water on the floor since it functions, in a sense, as a "dry mop."

When the task is completed, the public area attendant should reassemble his/her supplies and make one final check. Any unusual or offensive odors could

Swimming pools are a popular attraction in many types of properties. (Courtesy of Best Western Midway Motor Lodge, Grand Rapids, Michigan)

indicate an improperly functioning ventilation system. If the floor is still wet from mopping, an appropriate warning sign should be in place to alert guests until the floor is dry. As always, any condition requiring maintenance should be noted and reported to the appropriate supervisor.

Swimming Pool Areas

Swimming is perhaps the most popular of all recreational sports. Many properties—particularly resorts—cater to this interest by providing swimming facilities.

Pools can be either indoor or outdoor. Pool designs are as varied as hotel operations and range from the very basic to very elaborate settings. Some pool areas include whirlpools and saunas. In most hotels, the daily care and maintenance of the pool, sauna, or whirlpool is the responsibility of the engineering and maintenance department. However, the housekeeping staff does "make the rounds" to attend to specific tasks within the pool area.

Among the duties usually appointed to housekeeping staff in pool areas are:

- Collecting wet towels and dirty linen
- Restocking towels and linen
- Emptying and cleaning trash receptacles
- Emptying and cleaning ashtrays

Many guests appreciate the health clubs and fitness centers offered by some commercial hotels, resorts, and conference centers. (Courtesy of Stouffer Esmeralda Resort, Indian Wells, California)

- Cleaning wall areas
- Sweeping and mopping hard floor surfaces
- Caring for any carpeted areas
- Washing window or glass areas
- Cleaning and straightening lounge furniture

As with any public area, unsafe, unsanitary, or damaged conditions should be noted and reported to management staff.

Exercise Rooms

The rising interest in physical fitness and health has produced a long-term demand for health products and services. In response to this trend, many properties have made health facilities a part of their overall package.

Housekeeping's part in servicing these facilities will be largely determined by the size and scope of these areas and the equipment involved. Health and fitness services can range from simply a pool and a sauna to providing fully outfitted gymnasiums with a trained staff. Equipment for an exercise room might include universal gyms, stationary bicycles, rowing machines, floor mats, barbells, and dumbbells. The design of these facilities often incorporates special flooring or hardwood surfaces, mirrors, and special light fixtures. On a more elaborate scale, locker room and shower areas are provided.

The responsibility for maintaining the proper functioning of exercise equipment typically rests with the hotel's engineering staff. Housekeeping personnel, however, will play a role in ensuring that these facilities meet the same standards of cleanliness that the guest enjoys in other public areas. For the most part, an attendant will be assigned on a daily basis to perform such tasks as:

- Dusting equipment
- Cleaning mirrors and glass areas
- Sweeping and mopping floors
- Removing soiled linen
- Restocking clean linen
- Cleaning and straightening any furniture
- Dusting light fixtures
- Spot-cleaning walls

For safety reasons, it is extremely important for attendants to note the general condition of the equipment and report any suspected malfunctions to their supervisor. Attendants may also be responsible for cleaning shower and locker areas and replenishing appropriate guest amenities.

Other Functional Areas

In addition to typical public areas such as lobbies and restrooms, housekeeping staff may also be responsible for cleaning other functional areas in a property. In some of these areas, housekeeping staff will have limited responsibilities; in others, cleaning activities will be as elaborate as the area's counterpart in the front of the house. Included among these spaces are property dining rooms, banquet and meeting rooms, administration and sales offices, employee areas, and housekeeping office and work areas.

Dining Rooms

Cleanliness in dining rooms is important not only for image, but for safety and sanitation reasons. For the most part, the dining room staff will be responsible for keeping the area presentable during operating hours. This includes cleaning tables, changing linens, attending to on-the-spot spills, and light vacuuming or sweeping. In most properties, housekeeping comes in on a nightly, weekly, or monthly basis to assist with more thorough cleaning tasks.

Vacuuming is perhaps the largest task which housekeeping "piggybacks" with an operation's food service outlet. As in other front-of-the-house areas, vacuuming is performed at night or "after hours" to avoid disturbing the guest. To achieve the best results, the attendant should move all the chairs away from a table before vacuuming beneath it. If any food spots are noticed, the attendant should follow the recommended procedure for spot-cleaning the carpet. Another major housekeeping responsibility may involve collecting dirty table linen and replenishing clean linen supplies on a daily basis. Other scheduled housekeeping tasks might include:

- Cleaning phones
- Wiping down the hostess station
- Spot-cleaning walls
- Wiping windowsills
- Dusting and polishing furniture
- Cleaning upholstery
- Attending to light fixtures

Many properties establish schedules that designate when certain cleaning tasks should be performed.

Banquet and Meeting Rooms

Like cleaning dining areas, banquet and meeting rooms are often cleaned by another department—in this case, the banquet or convention service staff—with housekeeping's assistance. All meeting rooms should be cleaned immediately after a function. Nothing is more unappetizing to passing guests or prospective clients than viewing a meeting room that displays the dredges from the previous night's festivities. Furthermore, stains that are allowed to set into carpeting or other furnishings can become nearly impossible to remove the next day.

In some properties, housekeeping staff will be responsible for cleaning chairs, tables, furniture, and wall and floor areas after all special meeting and food service items have been removed. When cleaning furnishings, special attention should be given to removing food particles and stains from upholstered surfaces. Attendants should note the condition of each piece of furniture and report any damage or stains to the appropriate supervisor. Carpet care involves thorough vacuuming, stain removal, and shampooing on a frequent basis. In addition, attendants may have special cleaning tasks depending on the design and function of the room. Sometimes, too, the services of outside contractors may be needed—particularly in the case of high ceilings and ornate lighting fixtures such as chandeliers.

Administration and Sales Offices

The blueprint of most properties includes office areas for conducting various administrative and sales activities. Depending on size and service level, the operation may contain administrative offices for the management of all divisions (such as human resources, rooms, food and beverage, marketing and sales, etc.)—plus areas for supporting staff. Although technically considered back-of-the-house areas, administrative and sales offices periodically host important interactions between property staff and outside clients, vendors, business associates, and prospective employees.

To ensure that these areas look their best at all times, the housekeeping staff is called in to maintain their overall cleanliness and appearance. The scope of the tasks will vary from property to property. For the most part, attendants will dust, empty wastebaskets, spot-clean wall areas, and sweep or vacuum on a nightly basis. Other tasks such as washing windows are scheduled on a weekly or monthly basis. Deep cleaning tasks which require moving furniture are typically scheduled during low-volume business periods to minimize disruption. While doing any cleaning task in these areas, housekeeping attendants should avoid moving or rearranging any items on desks or work surfaces—particularly business papers and folders.

Employee Areas

Employee areas in the back of the house often constitute as much space—if not more—as front-of-the-house areas. Although technically "closed" to the guest, these areas deserve the same level of cleaning attention as do areas in the public eye. A property that ensures that employees have a safe, clean, and pleasant environment in which to work, eat, take breaks, and go to the bathroom will gain the respect and loyalty of those employees. In the long run, this will invariably show in the quality of products and services provided by employees.

Although each employee does his/her part in maintaining the neatness of the back-of-the-house areas, the responsibility for heavy-duty cleaning tasks rests with the housekeeping staff. Housekeeping attendants will see that employee areas are free of dirt, grime, and dust by adhering to a cleaning schedule that is as carefully devised as that for cleaning the front of the house. Typical areas for which housekeeping will be responsible include service corridors and elevators, employee dining areas, employee washrooms, loading docks, and storage areas. In properties with food service outlets, housekeeping may also be responsible for some limited tasks in kitchen areas such as maintaining floor, wall, and ceiling surfaces.

Back-of-the-house areas are generally designed to withstand more wear and tear than areas open to guests—which basically reflects their function. Floors may consist of such stain-resistant materials as painted concrete, tile, or flat, tightly woven carpets. Walls and ceilings, too, will usually be smooth in texture and void of complicated angles or hard-to-reach areas. Even though procedures for back-of-the-house cleaning resemble front-of-the-house practices, attendants may enjoy a less complicated routine since surfaces tend to be smoother and furniture and fixtures less ornate. Often, too, these areas lend themselves to cleaning with heavy duty equipment which can speed cleaning tasks.

Housekeeping Areas

Housekeeping areas should be spotless since they represent the headquarters of the property's professional cleaning crew. The size and composition of these areas is dependent on the size and service level of the property itself. Typically, housekeeping operations emanate from three basic areas: the housekeeping office, laundry, and linen room.

Like any administrative office, the housekeeping office should be cleaned and maintained to the same standards as a public area. Floors should be swept, mopped, and vacuumed; walls spot-cleaned; and wastebaskets cleaned and emptied. Baseboards should be free of dust and dirt; windows free of grime and streaks; and furniture free of dust.

The same applies to the laundry area. In addition, special care should be taken to keep the exterior surfaces of machines clean to prevent resoiling linen. The inside drums of washers, dryers, and extractors should be wiped down between cycles; buildup on lint filters and in detergent areas should also be removed. Distinct areas should be maintained for processing dirty and clean laundry items. Folding tables, irons, clips, hangers, and other laundry aids should also be frequently wiped and cleaned to prevent the contamination of clean linens. Finally, all shelves and storage areas should be periodically wiped and dusted. Any laundry supplies should be straightened and tidied for easy access and inventory purposes.

Linen rooms also contain rows of supply shelves which require frequent cleaning and dusting. It is extremely important to keep these areas well-organized—both for efficiency and for maintaining accurate inventories. Cleanliness, too, is critical since the property's laundered sheets, pillowcases, blankets, towels and washcloths, and food service linens are stored in this area. In addition to cleaning floors and walls, routine tasks in the linen room may include dusting shelves, straightening supplies, and positioning room attendant carts for ease in restocking.

Special Projects

Depending on the property's design and architectural features, housekeeping may be assigned projects that involve complicated cleaning techniques, special equipment, and team efforts. Special projects might include cleaning certain weaves and types of carpet, ornate light fixtures such as chandeliers, staircases and handrails, interior fountains, windows and window coverings, and other decorative features such as wall hangings. Most special cleaning projects require a great deal of expertise and planning in terms of assembling supplies and equipment.

The procedures section of this chapter provides sample guidelines for cleaning the more standard public areas.

Key Terms

back-of-the-house areas
frequency schedule
front-of-the-house areas
public areas
night cleaning

Discussion Questions

1. What message does a spotless and well-kept public area convey to guests?

2. Why do the housekeeping needs of public areas vary considerably from property to property?

3. Name five cleaning activities which are generally performed every 24 hours in a lobby area.

4. What is the minimum number of times that public washrooms should be cleaned each day?

5. What is the first cleaning task many properties recommend that the public area attendant do after he/she enters a public washroom? Why?

6. Name five duties usually assigned to housekeeping staff in swimming pool areas.

7. In hotels with health facilities, which department generally has the responsibility for maintaining the proper functioning of exercise equipment?

8. Name three cleaning tasks which housekeeping may perform in dining areas.

9. Why is it important to maintain the same standards of cleanliness for back-of-the-house areas as it is for front-of-the-house areas?

10. What are some of the special tasks involved in cleaning housekeeping areas? Why are they important?

Procedures:
Public Area and Other
Types of Cleaning

The procedures presented in this section are for illustrative purposes only and should not be construed as recommendations or standards. While these procedures are typical, readers should keep in mind that each property has its own procedures, equipment specifications, and policies regarding protective gear which are designed to fit individual needs.

Cleaning the Hotel Entrance

Position: Public Area Attendant

Reports to: Shift Supervisor

Equipment

- broom and dustpan
- mop and bucket
- all-purpose cleaner
- floor cleaner
- glass cleaner
- cloths and sponges

Procedures

Step 1

Swab any excess water from the floor using a dry mop, rag, or sponge.

Step 2

Sweep floor area, including mattings or runners.

Step 3

Mop floor area, including mattings or runners, if appropriate.

Step 4

Clean glass areas of doors on both sides, working from the top down.

Step 5

Clean non-glass areas of doors. Pay particular attention to fingerprints and smudges around handles and knob areas.

Step 6

Polish knobs or handles.

Step 7

Clean door tracks.

Step 8

Make sure all mats and runners are laying straight and flat. As a final check, open each door to ensure proper clearance along the bottom.

Cleaning the Lobby—Nightly Activities

Position: Public Area Attendant

Reports to: Shift Supervisor

Equipment

- clean ashtrays
- glass cleaner
- broom and dustpan
- mop and bucket
- all-purpose cleaner
- floor cleaner
- cloths and sponges
- dusting solution
- vacuum

Procedures

Step 1

Remove and replace dirty ashtrays.

Step 2

Pick up loose papers and trash. Report any items left behind by a guest to your supervisor.

Step 3

Empty trash containers. Replace wastebasket liners.

Step 4

Clean glass and window areas, including any glass tabletops.

Step 5

Dust furniture, fixtures, and lobby telephones.

Step 6

Polish drinking fountains. Wipe drinking area dry and polish operating buttons or knobs.

Step 7

Spot-clean walls and wall fixtures. Dust top and sides of any picture frames.

Step 8

Dust or polish hand railings.

Step 9

Clean hardwood or tile floor areas.

Step 10

Vacuum carpeted floor areas.

Step 11

Straighten furniture, including loose cushions on sofas and chairs.

Cleaning the Front Desk Area

Position: Public Area Attendant

Reports to: Shift Supervisor

Equipment

- clean ashtrays and matches
- cloths and sponges
- dusting solution
- vacuum
- broom and dustpan

Procedures

Step 1
Remove and replace dirty ashtrays. Restock with matches.

Step 2
Empty trash receptacles. Replace wastebasket liners.

Step 3
Dust light fixtures and decorative wall items.

Step 4
Dust and polish front desk surfaces. Work from the top down, paying particular attention to removing fingerprints, smudges, and scuff marks.

- Clean around any folders or paperwork. *Do not* move or throw anything away.
- Report any surface damage to your supervisor.

Step 5
Spot-clean wall areas. Check for smudges around switches and electrical outlets.

Step 6
Vacuum behind the front desk area. Cover all exposed areas of the carpet you can reach including those under any tables and chairs. Use a broom to get hard-to-reach areas and edges.

Cleaning Corridors

Position: Public Area Attendant

Reports to: Shift Supervisor

Equipment

- dust mop for dusting high or hard-to-reach areas
- dusting solution
- all-purpose cleaner
- wall cleaning solution
- cloths and sponges
- vacuum
- light bulbs

Procedures

Step 1
Dust air supply vents, sprinklers, and ceiling corners.

Step 2
Dust and polish light fixtures. Replace burned-out light bulbs.

Step 3
Spot-clean walls.

- Begin at one point in the corridor and work your way completely down one side then the other until you arrive back at your starting point.
- Check trim areas around guestroom doors and remove any fingerprints or smudges.

Step 4
Clean baseboards.

- Begin at one point in the corridor and work your way completely down one side then the other until you arrive back at your starting point.

Step 5
Clean both sides of all exit doors. Wipe down surrounding trim and door tracks.

Step 6
Vacuum carpet. Begin at one end of the corridor and work your way back to an exit door.

Cleaning Elevators

Position: Public Area Attendant

Reports to: Shift Supervisor

Equipment

- cloths and sponges
- all-purpose cleaner
- glass cleaner
- dusting solution
- dust mop for dusting high or hard-to-reach areas
- light bulbs
- portable canister vacuum

Procedures

Step 1
Wipe down exterior of elevator door. Clean smudges and fingerprints from outside controls and surrounding wall area.

Step 2
Empty and clean ashtray near elevator entrance.

Step 3
Enter the elevator and key or push the appropriate control on the interior control panel so the elevator remains stationary with the doors open.

Step 4
Dust the ceiling light. Replace any burned-out light bulbs.

Step 5
Wipe down interior surfaces. On each wall or mirrored surface, begin at the top right-hand corner and work your way across and down.

Step 6
Clean and polish hand rails.

Step 7
Wipe down control panel so it is free of fingerprints and smudges.

Step 8
Vacuum elevator carpet. Begin in far corner and work your way back toward the door.

Step 9
Vacuum and wipe elevator door tracks.

Step 10
Close elevator door and wipe down interior surface.

Step 11
Before leaving the elevator, key or push the appropriate control on the interior control panel so the elevator resumes normal operation.

Cleaning Public Restrooms

Position: Public Area Attendant

Reports to: Shift Supervisor

Equipment

- all-purpose cleaner
- cloths and sponges
- glass cleaner
- bowl brush
- supplies for paper, tissue, and soap dispensers
- light bulbs
- broom and dustpan
- mop and bucket

Procedures

Step 1
Check status of restroom.

- Knock firmly on the door and say "Housekeeping." If no answer is heard, knock again and repeat "Housekeeping."
- Wait a second time for a response. If you still do not receive an answer, open the door slightly and repeat "Housekeeping."
- If the washroom is occupied, excuse yourself and close the door.
- If the washroom is unoccupied, prop the door open with the doorstop and position the approved floor sign that indicates the room is being cleaned.

Step 2
Flush toilets and urinals. Apply cleaner around and under the lip of the bowl, and around and under the rim and drain holes of the urinal. Let cleaner stand while you attend to other tasks.

Step 3
Empty trash containers. Replace wastebasket liners.

Step 4
Empty and clean ashtrays in sink area and in individual stalls.

Step 5
Clean sinks and countertop areas.

- Run warm water into each sink. Add the correct amount of cleaner.
- Clean the countertop area.
- Clean the sink. Remove drain trap and clean.
- Clean sink fixtures. Polish dry to remove water spots.
- Wipe countertop areas dry.
- Dust and clean any exposed piping under the sink.
- Clean mirror with glass cleaner.

Step 6
Clean toilets and urinals.

- Scrub the inside of the toilet or urinal with the bowl brush. Flush.
- Using a clean rag for each unit, clean exterior surfaces from top to bottom.
- Wipe dry all exterior surfaces. Polish handles.

Step 7
Clean partitions between stalls.

Step 8
Clean washroom walls and fixtures.

- Spot-clean walls for fingerprints and smudges.
- Dust light fixtures. Replace any burned-out light bulbs.

Step 9
Restock dispensers for toilet paper, tissue, paper towels, and soap. Dust or polish dispensers to remove any fingerprints or smudges.

Step 10
Clean the floor.

- Sweep all exposed floor areas. Run edge of broom along baseboards.
- Mop floor with warm water and appropriate cleaning solution.
- Rinse floor using hot water; wring mop frequently.
- Dry-mop floor.

Cleaning Public Restrooms *(continued)*

Step 11
Make one final check. Visually scan all areas of the public restroom for areas you may have overlooked. Smell the air for any unusual odors. Reassemble cleaning supplies and close the door.

REVIEW QUIZ

When you feel you have covered all of the material in this chapter, answer these questions. Choose the *best* answer.

True (T) or False (F)

T   F   1. Within the lodging industry, public area cleaning procedures are more standardized than guestroom cleaning procedures.

T   F   2. The best time to clean the hotel lobby is late morning.

T   F   3. When cleaning the front desk area, attendants will often have occasion to move papers and unplug equipment.

T   F   4. Corridor floors should be vacuumed at least once a week.

T   F   5. The engineering and maintenance department is usually responsible for dusting air supply vents and sprinkler fixtures in corridor areas.

T   F   6. In most hotels, the housekeeping department is responsible for the daily care and maintenance of pool, sauna, or whirlpool areas.

T   F   7. Front-of-the-house areas are designed to withstand more wear and tear than back-of-the-house areas.

Alternate Choice

8. The frequency of cleaning hotel entrances mainly depends on:

    a. the occupancy level of the hotel.
    b. weather conditions.

9. The best time to thoroughly clean the surfaces and tracks of entrance doors is:

    a. early in the morning.
    b. mid-afternoon.

10. The responsibility for maintaining the proper functioning of exercise equipment typically rests with:

    a. the housekeeping department.
    b. the engineering and maintenance department.

Chapter Outline

Selection Considerations
    Flammability Considerations
    Acoustical Considerations
Types of Ceiling Surfaces and Wallcoverings
    Painted Surfaces
    Vinyl Surfaces
    Fabric Surfaces
    Ceiling and Wall Cleaning
    Window Coverings
Types of Furniture and Fixtures
    Public Areas
    Guestrooms
    Staff Areas
Care Considerations
    Seating
    Case Goods
    Bathroom Fixtures
    Lighting and Lamps

Learning Objectives

1. Explain the importance of following the cleaning procedures suggested by manufacturers of ceilings materials, wall coverings, furniture, and fixtures.

2. Identify flammability considerations important to the initial selection of hotel furnishings.

3. Explain how a noise reduction coefficient is used to measure the acoustical quality of ceiling and wall materials.

4. Identify common types of ceiling surfaces and wallcoverings.

5. List criteria used in selecting chemicals and equipment for cleaning ceilings and walls.

6. Cite the advantages of selecting curtains and blinds that operate by batons instead of cords.

7. Explain how chain-affiliated properties purchase guestroom furniture.

8. Explain how housekeepers can determine the proper cleaning method for upholstery fabrics.

9. Explain why housekeepers would prefer acrylic tub/shower units over enameled cast iron tubs and tile walls.

12

Ceilings, Walls, Furniture, and Fixtures

THE BEST RULE OF THUMB for the care of ceilings, walls, furniture, and fixtures is easy to remember: Always follow the manufacturer's suggested cleaning procedures. Failing to follow care recommendations from the manufacturer can mean wasting valuable time researching effective cleaning methods that could be just a phone call away. It may also lead to costly damage or even the total loss of expensive items. In addition, the wrong cleaning methods could negate warranties that protect the property against faulty products.

Executive housekeepers can make cleaning and maintaining ceilings, walls, and furnishings most efficient—and contribute to property safety—by keeping up with new products on the market and making sensible purchasing recommendations.

This chapter discusses some selection criteria for ceilings, walls, furniture, and fixtures; outlines some types of ceiling surfaces, wallcoverings, and appointments; and offers some general cleaning and care guidelines.

Selection Considerations

"If they had only asked me, I would have told them it was going to be impossible to keep up" is an all too common refrain heard in many hotels. Certainly it makes good business sense for hotels to choose items that will make a good impression on guests. But it makes equally good sense for those choices to be practical from a housekeeper's point of view. Surfaces and items that will not wear well or cannot be cleaned efficiently will not attract many guests, no matter how much they cost initially. Flammability and acoustical factors are also important considerations when selecting ceiling surfaces, wallcoverings, and furnishings.

Flammability Considerations

A number of well-publicized hotel fires during the last decade have made fire safety a primary concern of lodging property managers. In fact, some state and local governments require that ceiling and wall materials, upholstered furniture, and bedding meet certain flammability standards. Having flame-resistant furniture, wall, and ceiling materials that meet safety standards is an important part of any fire safety program. Purchasers should know what materials furnishings are made of and how flammable they are before they buy.

Many state and local building codes require that hotels use only Class A materials. Class A materials rate 0 to 25 on the **flame spread index**, a scale that measures how quickly flames will spread across a material's exposed finished surface.

Some manufacturers have developed wallcoverings that will trigger smoke detectors when heated to a certain temperature. BFGoodrich's Koroseal, Cosmos, and Cornerstone vinyl coverings, for example, emit an odorless, colorless vapor when heated to 300°F [150°C]. The vapor triggers the alarm on ionization smoke detectors. (Ionization detectors are the most common type in commercial operations.)

Furnishings may be made of inherently fire-resistant materials. Or they may be treated by the manufacturer with chemicals to make them fire-resistant. Many hotels, especially those operating in states which require properties to use fire-resistant furnishings, ask manufacturers to supply documentation stating that pieces have been treated with fire-retardant chemicals. Dry cleaning companies or other firms specializing in treating furnishings can re-treat items if necessary.

Furnishings made of inherently fire-resistant materials are somewhat expensive, but they remain fire-resistant when laundered or cleaned by other processes and never have to be re-treated. These materials are especially practical in buildings that cannot be evacuated quickly—for example, high-rise hotels.

Some furnishings are made of materials that produce toxic gasses when burned. The National Fire Protection Association (NFPA) sets standards on fire retardancy and toxicity. Information on standards can be obtained by writing the NFPA at Batterymarch Park, Quincy, Massachusetts 02269 or calling (617) 770-3000. Interior designers usually quote NFPA specifications when they submit their finished drawings.

Acoustical Considerations

Because privacy is an integral part of hotel accommodations, **acoustical** considerations are important when choosing wall and ceiling materials. Acoustical materials are measured by a **Noise Reduction Coefficient (NRC)**. For example, an NRC of .75 indicates that the panel absorbs 75% of the sound waves which hit it. Most materials available for commercial use have an NRC of .60 to .95. Recommending a specific NRC for use in hotels or other settings is difficult because many factors affect sound absorption. For example, floors covered with carpet offer additional sound absorption, which may reduce the NRC needed in wall or ceiling surfaces.

Acoustical ceiling materials and wallcoverings usually come in panels. The panels have a fiberglass or mineral fiber filling, which absorbs the sound, and are covered with vinyl or fabric. The standard wall panel size ranges from two to two and a half feet wide by nine to ten feet high. The panels are designed for easy installation, usually attached to **splines** (thin wooden strips), which are glued or bolted to the wall. Ceiling tiles are usually fitted together on a metal grid suspended from the ceiling.

Types of Ceiling Surfaces and Wallcoverings

Ceiling, wall, and window coverings date back to the Middle Ages, when the residents of castles hung intricately woven tapestries and bed curtains to brighten up

Housekeeping Success Tip

Cynthia A. Baudoin
General Manager Trainee/Food & Beverage
Holiday Inn
New Iberia-Avery Island
New Iberia, Louisiana

❝The climate here in Louisiana is very humid, and one of our biggest problems is mildew. Mildew tends to grow more freely in damp, hot weather. Just about any surface is susceptible to mildew growth. If left unattended, mildew can permanently stain or even rot any surface.

The best way to combat this situation is to conduct a general cleaning whenever necessary. It is best to treat for mildew on a regular basis, especially if mildew growth is visible.

Bleach, of course, is a good mildewcide, but it tends to stain carpets, wallpaper, and other surfaces, especially with frequent use. We've found that a vinegar and water solution is very effective and won't fade or stain surfaces to be cleaned.

General cleaning consists of spraying curtains, walls, ceilings, and any other problem areas with the vinegar solution, then wiping with a damp cloth and allowing the surface to dry. Spraying, wiping, and drying with this solution will remove mildew that has already appeared, as well as retard its growth. Windex or some other ammonia-based glass cleaner is helpful on the backs of curtains and on door frames or backs of doors. Baking soda works well, too.❞

rooms and to keep out drafts. Today, ceiling, wall, and window coverings are chosen more for their acoustical properties, safety, and appearance than as insulation against the cold.

There are a wide variety of ceiling surfaces and wallcoverings on the market today. Paint is by far the most common. However, vinyl manufacturers have introduced a wide variety of practical and attractive products in recent years, making vinyl a popular alternative to paint in properties of all types.

Ceiling surfaces and wallcoverings include various kinds of wood surfaces such as laminated plywood, veneer, and paneling; synthetics such as carpet, paneling, and spray-on textured coatings; wallpaper; and stone such as ceramic tiles or marble. The following sections discuss more common surface coverings—paint, vinyl, and fabric.

Painted Surfaces

Much of paint's appeal derives from the fact that it is relatively inexpensive to purchase and apply to both walls and ceilings. It can also be cleaned easily with mild soap and water. In recent years, manufacturers have greatly improved the durability and cleanability of paint by decreasing its porosity. In general, the less porous the paint, the more durable and stain-resistant.

Vinyl Surfaces

Vinyl is now widely used as a wallcovering and also as a surface for ceiling panels. Vinyl wallcoverings are made by laminating vinyl to a cotton or **polycotton** backing. Polycotton-backed vinyl is recommended more often because it is more durable and less flammable than cotton-backed vinyl.

Like wallpaper, vinyl comes in rolls and is applied with a special adhesive. Vinyl wallcoverings should be applied with adhesives that contain mildewcides—especially in areas with hot, humid climates. Mildew causes the wallcovering to loosen from the adhesive, creating ripples in the surface of the vinyl. Installing wallcovering is the job of maintenance personnel or an outside contractor. However, housekeeping staff may be faced with cleaning adhesive seepages off the wall, especially around seams. A solvent recommended by the manufacturer can be used to remove the adhesive.

Vinyl was once chosen strictly for practicality; it can be scrubbed with a brush and soap and water or harsher cleaning agents if necessary. Today, however, vinyl wallcoverings come in a wide assortment of colors and textures, which adds aesthetic appeal to practicality.

The federal government classifies vinyl wallcoverings into three types. **Type II vinyl** is the most practical for public areas because of its durability and appearance. It usually lasts about three times as long as paint. However, it is vulnerable to tears and rips from accidental bumping. Highly textured vinyls may not be scrubbable. Exhibit 1 shows the type of information available from manufacturers to help properties care for vinyl surfaces.

Fabric Surfaces

Fabric wallcoverings are considered among the most luxurious. They are also expensive, tricky to install, easily damaged, and hard to clean.

Linen, once the material of choice for fabric wallcoverings, is giving way to a wider variety of materials—cotton, wool, and silks. Sometimes two or more materials may be combined to make the texture of the covering more appealing.

Fabric wallcoverings may be paper- or acrylic-backed. Paper-backed wallcoverings ravel less at the seams and are easier to install than acrylic-backed coverings. But acrylic-backed coverings are less vulnerable to wrinkling and can be adjusted more easily on the wall during installation. All fabric wallcoverings should be vacuumed regularly. Stains and spots should be removed with a cleaner recommended by the manufacturer. Water should *never* be used on fabric wallcoverings because shrinking can occur.

Ceiling and Wall Cleaning

Cleaning walls and ceilings can be a major money-saving operation for the hotel compared to the alternatives—painting or replacement. Painting, for example,

Exhibit 1   Sample Manufacturer Care Recommendations for Vinyl Wallcoverings

BFGoodrich
WALLCOVERINGS
The BFGoodrich Co., 500 S. Main St., Akron, OH 44318

Koroseal®

Cleaning Instructions

Cleaning instructions for Koroseal® Vinyl wallcoverings

Stains should be removed as quickly as possible to eliminate any possible reaction between the staining agent and the wallcovering. (Not as critical with products coated with Tedlar* film.) Time is especially important for removing materials containing colors or solvents such as ball point ink, nail polish, lipstick, oil shampoo tints, paint, lacquer or enamel and some foodstuffs.

Precautions: Excess soiling materials such as chewing gum, asphalt crayon, paint, nail polish or tar should be carefully scraped off prior to other cleaning attempts.

It is desirable to start cleaning with mild ingredients such as soap — detergent and water. If necessary, stronger cleaners can be used such as liquid household cleaners (with or without ammonia), rubbing alcohol, and solutions up to 3% of hydrogen peroxide, turpentine, gasoline or kerosene. High strength detergents chlorine bleaches, abrasive household cleansers, rubbing alcohol, hydrogen peroxide, turpentine, gasoline and kerosene should first be tried on some inconspicuous portion of Koroseal wallcovering to make sure that there will not be any adverse effect on print (if any), color or gloss.

*DuPont registered trademark

Gasoline, kerosene, and turpentine are explosive and should be handled carefully. NEVER MIX CLEANING REAGENTS TOGETHER — VIOLENT REACTIONS MAY OCCUR WHICH COULD RESULT IN SERIOUS INJURY. OBSERVE ALL LABEL PRECAUTIONS WHEN USING THESE AND ANY CLEANING AGENTS.

Repeated use of stronger cleaners will extract plasticizer from vinyl wallcovering causing the wallcovering to lose its suppleness.

Reagents:

Normal dirt
This can be removed with a mild soap or detergent and warm water; allow to soak for a few minutes, then rub briskly with a cloth or sponge. Use a soft bristle brush on rough textured patterns, rinse with clear water, then wipe with a clean dry cloth. Repeat if necessary.

Nail polish, shellac, lacquer
Remove immediately with dry cloth and be careful not to spread the stain. Go over quickly with rubbing alcohol and then rinse with clear water.

Paint, shoe polish, rubber heel marks, car grease, tar — asphalt
Wipe off as much as possible, then clean with kerosene or turpentine. Rinse thoroughly with clear water.

Ball point ink
Must be removed immediately, using a cloth dampened in rubbing alcohol.

Chewing gum
Wipe off as much as possible (will come off easier if rubbed with ice cube), then rub lightly with rubbing alcohol. Can also use kerosene or naphtha.

Pencil, crayon
Scrape off excess crayon. Erase pencil marks. Wipe any remaining stains with rubbing alcohol.

Fecal, blood, urine
Remove these staining materials quickly; wash stained area using a strong solution of soap and household-type chlorine bleach, rinse with clear water.

FPD-88-DP-3020   ®BFGoodrich   ©1988, The BFGoodrich Company   Litho in U.S.A.

(continued)

Exhibit 1   *(continued)*

Agents to remove stains from BFGoodrich wallcovering with Tedlar* film

Table key:

0. Dry paper towel
1. Damp paper towel
2. Mild soap and water
3. High strength household detergent (full strength)
4. Solvent (Toluene)

Stains

Stain		Stain	
Acetic Acid (5%)	0	Ink (Stamp Pad)	1
Acetone	0	Jam, Jelly	1
Alcohol	0	Lard	0
Ammonia (10%)	0	Lipstick	3
Amyl Acetate	1	Lye Solution	1
Beet Juice	1	Methyl Purple	1
Bluing	1	Methyl Red	1
Bromocresol Green in Methyl Alcohol	1	Methylene Blue in Phenol Indicator	1
Carbon Tetrachloride	0	Mercurochrome	2
Catsup	2	Merthiolate	1
Cigarette Smoke	1	Milk	1
Citric Acid (10%)	1	Moth Spray	1
Chocolate Syrup	1	Motor Oil	2
Coffee	1	Mustard	1
Crayon (wax)	2	Nail Polish	4
Cold Cream	2	Nitric Acid (5%)	0
Dreft Detergent	1	Olive Oil	2
Dye (hair)	1	Pencil	1
Dye (clothes)	1	Phenol (5%)	1
Fluorescin Sodium	1	Phenol Red (1%)	1
Fly Spray (Flit)	2	Phenol Blue	1
Gasoline	0	Potassium Permanganate in water (10%)	1
Grease	2	Permanent Eyelash Darkener	1
Grape Juice	1	Rubber Scuff Marks	1
Hair Oil	2	Salad Dressing	1
Hand Soap	1	Shoe Polish	2
Hydrochloric Acid (5%)	0	Silver Nitrate	2
Hydrogen Peroxide (30%)	0	Silver Protein	1
Hypochlorite Bleach	1	Sodium Bisulfate	1
Insect Spray (Raid)	2	Sodium Bisulfite	1
Ink (Ball Pen)	3	Stainless Mercresin	0
Ink (Higgins Drawing)	1	Synthetic Perspiration	1
Ink (Marking Pen)	3	Sulfuric Acid (5%)	0
Ink (Permanent)	1	Tea	2
Ink (Washable)	1	Trisodium Phosphate	1
		Tomato Juice	2
		Turpentine	2
		Urea	1
		Urine (Canine)	1
		Vinegar	1
		"Vitalis" Hair Oil	2
		Water	0
		"Wright" Blood Stain	2

A soft bristle brush will aid in cleaning deeply embossed grains.

Agents to remove stains from BFGoodrich wallcovering without Tedlar* film

Table key:

1. Mild soap and warm water
2. High strength household detergent
3. Strong solution of soap and chlorine bleach
4. Ice cube
5. Cleaning fluid
6. Kerosene or turpentine
7. Rubbing alcohol

Stains

Stain	
Asphalt†	5-6
Automobile grease	5-6
Ball point ink†	7
Blood†	3
Catsup†	1
Chewing gum	4
Coffee	1
Crayon	1-7
Fecal matter†	3
Lacquer†	7
Motor Oil	1
Mustard†	1
Nail polish†	5
Normal dirt	1
Paint†	6
Pencil marks	7
Rubber heel marks	5-6
Shellac†	7
Shoe polish†	5-6
Tea	1
Tomato juice	1
Tincture of merthiolate†	7

A soft bristle brush will aid in cleaning deeply embossed grains.

†These items may impart permanent stain if not removed immediately.

Courtesy of BFGoodrich Company, Akron, Ohio

costs more than cleaning. It also takes longer, which may mean lost guestroom sales. Painting does not kill bacteria or get rid of dirt; it simply covers them. And painting leaves odors that guests may object to. Replacement may cost ten times as much as cleaning and still requires cleaning air vents, grids, fans, and the like. Moreover, without cleaning, replacement becomes more frequent.

As noted previously, the best rule of thumb for caring for ceilings and walls is to follow the manufacturer's instructions. In general, ceilings and walls fall into three categories: **porous**, **non-porous**, and **semi-porous**, which dictate how to clean the surface.

Porous surfaces are those which absorb moisture. They include latex paint, acoustical ceiling tiles, unsealed wood, and textured ceilings. Non-porous surfaces do not absorb moisture. They include enamel paint, sealed wood, metal, vinyl wallpaper, and plastic ceiling tiles. Excess moisture must be wiped off these surfaces. Semi-porous materials include brick and stone.

Tips for selecting cleaning chemicals and equipment include:

- The more kinds of surfaces (porous, non-porous, semi-porous) a product can clean, the more economical it is.

- Choose non-toxic, biodegradable, and odorless chemicals that are safe for humans and pets. Solutions should not contain chlorine or heavy oxidizers (oxygen bleaches) and should not damage carpets or furnishings that they may come in contact with.

- Make sure cleaners have spray tips that can be held comfortably by the operator, usually at waist level. Cleaners should be sprayed in a fan-like motion to ensure even coverage without wetting.

- Choose equipment that allows housekeeping personnel to clean without having to use ladders or scaffolding. This not only saves time, but can cut down on accidents.

- Tools should have extenders so that operators can hold them comfortably. This will allow cleaning to be done more efficiently.

Areas around ceiling air vents, light fixtures, and fans should be vacuumed before the ceiling is cleaned. Cobwebs and soot should also be vacuumed. The most efficient way to vacuum a ceiling is to use an extension attachment that can be held by operators while standing on the floor. It is important to let the suction do the work; don't try to "scrub" at soot or dust with a brush attachment because this could grind dirt more deeply into the surface.

Furniture and fixtures should be covered before cleaning ceilings and walls to prevent moisture damage. Housekeeping personnel should be warned to wipe floors frequently if dripping occurs in order to prevent slippery spots that could cause falls and injuries.

Window Coverings

Window coverings include everything from vinyl blinds to roller shades to fabric curtains and drapes. All window coverings—but especially drapes—attract dust, cigarette smoke, and other pollutants. Vacuuming window coverings (if not proscribed by the manufacturer) can reduce the number of major cleanings (laundering or dry cleaning for drapes; soap and water washing for blinds) that need to be performed.

Drapes and blinds with cords are vulnerable to breaking and damage. Often guests don't immediately see a cord, so they pull or tug on the drape or blind itself to see out the window. Choosing curtains and blinds with batons instead of cords and opting for headings with automatic releases can help prevent damage. It is also wise to choose hardware that can be repaired without having to remove the rod or drapes.

Drapes that are easy to remove and hang increase the efficiency of room attendants. Choosing the proper hardware for drapes can make drape removal and hanging easier. The housekeeping department should keep extra drape panels on hand so that they can replace those soiled by guests if necessary.

Cleaning Window Coverings. The most important thing to know when cleaning window coverings is the material they are made of. The material may require vacuuming, hand washing, spot cleaning, laundering, or dry cleaning. All cleaning methods should be tested on a small, inconspicuous area of the window covering to avoid dye bleeding, fading, shrinking, or some other problem. The manufacturer can offer the best information on how to clean specific types of window coverings.

A portable canister vacuum is easiest for vacuuming drapes, curtains, and sheers. Be sure to vacuum in the same direction that the fabric is sewn.

Cleaning draperies should be simple and inexpensive. Choosing draperies that can be laundered can save the hotel time and money. If your hotel is planning to buy drapes that can be laundered, consider drapes made of washable fabric without a crinoline backing or sewn-in lining. Drapes with backing or linings sewn in need to be dry cleaned because the drapery fabric or lining could shrink or stretch, giving the drapes a permanently rumpled appearance. Choosing drapes that can be used with a low-cost liner that hangs independently from the drapes offers some advantages. For example, it cuts down the bulk of the drapes themselves. Moreover, liners deteriorate when exposed to light, especially in west and south windows. Separate liners can be replaced much more inexpensively than liners attached to drapes.

Many hotels recommend sending draperies that must be dry cleaned to off-premises laundries which offer warranties in case damage occurs. Generally, drapes are cleaned on an as-needed basis or every two years, whichever comes first.

Types of Furniture and Fixtures

A quick look at a detailed job list for a room or lobby attendant in any hotel will give you an idea of the mind-boggling number of furnishings that housekeeping staff must clean and check at least once a day. Furniture and fixtures include everything from wastebaskets to bedside lamps to poolside chairs to telephones in the lobby.

The number of furnishings in the hotel and the materials from which they are made depend on the size and service level of the hotel. In general, different types of furniture and fixtures are found in three main areas of a hotel—public areas, guestrooms, and employee areas.

Public Areas

Hotels vary widely in the size and types of public areas they contain. In many economy properties, for example, a small lobby is the only public area. Its furnishings may consist of simple lighting, seating, and perhaps **occasional tables**. Lobbies in world-class properties, of course, will have more luxurious appointments such as chandeliers, fountains, and sculpture or other artwork.

Public restrooms are often found in hotel lobbies of all sizes. These will contain basic furnishings such as toilets, sinks, towel dispensers, and hand dryers. Restrooms may also contain changing tables for infants, powder rooms with special makeup lights, or smoking lounges furnished with seating, tables, and cigarette urns.

Lobby attendants are responsible for cleaning all furnishings in these areas. In addition, they must check some items to make sure they are in working order. For example, are soap dispensers and toilets in the restrooms working properly? Do any bulbs in lamps need to be replaced? Are there any fluorescent lights burned out that should be reported to maintenance? Are fire extinguishers dusted and dated for inspection? (Normally, maintenance is responsible for filling and replacing extinguishers. However, housekeeping is responsible for dusting and cleaning and notifying maintenance if they notice that inspection dates are missing from extinguishers or if inspection dates have expired.)

Besides the lobby and public restrooms, many mid-range and world-class properties have meeting and reception rooms, convention space, recreation areas, restaurants, and bars. Furnishings in these areas may include seating, tables, portable platforms, and portable screens or room dividers. Party and reception items might include portable bars, dance floors, and pianos. Meeting rooms might be furnished with chalk boards, easels, projector screens, projectors, and lecterns.

The housekeeping department, however, is not responsible for special equipment or setting up rooms for special events. Meeting and reception rooms, for example, are generally set up by the banquet setup department staff, who also clear away tables, chairs, and other items used for the occasion. Housekeeping is responsible for daily and scheduled cleaning projects such as cleaning chandeliers, walls, windows, and shampooing and vacuuming carpets. Likewise, convention center space may be handled through the convention services department while housekeeping cleans and maintains facilities on a daily cleaning schedule. Audiovisual equipment needed for conventions or meetings is cared for and provided by audiovisual staff, usually attached to the engineering or convention services departments. Recreation areas inside a property normally fall under the rooms division and are therefore handled by housekeeping. Outside recreation areas, however, are usually maintained by engineering/landscaping personnel. Restaurants and bars in large properties are maintained during serving hours by food and beverage staff, but by housekeeping during off hours.

Guestrooms

Many chain companies help member properties make wise furnishing choices by offering a catalog of items from which to choose. The items in the catalog will fit the quality standards mandated by the company as well as ensure that the property will have the same "look" as other properties in the chain. In addition, these furnishings are usually chosen because of their durability, safety qualities, and ease of maintenance. In some cases, manufacturers may build furnishings according to the company's specifications.

Exhibit 2   Sample Suite Furnishings

Luggage Bench
0054-2670
48W x 21D x 18H
70 lbs.

Desk
0054-2620
42W x 22D x 29H
75 lbs.

Double
4/8 Headboard
Wall Hung
0054-2637
56W x 2D x 16H
30 lbs.

Queen
5/0 Headboard
Wall Hung
0054-2638
60W x 2D x 16H
32 lbs.

King
6/8 Headboard
Wall Hung
0054-2635
80W x 2D x 16H
40 lbs.

Coffee Table
0054-2650
22W x 13D x 17H
51 lbs.

Mirror
0054-2640
27W x 45H
33 lbs.

Nightstand
0054-2645
22W x 18D x 23H
75 lbs.

4-Drawer Credenza
0054-2615
78W x 21D x 24H
178 lbs.

Corner Table
0054-2610
28W x 28D x 24H
60 lbs.

Hutch
0054-2617
37.5W x 21D x 40H
140 lbs.

6-Drawer Dresser
0054-2612
68W x 21D x 30H
185 lbs.

Activity Table
0054-2655
34Dia x 29H
43 lbs.

2-Drawer Chest
0054-2616
40.5W x 21D x 24H
95 lbs.

NOTE: *Dimensions are shown in inches.*
D=Depth, W=Width, H=Height, Dia=Diameter.
Cs. pk. : 1 ea. unless otherwise indicated.

Courtesy of Holiday Corporation, Memphis, Tennessee

Virtually all chain properties and many independents purchase furniture in **suites** to ensure a certain level of quality among pieces and to ensure that all pieces match. Some suites come with warranties, which further guarantee the workmanship of pieces. Exhibit 2 shows various pieces of suite furniture.

Essential furnishings in sleeping areas of guestrooms include beds, bureaus, nightstands, and some sort of lighting—typically overhead and bedside. Room attendants must clean these furnishings and make sure they are properly arranged. Making sure the hotel's literature and amenities are properly arranged on desks, bureaus, and tables is also part of the job.

The type of materials from which furnishings are made dictate care and cleaning procedures and affect durability and ease of maintenance. Items may be made of wood, metal, plastic, fabric (synthetic or natural), or from a wide assortment of synthetic materials.

The basic design of any piece of furniture can help or hinder maintenance efforts. For example, furniture equipped with glides or carpet protectors will prevent carpet damage. Casters can make heavy pieces easier to move. **Case goods** with recessed fronts and sides will collect less dust. Bureaus and cube-shaped tables should have sealed bottoms so that moisture from spills or carpet cleaning does not seep into the material and damage it. Table, desk, and bureau tops finished with a water-repellent material will not become water-damaged and are usually more resistant to stains.

In most properties, a color television set has become a standard item. Nearly 70% of the properties in the United States offer not only a television set, but some type of cable hook-up as well.[1] Remote control televisions are becoming increasingly popular. Radio/alarm clocks are also provided in about two-thirds of the properties in the nation.[2] And most rooms today have a telephone. Room attendants are responsible for cleaning these items and making sure they are in proper working order. Items that are not in working order must be reported to maintenance.

Bathroom areas may be as basic as a sink, toilet, towel rack, mirror, lighting, wastebasket, and shower stall, though most properties offer combination tubs and showers. Exhibit 3 shows some of the smaller bathroom furnishings. Often vanities and special shower massagers are provided. Some properties, recognizing the ever-increasing number of business women who travel, now offer such amenities as wall unit hair dryers (nearly 30% of the properties in the United States offer these) and special makeup mirrors.[3]

Bathrooms in some mid-range hotels and many world-class properties are extremely luxurious. Bidets, Jacuzzis, whirlpools, saunas, and powder rooms may be provided. All these items must be cleaned and checked by the room attendant. Bathroom amenities must also be arranged properly around the vanity.

Increases in business travel have led all types of properties to offer living/work space, either in the same room as the sleeping area or in a separate room, as in a suite. Living areas include seating, tables, and often desks. Coffee makers have appeared in guestrooms in nearly 20% of the properties in the United States.[4] Sometimes, food and beverage storage space is provided, including small refrigerators and/or wet bars. Exhibit 4 shows a suite of furniture that might be used in a guestroom with a living area.

Staff Areas

Staff areas consist of office space, lounges, and work areas. In many cases, the staff who work in various areas are responsible for cleaning their work stations and keeping them neat. Engineering and maintenance staff, for example, will clean their own shops. As noted previously, dining room servers bus and clean tables

Exhibit 3   Sample Bathroom Furnishings

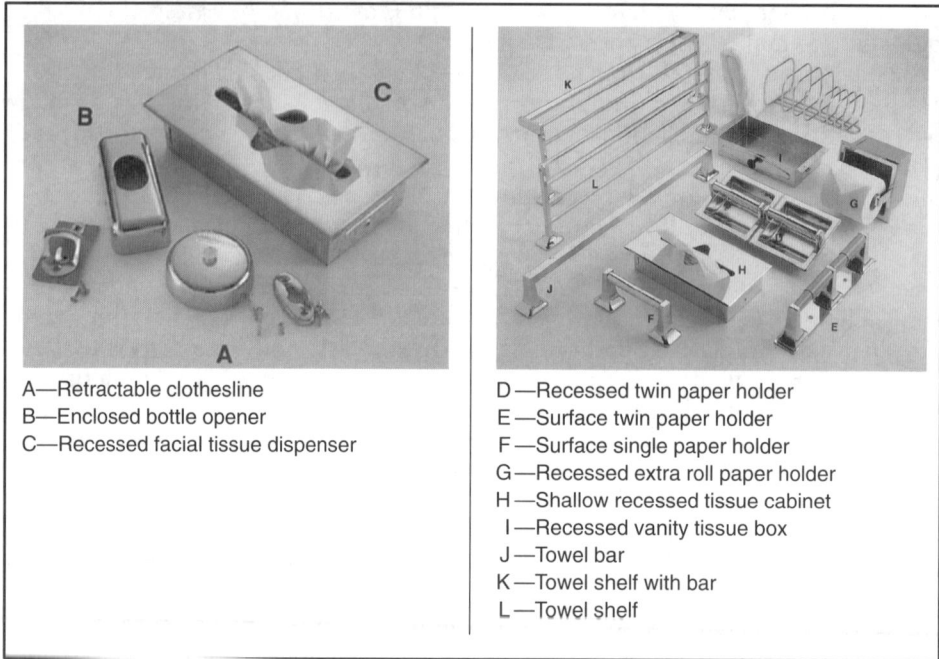

A—Retractable clothesline
B—Enclosed bottle opener
C—Recessed facial tissue dispenser

D —Recessed twin paper holder
E —Surface twin paper holder
F —Surface single paper holder
G —Recessed extra roll paper holder
H —Shallow recessed tissue cabinet
 I —Recessed vanity tissue box
J —Towel bar
K —Towel shelf with bar
L —Towel shelf

and other furnishings used in the dining areas. Office staff are usually responsible for keeping their own desks neat and clean. Housekeeping staff are only responsible for vacuuming, cleaning walls and ceilings, and emptying wastebaskets in these areas. Very large properties may even have a special office cleaning staff to clean offices daily.

Care Considerations

In general, major cleaning procedures include shampooing upholstered furnishings—usually about every six months or as needed—and cleaning washable furnishings with water and/or an appropriate cleaning solution.

Most major cleaning can be performed with very simple tools—buckets, rags, and a cleaning agent. Upholstery shampoos, however, usually require special shampooers. These machines are typically designed to be hand-held. A variety of attachments can be purchased to make cleaning easier. They can often be used to spot-clean carpets and carpeting on stairs and other hard-to-reach areas. The units usually work by feeding dry foam into a rotary brush with which soiled areas can be scrubbed. Some upholstery shampoos are very quick-drying and can even be used to clean furniture in occupied rooms.

Minor cleaning is performed more frequently than major cleaning. Minor cleaning includes such tasks as dusting and vacuuming lampshades and seat cushions and polishing metal fixtures. Paper dust cloths treated with furniture

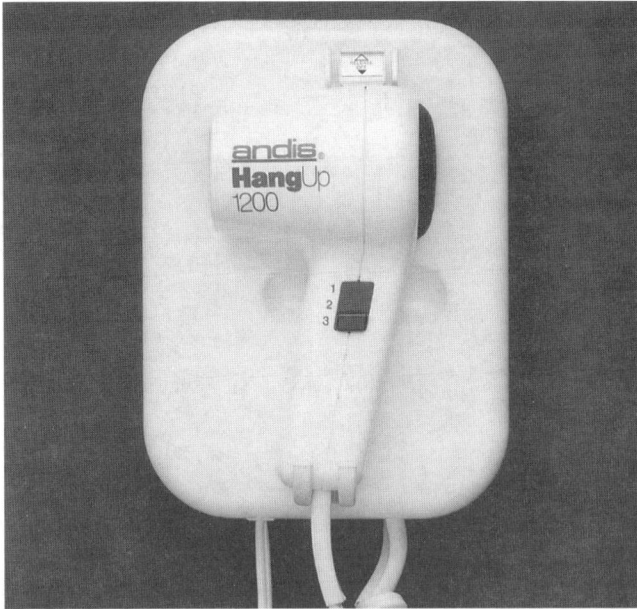

To help meet the ever-changing needs of today's travelers, many properties provide durable wall-mounted hair dryers as a guestroom convenience. (Courtesy of the Hotel Source, Boston, Massachusetts)

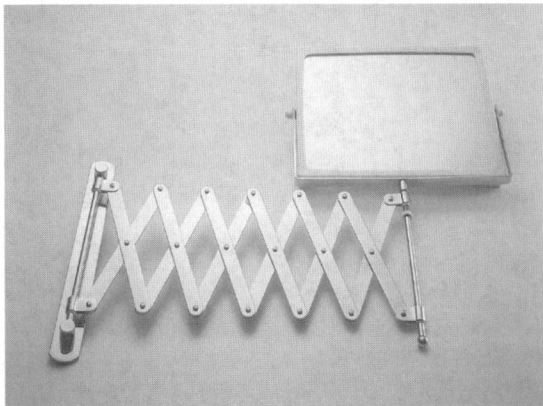

The increasing number of women business travelers has prompted some properties to offer a new line of guestroom convenience such as this extendable makeup mirror. (Courtesy of the Hotel Source, Boston, Massachusetts)

polish are generally used for dusting. Portable canister vacuums are easiest to use on upholstered furniture and other furnishings that should be vacuumed regularly. Care considerations for specific types of furniture and fixtures follow.

Exhibit 4   Sample Living Area Suite

Wet Bar Storage Unit
2 Compartment with
Doors and Shelves
2079-2020
20W x 18D x 48H
170 lbs.

Light Bridge
2079-2002
84W x 20D x 4H
90 lbs.

Wardrobe Unit
2079-2004
36W x 24D x 82H
350 lbs.

Wet Bar Unit
with Refrigerator
2079-2019
36W x 24D x 37H
542 lbs.

Wet Bar Unit with
Combo Refrigerator/
Ice Maker
2079-2018
36W x 24D x 37H
577 lbs.

Nightstand/
Wall Unit
2079-2006
12W x 24D x 82H
200 lbs

7/0 King Sloping
Headboard
2079-2000
84W x 12D x 33H
125 lbs.

Executive Desk
with Drawer
2079-2012
60W x 60D x 29H
190 lbs.

Connecting Desk
2079-2010
66W x 18D x 2H
125 lbs.

TV Cabinet
Mount
2079-2008
36W x 21D x 60H
175 lbs.

Curved Sofa - 3 Cushion
Add for Fabric or COM
2079-2026
72W x 72D x 30H
225 lbs.

Sofa Table
2079-2014
72W x 27D x 27H
70 lbs. ea. (x 2 sections)

NOTE: *Dimensions are shown in inches.*
D=Depth, W=Width, H=Height, Dia=Diameter.
Cs. pk. : 1 ea. unless otherwise indicated.

Courtesy of Holiday Corporation, Memphis, Tennessee

Seating

Seating comes in many sizes and shapes, from side chairs, **arm chairs**, and **lounge chairs** to **love seats**, sofas, sofa-sleepers, and ottomans. Well-made seating is relatively expensive, but it will pay off for any lodging property in the long run.

In general, kiln-dried hardwood frames will resist warping and loosening. Any wooden arms, legs, or backs that are not upholstered should be finished to resist stains, mars, and perspiration. Polyurethane finishes, for example, are impervious to water and other liquids.

Upholstered seating should also be chosen for durability. Foam pad cushions will wear longer and retain their shape better than shredded foam or cotton cushions. Many times manufacturers offer a wide range of fabrics that can be used with a particular seat frame.

Upholstered fabric needs to be vacuumed regularly and spot-cleaned. Always check with the manufacturer before cleaning pieces. Some manufacturers tag upholstered furniture with labels marked "S," "W," or "S-W." These codes mean that the piece can be cleaned with solvent only, with water only, or with either solvents or water. If a piece is marked "X," it cannot be wet-cleaned, only vacuumed. Many care experts recommend that all upholstery be pre-tested according to the following steps:

1. Pre-test water-based cleaning solutions by mixing a small amount of the solution with the shampoo or detergent you plan to use. Follow the instructions printed on the label and use the same temperature of water that you plan to clean with.

2. Thoroughly wet an inconspicuous section of the fabric with the solution. (An ideal place to test is on one-half of the zippered panel of a cushion. Wherever you test, make sure that all colors on the upholstery are included.) Blot the area with a white towel or rag. If color comes off the upholstered fabric, it is not wet cleanable. Allow the test section to air dry; do not hasten drying.

3. Inspect the area after drying. If one color in the fabric has bled into another, the fabric is not wet cleanable. If the test area has puckered, shrinking will occur with wet cleaning.

4. Use wet cleaning as a last resort if the fabric shows signs of brown rings in the wetted areas after drying. If the fabric has a nap (such as velvet), wet cleaning should only be attempted as a last resort *if pre-testing with water is successful.*

5. To pre-test solvents, apply a small amount on a white rag, blot an inconspicuous portion of the piece, and allow it to dry. Look for color loss or other damage before proceeding.

Once upholstery has been pre-tested, record the results on the piece for future reference. Remember that solvents are flammable; you should therefore purchase equipment suited for solvents. Although solvents dry quickly, they may leave odors. Solvents should be thoroughly evaluated before purchasing.

In general, woven nylon and good quality vinyl are the most stain-resistant and easily cleaned upholstery materials. Seating in heavy-use areas should be upholstered in vinyl, which needs only a daily wiping, or some other fabric that is especially designed for heavy use. Satiny or lightweight fabrics should be avoided because covers made from these materials tend to shift over the cushions and do not hold up well.

Exhibit 5   Case Goods Specifications

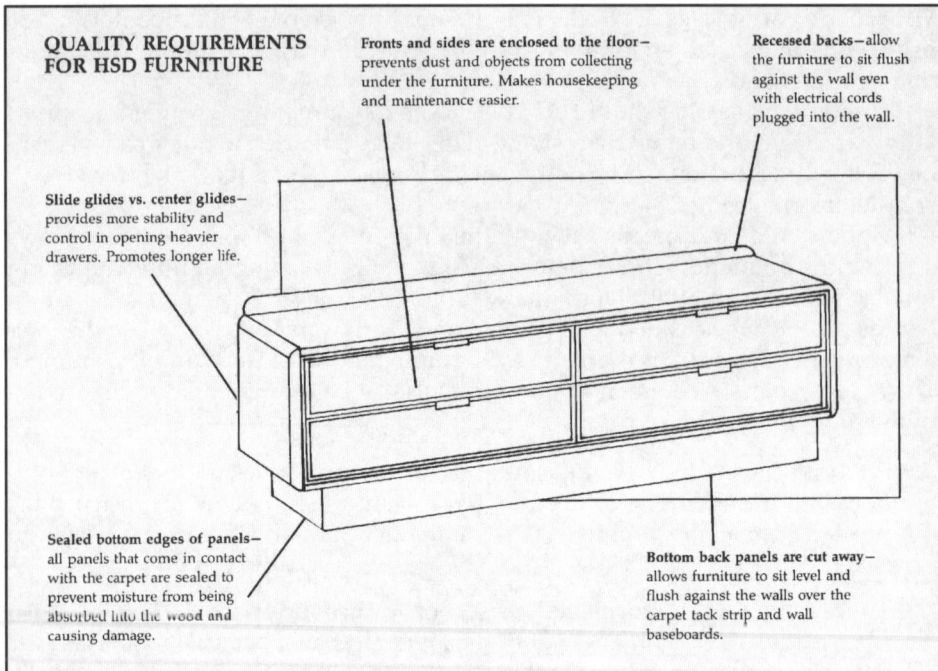

QUALITY REQUIREMENTS FOR HSD FURNITURE

Fronts and sides are enclosed to the floor—prevents dust and objects from collecting under the furniture. Makes housekeeping and maintenance easier.

Recessed backs—allow the furniture to sit flush against the wall even with electrical cords plugged into the wall.

Slide glides vs. center glides—provides more stability and control in opening heavier drawers. Promotes longer life.

Sealed bottom edges of panels—all panels that come in contact with the carpet are sealed to prevent moisture from being absorbed into the wood and causing damage.

Bottom back panels are cut away—allows furniture to sit level and flush against the walls over the carpet tack strip and wall baseboards.

Courtesy of Holiday Corporation, Memphis, Tennessee

Case Goods

Bureau drawers can pose a variety of safety threats and irritations to guests and housekeeping staff alike. Recessed handles will not snag clothes or protrude so that people bump into them. Nylon ball bearings and slide glides in drawers make opening and closing drawers quieter, reduce wear on the drawers, and eliminate jamming. Some drawers will even close all the way with a gentle push, which helps prevent people from bumping into open drawers. Better quality bureaus have a dust panel under the bottom drawer to keep floor dust and lint out of drawers. Items that have recessed backs can be placed flush to the wall, even though electrical plugs may be behind them. Exhibit 5 outlines some factors to consider when choosing case goods.

As with seating, case goods should be finished with some durable material. Polyurethane helps protect the furniture from stains and water damage. Some finer furniture, however, is finished with less durable materials. Waxes can protect these finishes and, to some extent, the wood below. Some hotels use cut glass on the tops of tables, bureaus, and other pieces of fine furniture for additional protection.

Bathroom Fixtures

Toilets are most often made of **vitreous china**. Models with elongated bowls often use space more efficiently. Because the bowl juts out farther, a shelf can be built behind the toilet.

Luxurious fixtures (note the bidet on the far left) and decor are standard features in some mid-range and many world-class properties. (Courtesy of Kohler Company, Kohler, Wisconsin)

Vanities (bathroom countertops) are usually made of marble or a synthetic material. Acrylics, polymers, and combinations of the two are the most popular synthetics. Marble is more expensive and more vulnerable to damage than acrylics or polymers. Synthetics are easier to care for; scratches or other damage can often be sanded out.

Bathtubs are most frequently made of **enameled cast iron** or **acrylic**. Acrylic models are often reinforced with fiberglass and may come in combination tub/shower units that include shower stall walls. Ceramic tile is frequently used around the enameled cast iron tubs and in the shower stall.

Many designers and owners select enameled tubs and tile walls. But housekeepers often prefer acrylic tubs or tub/shower units because they are cheaper, easier to care for, and at least as durable as enameled cast iron tubs. An inexpensive vinegar solution cleans acrylic tubs quickly and easily (though some hotels prefer to use a mild all-purpose cleaner to prevent the bathroom from smelling like a salad). Moreover, the tub/shower units do not require the attention that tile and grout do. Scratches or cigarette burns, too, can be sanded out without loss of color or finish. All in all, housekeepers say they can clean an acrylic unit in half the time it takes to clean a conventional enameled cast iron/tile tub and shower.

Lavatories most often come in vitreous china or enameled cast iron. However, very luxurious models may be made of marble or even teak, which is water-resistant and can be treated with a water-tight seal. Sinks can also be made in a variety of synthetics designed to look like natural materials such as granite or marble.

Faucets and handles on tubs, showers, and lavatories may be made of nickel, chrome, or, on the most elegant fixtures, brass. Brass and chrome should never be exposed to acidic cleaners.

Some modern baths have bathing areas with extra shoulder width at the reclining end. (Courtesy of Kohler Company, Kohler, Wisconsin)

Lighting and Lamps

Overhead lighting is the lighting of choice in public areas. There is a wide variety of overhead lighting to choose from, including elaborate chandeliers, recessed lighting, track lighting, and fluorescent lighting with diffuser panels. Wall lights on brackets may also be used.

A variety of lighting is used in guestrooms. Overhead lights are usually installed in the bathroom and, often, just inside the door of the room. In a room with a large work area or in a suite, there may be an overhead light fixture over the work table. The bathroom or powder room may also have a makeup mirror with wall lights.

The living room/bedroom is usually equipped with lamps on bureaus, nightstands, and near seating. Metal lamp bases are more break-resistant than ceramic. Superior quality metal bases do not have side seams. Colors in ceramic lamps will last longer if they are glazed into the surface instead of painted on. Some properties opt for lamps that can be bolted to tables and desks to eliminate breakage and theft.

Lampshades must be replaced more often than the lamp's base. Good quality plastic-lined shades will wear longer than fabric-lined shades. Switches on the lamp base will prevent guests and staff from having to grope around under the shade to find the switch, thus damaging the shade inadvertently. **Permanent assembly** prevents loosening around the socket. Exhibit 6 shows some factors to consider when choosing lamps.

Exhibit 6 Lamp Specifications

Lighting sets the mood in a room and is an important part of good decor. A versatile lighting scheme will let guests tailor their lighting usage to suit their needs and wants. Here are some pointers to help you make a wise investment in contract lighting:

Basic Lamp Construction

Material in a shade may be made from paper or from fabric that is laminated to paper or to plastic material (usually styrene). Plastic backed shades are more expensive, but are generally a better value because of their longer life.

One of the major problems with commercially used lamps is their tendency to become loose over a period of time. Be sure to specify lamps that have permanent assemblies. The pipe connecting the base with the socket is usually threaded at both ends and must be assembled with a strong adhesive to prevent the threaded joints from becoming loose.

Most lamps designed for contract use have an on-off switch at the base. This convenience will add to the life of shades since the guest never has to reach around a shade to locate a switch.

The base of a lamp can be made of ceramic, wood, metal or polyester resin. If you select a ceramic base, be sure that the colored glaze has been fired into the ceramic and is not just painted on. The total absence of any seams indicates better quality in a metal base.

Another way to provide stability and also discourage theft is the permanent mount. This optional feature lets you bolt the lamp to the table through a hole in the table's surface.

Courtesy of Holiday Corporation, Memphis, Tennessee

Endnotes

1. *The U.S. Lodging Industry 1988* (Philadelphia: Laventhol & Horwath, 1988), p. 11.
2. *The U.S. Lodging Industry 1988*, p. 12.
3. *The U.S. Lodging Industry 1988*, p. 12.
4. *The U.S. Lodging Industry 1988*, p. 12.

Key Terms

acoustics
acrylic
arm chair
case goods
enameled cast iron
flame spread index
lounge chair
love seat
non-porous
NRC scale

occasional table
permanent assembly
polycotton
porous
semi-porous
spline
suite
Type II vinyl
vitreous china

Discussion Questions

1. List the major types of wallcoverings and discuss their relative merits in terms of cost, care, and aesthetic appeal.

2. Why are acoustics important when considering ceiling materials and wallcoverings?

3. What is the NRC scale? What does it mean if a wallcovering has an NRC rating of .60?

4. Discuss how technology has decreased flammability in wallcoverings and ceiling materials.

5. Explain why cleaning ceilings and walls can be thought of as a cost-saving operation.

6. Why is the porosity of a ceiling or wall important to cleaning?

7. Discuss some of the types of window coverings and their cleaning considerations.

8. Explain how the hotel's size and service level affect the number and kind of furnishings and how these, in turn, affect the housekeeping department.

9. Discuss some of the various types of tables, seating, and case goods found in a hotel. Discuss where different types of these items might be found.

10. Discuss some of the materials from which various furnishings are constructed.

Procedures: Ceilings, Walls, Furniture, and Fixtures

The procedures presented in this section are for illustrative purposes only and should not be construed as recommendations or standards. While these procedures are typical, readers should keep in mind that each property has its own procedures, equipment specifications, and policies regarding protective gear which are designed to fit individual needs.

Ceiling Cleaning

Equipment

- cleaner
- waterproof tarp
- sprayer for cleaner (single- and double-headed attachment is best)
- extension pole
- sponge
- bucket
- clean, dry rags

Procedures

Step 1
Select a cleaner that is appropriate for the type of ceiling.

Step 2
Cover furnishings or fixtures that must be protected from moisture with a waterproof tarp. While many chemicals will not damage carpets, upholstery, or other furnishings, computers, paper (in offices), and other items should be protected from possible spills.

Step 3
Spray cleaning solution on the outer edges of the ceiling first. It is best to use a single-head sprayer for the edges of the ceiling. Hold the sprayer parallel to the ceiling, about six inches away from the wall and ten inches below the ceiling. Walk backward so you can check the amount of solution being applied to the surface. The surface should be evenly wet, not dripping. Wipe drips from walls as soon as possible to prevent streaks.

Step 4
Spray the rest of the ceiling with a double-head sprayer if available.

Step 5
Check the cleaner manufacturer's recommendations. Some porous ceilings should not be wiped; the cleaning solution will simply dissolve soils, and will evaporate.

Step 6
Wipe non-porous surfaces—including grids and ceiling fixtures—with an extension pole and sponge after the solution has emulsified soils. Rinse the sponge frequently.

Wall Cleaning

Equipment

- bucket
- cleaner
- rag or mop head
- extension pole
- clean, dry rags
- drop cloth

Procedures

Step 1
Select a cleaner that is appropriate for the type of wall surface.

Step 2
Place a drop cloth on the floor next to the wall to catch drips and spills.

Step 3
Use a rag or mop head on an extension pole to dust the wall before cleaning.
Move pole in an upward motion with gentle pressure. Do not dust in highly humid conditions.

Step 4
Dip rag or mop in cleaning solution. It should be wet, but not dripping; an overly wet rag or mop can cause streaking on the wall.

Step 5
Wash the wall from the bottom to the top. Use a sweeping motion to protect back muscles. Change rags frequently and clean up floor spills as often as necessary to prevent slips and injuries.

Step 6
Rinse with clean rag and water in a sweeping motion.

Cleaning Window Blinds

Equipment

- bucket
- clean, dry rags
- cleaner

Procedures

Step 1
Fill bucket with cleaner and water.

Step 2
Turn baton or pull cord so that the louvers of the blinds are parallel with the floor.

Step 3
Wet rag in cleaning solution, wring, and hold it between your thumb and index finger so that the louver of the blind is sandwiched between the rag.

Step 4
Beginning at the top, slide the rag across each louver.

Solvent- or Wet-Cleaning Upholstery

Equipment

- vacuum and attachments appropriate for upholstered furniture
- power rotary shampooer or hand bonnet
- carpet extractor with upholstery attachments
- bucket
- clean, dry rags
- shampoo
- defoamer (optional)
- pre- and post-spotting chemicals (optional)
- fan (optional)

Procedures

Step 1
Pre-test upholstery before cleaning.

Step 2
Fill shampooer with solution according to the manufacturer's directions. Add defoamer to prevent excess foam and spills. Solvents are used full-strength and do not require mixing. Avoid mixing chemicals in public areas.

Step 3
Vacuum upholstery.

Step 4
Clean.

- If using a rotary shampooer, run the shampooer over the upholstered area and use a carpet extractor with the appropriate attachments to extract excess shampoo.
- If using a hand bonnet, slip the bonnet over the hand and work the shampoo gently in the fabric. Turn the mitt frequently, and rinse and wring out as needed in the shampoo solution. Extract excess moisture.
- If using solvents, be sure to vent the exhaust from the extractor to an open, flame-free area. The fan may be used to help exhaust fumes.

Step 5
Allow upholstery to air dry.

Step 6
Vacuum upholstered surfaces after drying to remove shampoo residues.

Cleaning Leather (Not Suede)

Equipment

- bucket of warm water
- mild facial soap*
- clean, dry rags

Procedures

Step 1
Dust leather surface, particularly around button-tufted areas.

Step 2
Wash leather with soap and warm water.

Step 3
Rinse with clean dry rag.

*Today's leather does not require oiling, waxing, or saddle soap. Ammonia, abrasives, or bleaches should never be used. Aniline-dyed leather can mildew in damp areas. A gentle fungicidal soap can be used to kill and prevent mildew growth if warranted.

Cleaning Metal Desks and Cabinets

Equipment

- two buckets
- clean, dry rags
- cleaner

Procedures

Step 1

Fill two buckets, one with cleaner, one with water. Have a separate rag to use with each bucket.

Step 2

Soak one rag with cleaner and wipe all metal surfaces. Include the knee space (if any) and bottom surfaces. Rinse the rag frequently in the bucket of cleaner.

Step 3

Use the bucket of water and a clean rag to rinse all surfaces of the desk. Rinse the rag often in the bucket with clear water. Change water as necessary.

Cleaning Wood Furniture

Equipment

- canister vacuum sweeper
- cleaning solution
- bucket
- clean, dry rags
- treated dust cloths or furniture polish and rags

Procedures

Step 1

Vacuum upholstered portions of the furniture (if any) first. This will prevent dust from upholstered areas from settling on wood after it has been cleaned.

Step 2

Mix cleaning solution in a bucket. Avoid mixing chemicals in public areas.

Step 3

Wipe down all wood surfaces with the solution, stroking along the grain of the wood.

Step 4

Dry wood thoroughly.

Step 5

Polish with treated dust cloth or furniture polish and rag according to the manufacturer's directions.

Polishing Brass Fixtures

Equipment

- drop cloth
- brass cleaner (do not use cleaning fluid or abrasives on lacquered brass)
- clean, dry rags

Procedures

Step 1
Cover surface or floor below brass items with a drop cloth to protect it from polish. If possible, take fixtures to a cleaning area and put a drop cloth over the work surface to protect it.

Step 2
Read the directions on the cleaner container carefully and follow all directions.

Step 3
Apply brass cleaner to fixture surface as cleaner directions indicate.

Step 4
Rinse and/or buff according to cleaner directions.

Step 5
Use paste wax on lacquered brass every six months or so to help maintain the finish and prevent finger markings.

Polishing Stainless Steel Fixtures

Equipment

- drop cloth
- clean, dry rags
- polish

Procedures

Step 1
Cover surface or floor below items with a drop cloth to protect it from polish. If possible, take fixtures to a cleaning area and put a drop cloth over the work surface to protect it.

Step 2
Read the directions on the polish container carefully and follow all directions.

Step 3
Apply polish to fixture surface as the directions indicate. Rub in a circular motion.

Step 4
Buff fixture according to polish directions.

Cleaning Mirrors

Equipment

- bucket
- glass cleaner or vinegar and water solution
- treated dust cloths or furniture polish and rags
- clean, dry rags
- old newspapers

Procedures

Step 1
Fill a bucket with vinegar and water solution.

Step 2
If the mirror has a wooden frame, clean this first with treated dust cloth or furniture polish and a clean rag.

Step 3
Soak rag in solution and wring out. Wipe the mirror with a horizontal motion from top to bottom. For large mirrors, clean one section at a time.

Step 4
Wad up a section of newspaper and dry mirror with a horizontal motion.

Cleaning Picture Frames and Glass

Equipment

- clean, dry rags
- glass cleaner or vinegar and water solution in spray bottle

Procedures

Step 1
Clean frame with side-to-side strokes using a damp rag. Do *not* use glass cleaner or vinegar solution on the frame.

Step 2
Spray glass cleaner on rag and clean with up-and-down strokes.

Step 3
Buff glass with dry rag.

Cleaning Sconce Lights and Chandeliers

Equipment

- glass cleaner
- bucket
- clean, dry rags

Procedures

Step 1
Mix glass cleaner in bucket according to manufacturer's directions. Avoid mixing chemicals in public areas.

Step 2
Turn off the light fixture and allow glass and bulbs to cool before cleaning.

Step 3
Remove glass globes from sconces or chandelier fixture and put in bucket of glass cleaner.

Step 4
Wipe sconces or chandelier fixture with a rag dampened in cleaner solution. Remove bulbs and wipe with a dry rag.

Step 5
Put bulbs back in sconces or chandelier fixture. Turn on fixture and replace any burned-out bulbs.

Step 6
Remove globes from bucket and dry with clean rags. Put globes back in sconces; replace any broken or chipped globes.

Cleaning Recessed Lights

Equipment

- duster pole
- clean, dry rags
- rubber bands
- dusting spray

Procedures

Step 1
Attach a rag to the end of a duster pole with rubber bands.

Step 2
Spray rag with dusting spray.

Step 3
Move pole around light recess, staying below the light bulb.

REVIEW QUIZ

When you feel you have covered all of the material in this chapter, answer these questions. Choose the *best* answer.

True (T) or False (F)

T   F   1.   The best rule of thumb for the care of upholstered fabrics is to pre-test all cleaning solutions.

T   F   2.   Ceiling, wall, and furniture items classified as Class A materials would have high flame spread index values.

T   F   3.   Water should never be used to clean fabric wallcoverings.

T   F   4.   Chlorine or oxygen bleaches should not be used to clean guestroom ceilings or walls.

T   F   5.   Enamel paint is a good example of a porous wallcovering.

T   F   6.   Chain-affiliated properties often purchase guestroom furniture items as part of an entire suite package.

T   F   7.   Acrylic bathroom countertops are less vulnerable to damage than marble vanities.

Alternate/Multiple Choice

8.   Upholstered fabrics coded by a manufacturer with an "X" indicates that the items cannot be:

    a.   vacuumed.
    b.   wet-cleaned.

9.   Toilets are most often made of:

    a.   enameled cast iron.
    b.   vitreous china.

10.   Manufacturers have made paint more durable and stain-resistant by decreasing its:

    a.   latex content.
    b.   porosity.
    c.   solubility.
    d.   thickness.

Chapter Outline

Beds
    Springs
    Mattresses
    Frames
    Selection of Beds
    Maintenance of Beds
Linens
    Types of Linens
    Sizes of Linens
    Linen Care, Reuse, and Replacement
    Linen Selection Considerations
Uniforms
    Identifying Uniform Needs
    Selecting Uniforms

Learning Objectives

1. Describe three types of bed spring construction.

2. Identify three types of bed mattresses.

3. Explain the function of a linen committee.

4. Identify criteria for selecting linens.

5. Identify ways in which discarded linen can be reused.

6. Describe three types of fabric material to evaluate when selecting linens.

7. Distinguish between warp yarns and fill yarns.

8. Describe three types of fabric weaves.

9. Explain the importance of carefully selecting employee uniforms.

───13

Beds, Linens, and Uniforms

O VER THE YEARS, innovative innkeepers and hoteliers have tried to attract guests with the latest amenities—color television, air conditioning, drinking and dining facilities, pools, entertainment, and so on. But one item remains the hotel's biggest draw: the bed.

This chapter discusses the selection and maintenance of beds as well as various types of linens. Uniforms are included in this chapter because many of the selection and care criteria for linens apply to uniforms as well.

Beds

Beds, as a class, include conventional guestroom beds, cribs, and rollaways. All these beds will be discussed in this section of the chapter. Exhibit 1 lists standard bed sizes.

Most beds consist of springs, which provide resiliency and support; the mattress, which lies on top of the springs and provides extra padding; and the frame on which the springs and mattress rest. When carefully chosen, these items work together to provide a durable, comfortable bed that can be maintained and changed easily. When beds are poorly chosen, the hotel is stuck with beds that sag, are difficult to change, must be replaced often, and that guests will complain about.

In most hotels, headboards are not part of the bed. They are typically mounted on the wall behind the bed, not on the frame. Headboards are often included with guestroom suites and designed to match other pieces of furniture.

Springs

Springs add resiliency and durability to the bed. In general, springs are made by joining wire springs or coils together and covering them with padding. There are three basic types of spring construction: **box springs**, **metal coil springs**, and **flat bed springs**.

Box springs are mounted on a wood frame and covered with a pad. A sturdy cloth called **ticking** covers the springs and pad.

Metal coil springs may be arranged in two layers. The springs on the bottom are tightly coiled for good support. The top springs are more loosely coiled for resiliency. Metal coil springs also come in a single layer with metal bands criss-crossing the surface or extra wire at the top of the springs to form a semi-closed surface on which the padding lies.

Flat bed springs are simply lengthwise strips of metal attached to a frame with **helical hooks**. Helical hooks are small coils with hooks at both ends. Flat bed

Exhibit 1   Standard Bed Sizes

Crib	28 × 52 inches
Rollaway	39 × 75 inches
Twin	39 or 42 × 76 inches
Three-quarter	48 × 76 inches
Double	54 × 76 inches
Queen	60 × 80 inches
King	78 × 80 inches

Exhibit 2   Rollaway Bed

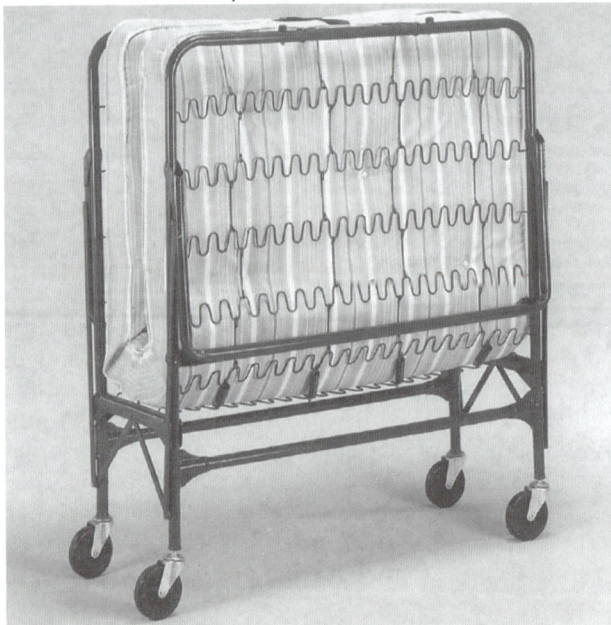

Source: The Hotel Source (undated catalog), p. 75.

springs are most frequently found on rollaway beds, such as the one shown in Exhibit 2.

Mattresses

There are three main types of mattresses: **innerspring, latex** (foam rubber), and **solid**. Most hotels use innerspring or latex mattresses because they are easier to clean and more durable than solid mattresses.

As the name implies, an innerspring mattress has an inner layer of springs between layers of insulation and padding. The springs may be tied together with wire or helical hooks. This is called the *Bonnell, Hager,* or *Karr* construction. The springs may also be individually encased in cloth (*Marshall construction*).

Most metal frames are easy to assemble and sturdy enough to support larger-size box springs and mattresses. (Courtesy of The Hotel Source, Boston, Massachusetts)

Latex mattresses are made from synthetic rubber which is whipped into a foam while in a semi-liquid state and poured into a mold.

Solid mattresses are made by filling a tick with padding—horse or other animal hair, cotton, or **kapok** (a type of plant fiber).

Frames

The bed frame supports the springs and mattress. The frame consists of four metal bars joined at the corners to make a square frame that the box spring and mattress sit in. An extra bar is placed in the center of queen- and king-size mattresses for added support.

Some properties opt for box or platform frames. A box frame consists of a box of solid wood or steel supporting steel bars. The bars support the mattress. Box frames sit tight to the floor, which means that the area under them does not have to be vacuumed or cleared of trash. These frames, however, are subject to scuffs and dents, which must be cleaned and repaired.

Selection of Beds

Box springs and metal coil springs are most often recommended for use in hotels and other institutional properties. Better quality box springs have coils that are tied together and are installed on hardwood frames. Innerspring mattresses may have to be used with box springs in order to bring the bed to standard height. Metal coil bedsprings can accommodate most mattresses. Flat bed springs are less expensive than the other types, but also the least durable. Helical hooks used to connect the metal strips to each other can help improve the quality of flat bed springs. Flat springs are acceptable for rollaways that are used infrequently.

In general, tufted mattresses—innerspring, latex, or solid—are more durable than those that are quilted or tuftless. While buttons are often used to tuft mattresses, guests may complain that a button-tufted mattress is uncomfortable.

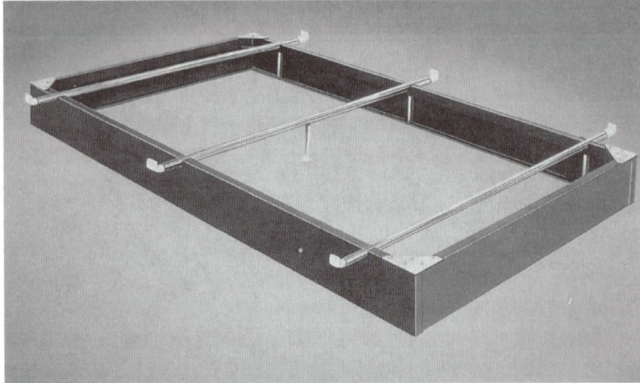

Some box frames are all-steel construction with sturdy, center-supported bars. (Courtesy of The Hotel Source, Boston, Massachusetts)

Mattress ticking should be sturdy—at least six ounces per inch, according to some manufacturers—and the seams around the edges of the mattress should be rolled or reinforced in some way. Handles on the sides of the mattress make moving and turning mattresses easier.

Some properties prefer latex mattresses because they are lighter and easier to lift when making beds and performing mattress maintenance. They can also be cleaned more easily than other types. (Some sample mattress cleaning procedures are included in the procedures section at the end of this chapter.) However, foam mattresses do have drawbacks. They are often more easily torn by springs sticking out of the box springs. And unless the ticking on the mattress is bonded to the foam itself, ticking can shift around. Better quality latex mattresses are at least four and a half inches thick.

Among solid mattresses, those with horse hair stuffing are the most resilient and durable. Long staple cotton is also a good filling, but mattresses stuffed with cotton **linters** (short fibers and waste cotton) become lumpy quickly. Cotton linters and kapok stuffed mattresses cannot be rebuilt. A solid mattress is also more durable if it is made of separate stuffed compartments.

Hotels concerned with fire hazards frequently purchase mattresses wrapped in flame-retardant polyurethane foam to reduce flame spread. It is often a good idea to purchase sample mattresses to actually test the fire retardancy of the mattress material.

Maintenance of Beds

Turning a mattress and springs is a simple, easy maintenance task that can add as many as three years to the useful life of the bed. Many properties recommend turning the mattress four times per year. Mattresses can also be cleaned with a hand-held vacuum attachment when turned. Some properties check beds for wear and sagging when they are turned. Exhibit 3 offers a bed inspection checklist.

Exhibit 3   Bed Inspection Checklist

Mattress
  Ticking
    ☐ Check for tears
    ☐ Check for soil
  General Condition
    ☐ Check center
    ☐ Check edges
    ☐ Check for lumps
    ☐ Check to see if handles are in good repair
Springs
  Ticking
    ☐ Check for tears
  General Condition
    ☐ Check edges for firmness
    ☐ Check corners for frayed fabric
    ☐ Check for broken springs
Bed Frame (metal frame)
    ☐ Check casters or furniture glides (if any)
    ☐ Check joints
    ☐ Check for crossbar support on queen- and king-size frames
Bed Frame (box)
    ☐ Check for scuffs and dents in the box frame

Linens

Purchasing too many, too few, or the wrong kind of linens can be a costly mistake. Moreover, an inadequate supply of linens and bedding can disrupt the operation of the entire hotel if beds and dining tables are not ready when guests want them.

The linen supply is generally discussed in terms of **par**. A par is one complete set of linen for all guestrooms and food and beverage outlets in the hotel. How many par the property must keep on hand depends on a number of things:

- What is the delivery schedule of the outside laundry service or how efficient is the on-premises laundry?

- How effective are the property's measures to control linen loss?

- Does the property do a great deal of banquet or group business that frequently generates unexpectedly high levels of linen use?

Besides having linens on hand, properties must consider fabric material, construction, and finishing of linens to determine durability. All linens should "rest" on shelves at least 24 hours after being washed to reduce wear and tear. How attractive and comfortable linens and bedding are is important to guest satisfaction. The way in which items must be washed can affect the amount of equipment required in the on-premises laundry. And all these factors will ultimately affect the hotel's expenses and profits.

Because having the right amount and type of linens and bedding is so important, many hotels form a linen committee to help choose and review the current types, sizes, and uses of linens.

The linen committee helps all departments make their linen needs heard. At a large hotel, this committee might include the executive housekeeper, linen room manager, laundry manager, head of the maintenance department, dining room managers, and the hotel's general manager. Other staff whose tasks are affected by the supply of linens should also be included. At smaller hotels, the linen committee may consist simply of the executive housekeeper and general manager or owners.

Effective communication between housekeeping and other departments within the hotel is important for purchasing and controlling the supply of linens. For example, the dining room manager is probably the best person to determine how many par of tablecloths and napkins are needed for effective operation, to monitor the performance of table linens, and to measure guest satisfaction with the product.

Effective communication helps to pinpoint where linen loss is occurring since everyone knows the procedures of other departments. Similarly, the source of linen damage can be found more easily if all staff who handle linens are in close contact with the housekeeping department. Good communication with the front desk and reservations department will alert the housekeeping department when extra linens may be required for banquets, parties, meetings, and other special functions.

Types of Linens

Linens can be classified by where they are used: on beds, in bathrooms, or in dining rooms.

Imagine the effect on a hotel's business if a guest pulled back the blankets and bedspread to find worn, stained, and wrinkled sheets. Hotel sheets and pillowcases must not only be clean; they must *look* clean, crisp, and new. In addition, sheets and pillowcases must be comfortable. Sheets are available in muslin or percale. Percale is the better grade of fabric.

Many properties use plain white sheets and pillowcases. Some properties color-coordinate sheets and pillowcases with the bedspread and other room decor to add a touch of elegance. World-class properties may keep a special supply of monogrammed sheets and pillowcases in some luxurious fabric such as Egyptian cotton or satin.

Like sheets and pillowcases, blankets need to look clean and new and feel good. Blankets may also add to the elegance of the property. Climate is an important consideration in choosing blankets, and hotels in very cold or unpredictable climates may stock guestrooms with extra blankets.

Mattress pads protect mattresses. They may be made of a woven, quilted fabric or of felt. Because guests rarely see mattress pads, properties typically choose those that provide the best protection for the mattress at the best price. Felt pads are generally the least expensive, but do not hold up well under repeated washings. Other types of pads include cotton and synthetic blends or 100% polyester. The most expensive pads are the blends.

Bedspreads and pillows are usually purchased in new hotels as specified by the interior designers. It is best to follow manufacturers' specifications for cleaning and care.

Pillows can be feather, acrylic fibers, or hypoallergenic foam. Feather is more luxurious and costly. Acrylic or foam are less expensive and more durable.

Terry cloth is the most common fabric used for bath linens. Velvet towels may have a smoother **hand** (meaning feel), but they are less absorbent. Better quality towels have **selvaged** edges—that is, edges that are woven, not hemmed. Some properties recommend buying towels with hemmed selvages for extra strength. Selvaged towels last longer; they do not unravel as quickly as non-selvaged towels after repeated washing and drying. Loops should be one-eighth of an inch high.

Bath towels often come with the hotel's logo or initials woven into the fabric. Extra-large towels (called **bath blankets** or **bath sheets**) may also be stocked. Many properties see bath blankets as a luxurious amenity. And they are—for large or tall people. Some guests, however, find bath blankets difficult to handle and too heavy to manage easily. Many properties now provide bathrobes to guests as a bath amenity.

Shower curtains should be washable and able to be sent through the ironer. Bath mats generally have the same basic characteristics as other terry items, but they are usually heavier.

Table linens have both practical and aesthetic uses. Practically speaking, a tablecloth, place mats, or runners provide a sanitary eating surface, and napkins help guests stay neat while they eat. Aesthetically, a table set with crisp, fresh linens and fancy folded napkins lends an air of elegance to the dining room.

A hotel that offers a dining room and banquet service needs a large assortment of tablecloths. **Table skirts,** which fit under the cloths, are often used for banquets. **Silence cloths** may be used under tablecloths to protect the table surface and to absorb noise. Silence cloths are generally cotton felt or oil cloth backed with polyurethane foam.

Runners and place mats can make inexpensive and attractive alternatives to tablecloths. These items come in a variety of styles and weaves, from elegant to homespun.

Cotton napery is recommended most frequently for restaurants and dining rooms because it is more absorbent and can be starched to retain its shape. This is especially important when napkins are folded into fancy shapes.

Sizes of Linens

Sheets, blankets, tablecloths, etc., have to be sized according to the sizes of the mattresses and tables. Other items can be chosen on the basis of appearance and price. Exhibit 4 shows standard linen sizes for various bed and table sizes.

Tablecloths come in a wide variety of sizes. To make an attractive presentation, the edges of a tablecloth should have a sufficient **corner drop** off the end of the table. Exhibit 5 shows how to calculate drop to determine the right size tablecloth for any size table.

Purchasing many different sizes of sheets, sorting, and separating them is expensive. The careful selection of standard sizes makes purchasing, counting,

Exhibit 4   Standard Linen Sizes

Bed Items	Size in Inches
Sheets	
Twin	66 × 104
Double	81 × 104
Queen	90 × 110
King	108 × 110
Pillowcases	
Standard	20 × 30
King	20 × 40
Pillows	
Standard	20 × 26
King	20 × 36
Bath Items	**Size in Inches**
Towels	
Bath Sheets	36 × 70
Bath	20 × 40
	22 × 44
	24 × 50
	27 × 50
Hand	16 × 26
	16 × 30
Washcloth	12 × 12
	13 × 13
Bath mat	18 × 24
	20 × 30
Napery Items	**Size in Inches**
Napkins	17 × 17
	22 × 22
Tablecloths	45 × 45
	54 × 54
	64 × 64
	54 × 110
Place mats	12 × 18
	14 × 20
Runners	17 × variable lengths

storing, and maintaining inventories much easier. Sizes can be color-coded for easier sorting. Sheets are usually available with color-coded hem threads.

Linen Care, Reuse, and Replacement

Because linen is a major investment, it is particularly important to minimize its disappearance or **shrink** as it is sometimes called. Shrink may occur from wear, improper use, and theft. The cost of linen can be reduced by minimizing wear. One important way to reduce wear is by properly laundering items. Fabrics laundered the wrong way wear much faster because of the damage done to them.

Exhibit 5   Formula for Calculating Corner Drops on a Tablecloth

To find the corner drop of a 72″ square cloth on a 54″ square table:

Cloth size 72″ square × .707	= 50.90″
Table size 54″ square × .707	= 38.18″
Difference is drop at corner of the cloth	12.72″

To find the corner drop of a 72″ square cloth on a 54″ (in diameter) round table:

Cloth size 72″ square × .707	= 50.90″
The radius (half the diameter) of the table	= 27.00″
Difference is drop at corner of the cloth	23.90″

Source: Anthony M. Rey and Ferdinand Wieland, *Managing Service in Food and Beverage Operations* (East Lansing, Mich.: The Educational Institute of the American Hotel & Motel Association), 1985, p. 380.

Improper use of linen—using a guest towel to mop up spills, for example—can cause permanent damage and increase linen costs. Many properties color-code linen to reduce improper use. For example, sheets, spreads, and blankets might be white; table linens yellow; and cleaning rags blue. When items are color-coded, supervisors can easily spot an item being used improperly as a rag.

Housekeeping staff may repair linens that are not beyond repair. Blankets and bath mats, for example, can often be patched. Sheets may be saved by rehemming. And, depending on their construction, bedspreads can sometimes be pieced together. However, at some point it becomes more economical to buy new linen than to repair old.

Linen reuse or recycling can save properties a great deal of money. Turning discarded items into rags is probably the most simple and common type of recycling. Discarded sheets can also be used to replace torn or worn dustcovers on the bottom of box springs. Large sheets can be cut down for crib sheets, aprons, and other items. Tablecloths can be cut into ironing board covers. Some properties, in a final attempt to recoup linen costs, sell discarded linens to staff at a reasonable price. Besides generating revenue that can help replace linens, this policy may significantly reduce employee theft. Other properties donate used linen to charities.

Linen Selection Considerations

Getting linens to the hotel is a long process. It begins in cotton fields, on sheep farms, and in chemical factories where the raw materials used to make linens are produced. From there, the raw materials are shipped to textile mills where they are spun and woven according to a variety of methods. The process continues in finishing plants where various techniques are used to dye, cut, and sew the final products. And the final products themselves are tested at hundreds of sites in mills, factories, and laboratories by manufacturers, professional groups, consumer organizations, and government agencies.

Anyone responsible for purchasing linens and other textiles should be aware that the American Standards Association has developed and issued Minimum Performance Requirements for Institutional Textiles since 1956. The standards cover breaking strength, shrinkage, colorfastness, permanency of finish, seam strength, chlorine retention, components (that is, zippers, grommets, snaps, and other fasteners), thickness and resiliency of blankets, weathering resistance, shape retention after laundering, resistance to mildew and rot, resistance to wetting, and yarn distortion. Standards are available by writing to the American National Standards Institute, 1430 Broadway, New York, New York 10018, or calling (212) 354-3300.

While this chapter cannot cover every aspect of linen manufacturing, it can offer some practical information that will help hotels select the best materials for their guests. This section will cover fabric materials, fabric construction, and finishing.

Fabric Materials. All fabrics begin with raw materials that are spun into long strands of fibers called **yarns** that are woven or knitted into cloth.

A large number of synthetic fibers were developed during World War II. These fibers were frequently stronger than natural fibers and could often be spun to resemble luxurious materials like silk. To make fabrics easier to identify, the U.S. government passed a law in 1960 requiring that all textile products carry labels stating their fiber content.

Today, yarns can be made from dozens of different materials that fall into three basic groups—natural, synthetics, or blends.

Natural fibers. Linens generally come in one of three natural fibers: cotton, wool, or linen. Cotton is by far the most common of these fibers.

Before synthetics and blends became widely used, most of the linens in hotels were made of cotton. Cotton is strong, absorbent, and available in a wide range of grades. Blends and synthetics have replaced the bulk of cotton linens in most hotels. But this trend may be reversing. Natural fibers have enjoyed a renaissance in the clothing industry, and consumers are expressing a preference for other natural fiber items.

Despite the extensive use of synthetics in some items, cotton has continued to be the fabric of choice for napery (table linens) and towels. Cotton's superior absorbency makes it a good choice for napkins and bath towels. It can also be starched (synthetics cannot), which helps it retain a crisp appearance and makes napkins easier to fold into fancy shapes. **Mercerized** cotton, while more expensive, reduces cotton's tendency to produce lint. Cotton blends combine many of the advantages of cotton with the durability of synthetics (mostly polyester).

Wool, once the fabric of choice for blankets, is not as soft, does not wear as well, or wash as easily as synthetic fabrics. It also has a tendency to **felt**, which means that its surface fibers mat together. As a result, most blankets are now made from a variety of synthetic materials.

Cotton and wool can be either **carded** or **combed** and then spun. Combed fibers generally make stronger, more lustrous yarns and better quality fabrics (for example, cotton **percales**). Percales are made of polyester and cotton. Because they are softer to the touch, percales are often used in bed linen. Carded fibers are

Exhibit 6   Synthetic Fiber Generic Names and Some Common Trade Names

Generic	Trade Names
Acetate	Celanese, Celaperm
Acrylic	Acrilan, Creslan, Orlon
Polyester	Dacron, Fortrel, Kodel
Spandex	Lycra
Nylon	
Rayon	
Vinal	Vinylon

coarser and shorter, rougher to the touch, and yield duller fabrics (for example, cotton **muslins**) that produce more lint and pills.

Linen, made from the fibers of the flax plant, is another natural material. It is generally used only for napery today. Linen is smooth, durable, lintless, fast-drying, and absorbent. It is also expensive. A less costly natural fiber can be made by combining linen and cotton, which resembles 100% linen.

Synthetics. Blankets, bedspreads, and shower curtains are most frequently made from all-synthetic fabrics. Synthetics may be less absorbent than cotton or actually moisture-repellent. This makes them particularly attractive fabrics for shower curtains and bedspreads. Synthetics also have good thermal qualities which make them a good choice for blankets. And many synthetic fabrics are stronger than natural fabrics. As a result, some uniform items are made from woven or knit synthetics. There are dozens of synthetic fibers on the market today. A list of some of them is offered in Exhibit 6.

Blends. In the last two decades, many hotels have purchased "no-iron" sheets and pillowcases made of cotton/synthetic (usually polyester) blends. Whether these products can truly be called no-iron, however, is debatable. Usually, laundering erodes the wrinkle-resistance of these linens about halfway through their useful life. Moreover, if no-iron sheets and pillowcases are not removed from dryers and folded immediately, wrinkles generally set in.

Nevertheless, no-iron linens are often stronger than those made of 100% cotton—and they get stronger after more washings. A blended fabric can last for more than 500 washings; a 100% cotton fabric can last for only 150 to 200 washings. This fact alone represents a considerable savings to the hotel. Also, if ironing can be eliminated when the linens are new, then the hotel need not purchase as many ironers as it might if it were using all-cotton linens. No-iron linens can represent considerable labor savings for the property as well.

Some hotels have also found that using a small percentage of polyester fibers in the **base** of bath linens and all-cotton nap on the **face** offers a product with all the absorbency of cotton and some of the added strength of synthetics. Doing so also yields a fabric that tends not to shrink as much as 100% cotton.

Napery made of polycotton blends is widely available. These napkins and tablecloths initially have some of the absorbency of cotton and the easy care characteristics of polyester. With repeated washings, however, the cotton wears out, reducing the absorbency of the item.

Fabric Construction. Some linen items may not be made of woven fabric. Blankets, for example, may be bonded or made by a process known as **fiberlock**. Nylon fibers are flocked to a foam backing to make a bonded blanket. These blankets wash well and have a velvety look and hand.

Woven fabrics. Woven fabrics have two kinds of yarns. The yarns that run the length of the fabric are called **warp yarns**, and the yarns that run sideways are called **fill** (or **weft**) **yarns**. The strength and durability of the fabric depend not only on the material from which the yarn is made but on how thick the yarns are and how closely together the yarns are placed on the loom. When yarns are placed closely together, the fabric is stronger, heavier, and stiffer; when yarns are placed further apart, the fabric is weaker, looser, and more limp.

Balance of warp and **fill yarns** is an important indicator of fabric quality. Better quality fabrics are well-balanced, that is, they have approximately the same number of warp yarns (no more than 10) as fill yarns per square inch. Balance determines how well the fabric will stand up under repeated stretching through a flatwork ironer.

The number of yarns per square inch of fabric is the fabric's **thread count**. Thread count may be written to reveal the number of both warp and fill yarns, for example: 80 × 76. Or the number of warp and fill yarns per square inch can be added together to yield the fabric's total thread count per square inch, for example: T120. (The latter way of expressing thread count, of course, does not tell you whether the fabric is well-balanced. You may have to go directly to the manufacturer or sales representative for more information.) The most common thread count used in hotel bed linen is 180 threads per square inch.

Thread count is a good indicator of fabric durability only if you are comparing linens made from the same kind of fabric. Different fabrics can have the same thread count but different weights. If you are comparing two different fabrics, weight per square inch is a better way to determine which fabric is more durable. Towels are sometimes measured by the number of pounds per dozen.

Yarns can be woven into three basic fabric types called **weaves**: **plain weave**, **twill weave**, and **satin weave**.

Fill yarns in plain weave are simply woven under and over the warp yarns in a crisscross pattern. Twill weaves, somewhat more durable than plain weaves, are woven so that a diagonal pattern of yarns emerges. Most sheets, pillowcases, towels, tablecloths, and napkins are made in plain weave fabrics. However, some properties opt to purchase some of these items in more luxurious—and expensive—satin weaves. In satin weaves, the warp and filling threads interlace to produce a smooth-faced fabric.

Terry cloth is made with a plain or twill weave **base** with extra warp yarns pulled up on either side of the base to form the loops on the **face** of the towel. The face loops may be sheared to make a **velvet**. **Jacquard** towels are those in which a raised terry or velvet pattern is woven into the fabric for a sculpted effect.

Napery (table linens) may be made of plain weave fabric (also called **crash cloth**). **Dobby cloth** is another kind of plain weave fabric into which repeating geometric patterns are woven at regular intervals. **Momie cloth** is a type of dobby. **Damask** is a patterned cloth in which the pattern emerges in a twill weave; the background or **field** is made by passing warp yarns over several fill yarns for a satin effect. While the effect is elegant, the more fill yarns the warp thread passes over, the weaker the fabric. In **single damask** construction (as opposed to **double damask** construction) fewer fill yarns are passed over. This is why most properties prefer to use single damask linens.

Woven blankets come in plain weaves for summer and winter. Thermal weave, a type of plain weave, is designed for cold weather. The blanket surface has small depressions, giving the blanket a waffled texture, which trap and circulate warm air from the body.

Fabric Finishing. Other factors will affect a fabric's quality besides weave. A percale sheet with a high thread count and good balance, for example, is not worth buying if its colors fade quickly or run during washings or if its hems will not hold. Finishing, dyeing, and sewing are therefore important considerations.

Dyeing. Color coordinating linens with a guestroom or dining room's decor often seems like a good way to enhance the appearance of the property. However, colored linens complicate purchasing, laundering, and inventory procedures.

Purchasers should find out how a particular item has been dyed before buying it. Linens that have been **vat-dyed** in the yarn stage (before being woven) are the most colorfast fabrics available. In addition, the purchaser should make sure to purchase linens in the same **dye lot**. Items should be from the same dye lot to avoid minor color differences between pieces. Long-term replacement of these items may become a problem when the dye lot becomes obsolete.

While vat dyed linens retain colors better, all dyed natural fibers will fade after several washings. This fading will be more immediately noticeable in bright colors than in pastels. Moreover, chlorine bleach, used frequently to remove stains, can erode the color further. Laundering procedures must be established and carefully followed to help linens retain their original color.

Because linens will fade with washing, housekeeping staff in charge of storage and inventory should rotate colored linens carefully so that all pieces fade at the same rate. When linens become too faded to use, they will have to be discarded and new linens purchased.

Sewing. Most linens are woven in standard widths on the loom so that the piece need only be hemmed at the ends to prevent unraveling. However, for appearance' sake, it is a good idea to purchase napery that has been hemmed on all sides.

In general, hem threads should be made from some type of fabric that will ← ✳ shrink at the same rate as the piece itself. Otherwise, the piece will pucker after laundering. Hem stitches should be close together so that the hem lasts for the life of the item. Hems that pull out must be repaired by housekeeping staff, and if this happens frequently, the cost of rehemming items can become excessive.

Uniforms

Hotel staff may use many different types of uniforms. Door attendants, parking attendants, guest hosts and hostesses, male and female front desk personnel, bell attendants, chefs and kitchen personnel, waitstaff, banquet servers, maintenance personnel, room attendants, laundry workers, janitorial staff, and others may have their own special uniforms, perhaps even with seasonal variations. Each uniform may have a number of different components. A door attendant may need an overcoat, a summer jacket, pants, hat, and ascot. A hostess may require a skirt, blouse, vest and/or jacket, and scarf. In many hotels, name tags are also considered part of all uniforms.

Identifying Uniform Needs

Identifying uniform needs is not unlike determining linen needs. Staff members may be canvassed by managers to find out whether they wish to wear uniforms and what styles they would prefer. In most cases, especially in chain properties, corporate management decides which staff members will wear uniforms and what styles they may choose from. Staff who wear uniforms need to help track the quality of the items. The hotel must also decide who will pay for uniforms—the employee or the property—and how they will be maintained.

Par levels must also be determined and, again like linen, a number of factors will affect par: Will the hotel's laundry service or on-premises laundry take care of cleaning uniforms? Or will employees be responsible for laundering their own? Is turnover high among employees that wear uniforms? How effective are the property's measures to control uniform loss and damage? What kinds of work do staff members wearing uniforms perform that could damage or ruin uniforms?

Selecting Uniforms

Hotel managers like employees to wear uniforms because it gives them more control over how employees dress. Staff uniforms also allow managers to create an image for the property in the kind of uniforms they choose. Guests like uniforms because they identify staff members who can offer assistance and information. And many staff members like wearing uniforms because it eliminates the need for employees to choose, buy, and, sometimes, care for their own work clothes.

Uniforms must be chosen with caution, however. Food and beverage servers may refuse to wear revealing uniforms that could attract unwanted attention from guests. Staff members may balk at wearing uniforms that are unfashionable, uncomfortable, or ill-fitting. Managers must remember that uniforms should make staff feel well-groomed, neat, and confident about meeting the public. When employees are unhappy with what they are wearing, their dissatisfaction will be transmitted to guests.

Fortunately, the marketplace offers a wide variety of uniforms in a range of styles, colors, and fabrics. Most of today's uniforms are made of polycotton that is durable, is easy to care for, holds its color well, and offers much of the comfort of all-cotton fabrics. Polyester or other synthetic fabric is often used for jackets, coats, scarves, vests, ties, and other accessories. In some areas of the hotel, however, all-cotton uniform items are still preferred. All-cotton kitchen aprons, for example, are more absorbent and easily cleaned than those made of synthetics or blends.

Key Terms

box springs
fill yarns
flat bed springs
innerspring mattress
latex mattress
metal coil springs
napery
par
plain weave

satin weave
shrink
solid mattress
terry cloth
thread count
ticking
twill weave
warp yarns

Discussion Questions

1. Name the main parts of a bed.

2. Name the principal kinds of mattresses and springs and discuss the advantages and disadvantages of each. Which kinds would be most suitable for your property and why?

3. Review some of the factors that affect the quantity of linen that must be kept on hand. What specific factors affect the supply of linens at your property?

4. How does good communication between housekeeping and other departments affect the purchase and control of linens?

5. Discuss some of the main types of fabrics for linens, their advantages, and appropriate/inappropriate uses for each type.

6. What is linen reuse and why is it important? How are linens reused at your property?

7. What are some of the advantages of uniforms?

8. Discuss how uniforms are related to employee morale.

Procedures:
Beds, Linens,
and Uniforms

The procedures presented in this section are for illustrative purposes only and should not be construed as recommendations or standards. While these procedures are typical, readers should keep in mind that each property has its own procedures, equipment specifications, and policies regarding protective gear which are designed to fit individual needs.

Repairing Dustcovers on Box Springs

Equipment

- staple gun and staples
- extension cord (if necessary for stapler)
- discarded sheet
- scissors

Procedures

Step 1
Remove all bed linen from the bed, including dust ruffle (if any).

Step 2
Remove mattress and turn the box spring over on the bed frame so the dustcover side is facing you.

Step 3
If the dustcover is not torn, pull the cover back into place and staple onto the frame with the stapler. Staples should be about three inches apart.

Step 4
If the dustcover is torn:

- Spread the sheet over the box spring to cover the entire bottom. Allow a four-inch overlap on all sides. Trim off excess.
- Fold the top edge of the trimmed sheet under four inches and staple in place. Staples should be about three inches apart. Continue around the frame. Be sure to keep the edges neat and stretched tightly as you staple.

Step 5
Remake the bed.

Turning and Checking a Mattress and Springs

Equipment

- none

Procedures

Step 1
Consult your department's records to find out when the mattress was turned last. (Many properties mark the mattress itself with the date for easy reference.)

Step 2
Strip the bed of all linens.

Step 3
Remove the mattress, noting which side was facing up and which edge was on the right side of the bed. Check the mattress for tears and soil spots.

Step 4
Check the springs for tears in the ticking, broken springs, and tears in the dustcover. Turn the springs so that the right side is now on the left side of the bed.

Step 5
Replace the mattress so that the bottom side is now facing up and the right side of the bed is now on the left.

Step 6
Mark the mattress or note in the department's records that the mattress and springs were turned. Report any damages to your supervisor.

Step 7
If no damages are noted, remake the bed.

Cleaning an Innerspring Mattress

NOTE: Some properties schedule mattress cleaning at the same time as mattress and spring turning.

Equipment

- tank vacuum sweeper
- cleaning brush attachment
- dry foam upholstery cleaner (if necessary)
- sponge
- clean, dry cleaning cloths

Procedures

Step 1
Strip the bed of linens.

Step 2
Using a tank vacuum sweeper and cleaner brush attachment, vacuum the surface of the mattress. Pay special attention to tufting and the edges of the mattress where dirt is most likely to collect.

Step 3
Remove stains by applying dry foam to a sponge and rubbing the mattress from the edge of the mattress inward toward the stain. Remove excess moisture with a clean cloth.

Step 4
When the mattress is completely dry, remake the bed.

Cleaning a Latex Mattress

Equipment

- tank vacuum cleaner
- cleaning brush attachment
- cleaning solution and applicators

Procedures

Step 1
Strip the bed of all linens.

Step 2
Vacuum the entire surface of the mattress with a tank vacuum and cleaning brush attachment. Pay special attention to tufts and reinforced edges where dirt is most likely to collect.

Step 3
Clean the mattress by spraying, sponging, or soaking it with mild antiseptic solution.
CAUTION: Do not use solutions over 150°F [66°C]. Do not use any type of steam cleaning.

Step 4
Remake the bed when the mattress is completely dry.

Making a Crib Sheet

Equipment

- discarded sheets
- scissors
- sewing machine
- measuring stick

Procedures

Step 1

Examine the discarded sheet. You will need a usable section (not worn, torn, or stained) that is 29 inches by 92 inches.

Step 2

Cut the discarded sheet into 29 × 92-inch rectangles.

NOTE: Be sure to cut these rectangles along the straight grain of the fabric. The sheet will be skewed if you cut along the bias.

Step 3

Make a four-inch hem on one end of the sheet with a five-inch turndown.

Step 4

Make a shirttail hem on the other end of the sheet.

Step 5

Fold the sheet in half to make the length 41 inches.

Step 6

Serge the sides and turn.

REVIEW QUIZ

When you feel you have covered all of the material in this chapter, answer these questions. Choose the *best* answer.

True (T) or False (F)

T F   1. Innerspring mattresses are more durable than solid mattresses.

T F   2. Mattresses should be turned once a year.   *4 x a year*

T F   3. Feather pillows are more expensive than acrylic pillows.

T F   4. Silence cloths are used under tablecloths to protect the table surface.

T F   5. Linen is a fabric material made from a synthetic fiber.

T F   6. Linens made from combed fibers are courser, shorter, and rougher to the touch than linens made from carded fibers.

T F   7. Blankets and bedspreads are most often made of synthetic materials.

T F   8. Fabric with a twill weave construction displays a diagonal pattern of yarns.

T F   9. Damask is a type of dobby cloth.

T F   10. To prevent the edges of linens from puckering after laundering, hem threads should be made from the same fabric as the linen item.

Alternate/Multiple Choice

11. The most common type of spring construction used for rollaway beds are:

   a. metal coil springs.
   b. flat bed springs.

12. A thread count of 80 × 70 indicates:

   a. there are 80 fill yarns per square inch.
   b. there are 80 warp yarns per square inch.

13. Selvaged edges of bath towels are:

   a. woven.
   b. hemmed.

14. The most common weave used in the construction of sheets, pillowcases, towels, tablecloths, and napkins is:

   a. satin weave.
   b. plain weave.
   c. twill weave.
   d. jacquard weave.

15. Minimum Performance Requirements for Institutional Textiles are developed by the:

    a. National Textile Association.
    b. OSHA.
    c. American Standards Association.
    d. Federal Standards Board.

Chapter Outline

Carpet Construction
    Tufted Carpet
    Woven Carpet
    Face Fibers
Carpet Problems
    Pile Distortion
    Shading
    Fading
    Wicking
    Mildew
    Shedding/Pilling
Carpet Maintenance
    Routine Inspection
    Preventive Maintenance
    Routing Maintenance
Carpet and Floor Care Equipment
    Wet Vacuums
    Wet Extractors
    Rotary Floor Machines
Carpet Cleaning Methods
    Vacuuming
    Dry Powder Cleaning
    Dry Foam Cleaning
    Bonnet Spin Pad Cleaning
    Rotary Shampoos
    Water Extraction
Special Carpet Treatments
    Antimicrobial Treatment
    Electro-Static Dissipation
Types of Floors
    Resilient Floors
    Wood Floors
    Hard Floors
General Floor Maintenance
Floor Cleaning Methods
    Mopping
    Buffing and Burnishing
    Scrubbing
    Stripping and Refinishing

Learning Objectives

1. Identify the factors that affect a carpet's durability, texture retention, and serviceability.

2. Distinguish between tufted and woven carpet construction.

3. Identify characteristics of face fibers used in carpet construction.

4. Describe common carpet problems and how to remedy them.

5. Explain how carpet and floor cleaning schedules are established.

6. Explain the importance of developing a procedures manual for removing spots and stains on carpets.

7. Describe the carpet and floor care functions of wet vacuums, wet extractors, and rotary floor machines.

8. Describe several carpet cleaning methods.

9. Identify types of resilient floors.

10. Identify type of hard floors.

11. Describe several floor cleaning methods.

14

Carpets and Floors

The carpet sections of this chapter were written and contributed by
James H. Simpson, Vice President of Business Development,
Flagship Cleaning Services, Newton Square, Pennsylvania.

Until human beings learn to defy the law of gravity, carpets and floors will be walked on, spilled on, tracked in on, crushed, and eventually worn down. In a lodging operation, carpets and floors are walked on by thousands of feet every day. As a result, they can become worn and dirty very quickly. A soiled, stained, or faded carpet or floor creates one impression in the minds of guests: poor care and maintenance. No wonder, then, that most lodging properties rate durability, appearance, and ease of maintenance as the major concerns when choosing carpets and floors.

New kinds of floor coverings, cleaning solutions, and maintenance equipment appear on the market every year. Executive housekeepers must keep up with these advances to develop effective cleaning procedures and make wise recommendations about purchasing equipment or contracting carpet or floor cleaning services. For example, many properties can save money by purchasing equipment that can be used on both carpets and floors.

In this text, the term floors is used to refer to all floor surfaces other than carpeting. After a brief discussion of general carpet and floor care considerations, this chapter focuses on the types of carpets and floors found in lodging properties, common preventive and routine carpet and floor maintenance procedures and equipment, and typical carpet and floor cleaning methods. This chapter also includes a brief discussion of special carpet treatments such as antimicrobial treatment and electro-static dissipation.

Carpet Construction

Carpeting offers a number of benefits over other types of floor coverings. Carpeting reduces noise in halls and guestrooms, prevents slipping, and keeps floors and rooms warmer. Carpeting is also easier to maintain than many other floor coverings. Most lodging properties use commercial grade carpet specially designed to withstand more wear and tear than the retail (or consumer) grade carpets people install in their homes.

In general, carpets have three components: the **face**, the **primary backing**, and the **secondary backing**. Exhibit 1 shows a cross section of these components.

Exhibit 1   Cross Section of Carpet Components

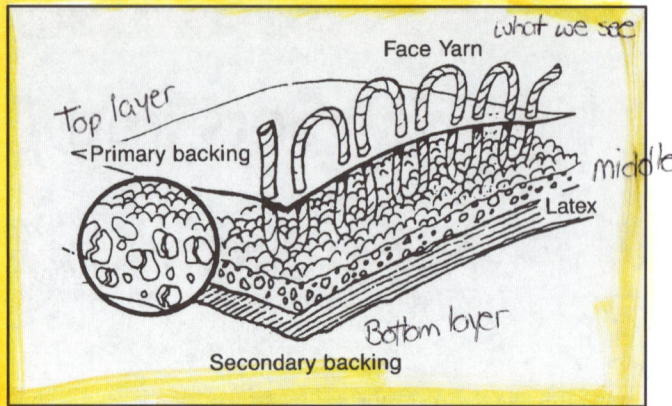

The face or **pile** of the carpet is the part you see and walk on. The face may be made of synthetic fibers or yarns such as polyester, acrylic, polypropylene (olefin), or nylon. The face may also be made of such natural fibers as wool or cotton, though cotton is seldom used as a face fiber today. Some carpets are made of blends of synthetics and natural fibers, or blends of different kinds of synthetics. The carpet's face fibers, as well as its density, height, twist, and weave, will affect the carpet's durability, texture retention, and serviceability.

The density of the carpet's face fibers is the best indicator of durability. In general, the greater the density, the better grade of carpet. Dense carpets retain their shape longer and resist matting and crushing. They also keep stains and dirt at the top of the fibers, preventing deeply embedded soiling. To determine how dense a carpet is, bend a corner of the carpet and see how much backing shows underneath the pile. The less backing that shows, the denser the carpet.

In carpets of equal density, the one with the higher pile and tighter twist will generally be the better product. Carpet that is more tightly twisted is more resilient and will retain its appearance better. When examining a carpet, you should be able to see the twist. The tips of the fibers should not be flared or open. Good quality cut pile carpets have a heat-set twist.

Pile weight, while not as important as density, can affect the carpet's durability. Pile weight is measured in **face weight**—the weight of the face fibers in one square yard of carpet. The greater the weight, the more durable the carpet.

Face fibers are attached to a primary backing which holds the fibers in place. This backing may be made of natural material (typically jute) or synthetic material such as polypropylene. Jute backings are durable and resilient but may mildew under damp conditions. Polypropylene has most of the advantages of jute and is mildew-resistant. Both jute and polypropylene are suitable for tufted or woven carpets.

Usually, the primary backing of the carpet has a backsize. A backsize is a bonding material made of plastic, rubber, latex, or other adhesive that holds the fibers in place. This material is spread in a thin layer over the back of the primary backing and prevents the carpet tufts or loops from shifting or loosening after installation. Some carpets have a secondary backing that is laminated to the primary backing to provide additional stability and more secure installation.

In the past, all carpeting was installed over a separate pad. Today, carpeting may be glued directly to the floor or installed over one of a variety of pads. Sometimes, particularly with carpet tiles, a pad may be bonded directly to the backing in the manufacturing process.

In general, pads should be chosen with as careful an eye toward quality as toward the carpet itself. A cheap pad will reduce the life of the carpet as well as its insulating, sound-absorbing, and cushioning abilities. A thick pad will prevent the carpet from shifting *unless the carpet is installed in an area where heavy equipment will be rolled over it frequently;* in this case, a thinner pad is preferred.

Tufted Carpet

Non-woven or tufted carpet is constructed with either **staple** or **bulk continuous filament (BCF)** fibers. Staple fibers are short (approximately seven to ten inches long) and are twisted together to form long strands. BCF fibers are one continuous strand. Of course, wool and other natural fibers are only available as staple fibers. The reason some carpets shed or pill (pills are small, round fibers appearing at the tips of the tufts or loops) is that not all the fibers in a staple construction are attached to the primary backing as well as the fibers in a BCF construction. Carpets that do not shed are almost always made of BCF fibers.

In tufted construction, needles on a large machine pull the face fibers through the carpet's backing to form tufts or loops. These tufts or loops form a thick pile or plush. Cut pile may be long, short, or cut in various lengths to provide a sculpted effect. Or, the tufts may be pulled to different lengths and left uncut to give a pattern to the finished carpet. Sometimes, both methods are used to create a cut and loop effect.

Berber carpets have short, nubby tufts, and are available in a variety of textures. Level loops are the most common commercial carpets, usually tufted in short continuous rows. Lodging properties typically use a level cut pile carpet in guestrooms to approximate the appearance of residential carpet. Other types of carpeting may be used in public areas, depending on the individual design requirements of the property.

Woven Carpet

In a woven carpet, a machine or loom weaves the face fibers and backing together as the carpet is being made. Generally, woven carpets are available only in narrow widths or strips that are attached or seamed together. Woven carpets do not have secondary backings, but they can perform as well as or better than tufted carpets if properly installed and maintained.

The weaving consists of warp (lengthwise) and weft (widthwise) yarns interwoven to form the face pile and backing at the same time. Different weaves include Velvet, Wilton, and Axminster. Many variations are available with the Velvet weaving method, including plushes, loop pile, multilevel loop, and cut and loop styles. Wilton refers to the special loom used to produce intricate patterns (sometimes

Exhibit 2   Typical Carpet Types and Characteristics

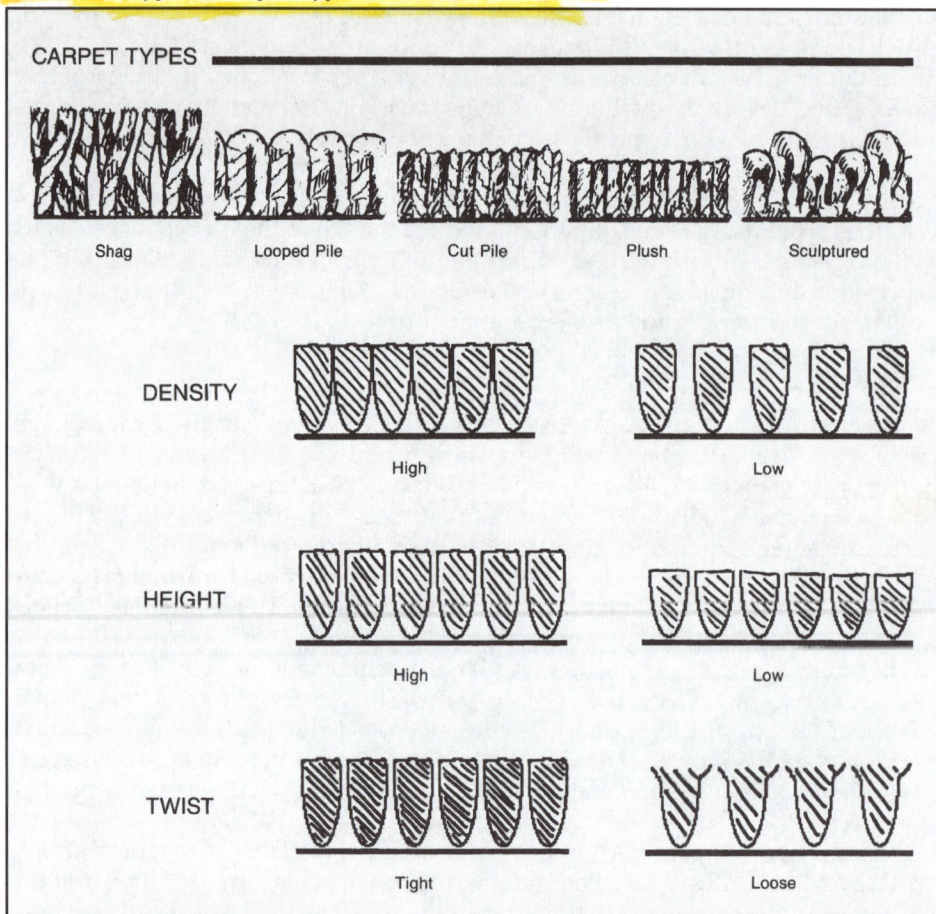

CARPET TYPES

Shag	Looped Pile	Cut Pile	Plush	Sculptured

DENSITY

High   Low

HEIGHT

High   Low

TWIST

Tight   Loose

multicolored) using perforated pattern cards. Axminster weaves are made from prearranged spools of different colored yarns which are fed into a mechanical patterning device. This method places most of the pile yarn on the surface and leaves a ribbed back.

Different types of carpets and various characteristics are shown in Exhibit 2.

Face Fibers

In general, synthetic fibers are more durable, more sanitary, and less expensive than natural fibers. These advantages help explain why synthetic fibers make up 90% of the carpets used in commercial operations. Face fibers are judged on their appearance, springiness and texture retention (ability to hold their shape), resistance to wear, resistance to soil, and cleanability. Typical face fibers include wool and other natural fibers and nylon and other synthetics.

Wool and Other Natural Fibers. People who buy and sell carpets agree that wool is good-looking, resilient, durable, and easily cleaned. It is also expensive.

Despite the cost, wool is especially well-suited for lodging properties because of its natural resistance to flame and its ability to shed soil. Indentations caused by furniture legs, sometimes permanent on synthetic carpet, can be removed easily from wool by applications of moisture and low heat.

Wool fibers are also water-loving, which means they are responsive to wet cleaning. Unfortunately, this also means that they provide a better breeding ground for microorganisms than do synthetic fibers. Molds, mildew, bacteria, and other growths can mar the carpet and/or cause odors. Cleaning solutions for wool should be chosen carefully; ammonia, salts, alkaline soaps, chlorine bleach, or strong detergents can damage the fabric.

Other natural fibers available but rarely used today are cotton, sisal (hemp), and silk.

Nylon. More than 80% of all carpets manufactured in the United States are nylon. Nylon holds its shape and color well, cleans easily, and costs much less than wool. When properly maintained, nylon carpet fibers are less likely to promote bacterial growth; they are easily treated for resistance to mold, mildew, and other organic growth. Nylon is particularly attractive because of its durability and its flexibility in construction and design. It is also comfortable under foot and is more resistant to stains and soil.

Nylon fibers normally have a shiny appearance. However, a "baking" process is now available that gives the carpet a duller or **delustered** finish that looks more like wool. Delustered carpets have the added advantage of soiling less quickly.

Other Synthetics. Acrylic fibers were developed in the 1950s to approximate the appearance and durability of wool. Generally, acrylic carpet is not as easily cleaned or as resilient as other synthetics. It can also turn brown during cleaning. It has poor resilience and a tendency to pill and fuzz. Oil may leave permanent stains if not treated quickly. Acrylic does, however, resist most acids and solvents. **Mod-acrylic** is similar to acrylic, but has less resistance to stains and abrasions.

Olefin (or **polypropylene**) fibers wear very well. They can be cleaned very aggressively without damage and are not as susceptible to sun fading as nylons and wools. Olefin carpets are solution-dyed, which means that color is added to the olefin in its liquid state. Olefin resists acids, solvents, and static electricity buildup. Olefin is, however, susceptible to heat or friction damage and is not as comfortable to walk on as other fibers.

Polyesters offer an appearance similar to wool. They are also very durable and clean easily. They do, however, have a tendency to mat under heavy traffic and are not very resilient.

Acetate is a low-cost silky fiber. It is colorfast and resistant to mildew, but it is relatively easy to soil and is vulnerable to abrasion. Spot-cleaning should be handled with care because acetate may dissolve when dry-cleaning fluid or solvents are applied.

Rayon has many of the same characteristics as acetate—poor resistance to soil and abrasion but good color retention and resistance to mildew. Dense, high-grade rayons have adequate resilience. Rayon is vulnerable to oil stains.

Carpet Problems

To keep carpets as attractive and clean as possible, housekeeping staff in charge of carpet cleaning should learn to recognize and remedy the following common carpet problems:

- Pile distortion
- Shading
- Fading
- Wicking
- Mildew
- Shedding/pilling

Pile Distortion

Pile distortion is a general term for a number of problems with the carpet's face fibers. Fibers can become twisted, pilled, crushed, or flared and matted. Pile distortion occurs when the carpet receives heavy foot or equipment traffic. Improper cleaning methods can also cause pile distortion. For example, the pile can lose its twist when it is subjected to cleaning solutions that are too hot or methods that are too aggressive.

Pile distortion is hard to remedy. It may be impossible to remedy in high-traffic areas. Mats, runners, and furniture glides can help prevent crushing. Regularly vacuuming or using a pile lifter or pile brush on high traffic areas will help to remove dry soil which can wear on fibers and cause pile distortion. A pile lifter will help pick up crushed pile while removing gritty soils which can damage the carpet. Pile distortion can also be prevented by grooming high traffic areas with a carpet rake. Using a pile lifter or carpet rake before corrective cleaning will improve the effect of the cleaning.

Shading

Shading occurs when the pile in a carpet is brushed in two different directions so that dark and light areas appear. Shading is a normal feature of almost all carpets. Vacuuming or pile lifting the carpet in one direction can help to reduce a shading problem, but will probably not eliminate it. Some properties instruct room attendants to leave shading marks purposely in the carpet while vacuuming. The notion behind this practice is that guests will see the vacuuming pattern and feel that the room has been properly cleaned.

Fading

Every carpet will fade with time. Sunlight, wear, cleaning, and natural aging can combine to accelerate color loss. Premature fading may occur if the carpet is improperly cleaned. Some professional carpet service companies can dye carpets that have faded prematurely. Improper cleaning or spot removal can actually do more damage than some permanent stains. Always pre-test carpets before using aggressive spot-removal techniques.

Wicking

Wicking (sometimes called browning) occurs when the backing of the carpet becomes wet and the face yarns draw or wick the moisture and color of the backing to the surface of the carpet. Wicking can often be prevented by promptly attending to spills and by following proper cleaning procedures that avoid overwetting the carpet.

Wicking occurs most frequently in jute-backed carpet that has a light color face fiber. Vinegar or synthetic citric acid solutions used in post-cleaning treatments or added to certain cleaning chemicals can help prevent or cure browning problems. As always, check with the manufacturer and/or pre-test the application before proceeding with an anti-browning treatment.

Mildew

Mildew forms when moisture allows molds in the carpet to grow. Mildew can cause staining, odor, and rotting. Natural fibers are especially prone to mildew, but all carpets should be kept dry and/or treated with an anti bacterial to prevent the problem. Proper cleaning procedures that avoid overwetting the carpet can help prevent mildew from forming.

Shedding/Pilling

Short pieces of face fibers are often trapped in the carpet when it is manufactured. As the new carpet is walked upon, these pieces work themselves to the surface of the carpet and can make a new carpet look littered and unkempt. Shedding will eventually stop. In the meantime, frequent vacuuming will prevent the carpet from looking littered. Pilling, often the result of cleaning, can be removed by heavy vacuuming or by gently cutting loose fibers from the carpet with scissors.

Carpet Maintenance

The aim of any carpet maintenance program is to keep the carpet clean and like new for as long as possible. To some extent, the carpet's fiber, backing, and construction will dictate the maintenance program required to keep it looking good. Most floor care experts encourage executive housekeepers to develop a cleaning schedule for all carpet and floor areas. The most efficient cleaning schedules are based on studies of the amount of traffic in the various areas of the property.

Heavy soiling typically occurs in high-traffic public areas, track-off areas, and funnel areas. Track-off areas are those directly in front of doors leading to the outside. Funnel areas, identified by their distinctive funnel-shaped soil spots, are those in which traffic converges into a narrower space. Funnel areas typically occur around elevators, stairways, and in front of vending machines.

High-traffic and heavy-soil areas can be indicated in colors or shades on a property floor plan. For example, one color or shade could indicate high-traffic areas—those that need to be cleaned at least once a day. Areas with less traffic that soil more slowly can be depicted through other colors or shades to indicate weekly, monthly, or quarterly cleaning. Exhibit 3 shows such a floor plan for a hospitality operation.

Once the floor plan is completed, a calendar plan can be devised. The calendar plan should list the cleaning tasks—vacuuming, spot-cleaning, deep cleaning—to be performed on specific days and the time required for each task. Implementing a regular cleaning schedule has a number of advantages:

- Housekeepers can accurately forecast monthly and yearly cleaning costs.

- Regular maintenance will prevent major problems from occurring and will extend the life of the carpet.

- Regularly scheduled carpet and floor cleaning allows the executive housekeeper to budget time for other major department projects.

Before implementing any maintenance programs or purchasing equipment or cleaning chemicals, the executive housekeeper should consult with the carpet supplier or manufacturer and follow the suggested cleaning procedures.

Routine Inspection

Inspection is an important part of all carpet and floor care programs. Housekeeping staff generally inspect carpets and floors in all areas of the property each day. All hotel employees should be instructed to help preserve carpets and floors by promptly reporting spots and spills to the housekeeping department. Good maintenance depends on immediate removal of spots and spills.

Housekeeping supervisors should routinely review the property's carpet cleaning procedures and ensure that employees follow these procedures properly. At many properties, supervisors routinely inspect cleaning equipment to make sure that all items function safely and efficiently.

Preventive Maintenance

Lodging properties can prevent carpet soiling and damage with frequently changed mats and runners in high-traffic, track-off, and funnel areas. Furniture glides on the bottoms of chairs and tables can help reduce pile distortion or tearing. Damage from food and beverage spills around self-serve bars in dining rooms or near vending machines can be reduced by using waterproof plastic carpet protectors or, if preferred, regularly cleaned customized mats. Carpet tiles, which can be easily replaced or rotated in high-use areas, can be used to reduce pile distortion.

Routine Maintenance

Most housekeeping departments vacuum all carpets once a day—or even more often—as part of routine maintenance. Routine maintenance also includes periodic deep cleaning (shampoos and hot or cold water extraction), spot-cleaning, and stain removal as necessary. Spots should be removed quickly before they set into the fabric as more-difficult-to-remove stains.

Exhibit 3   Sample Floor Plan of Carpet Traffic Areas

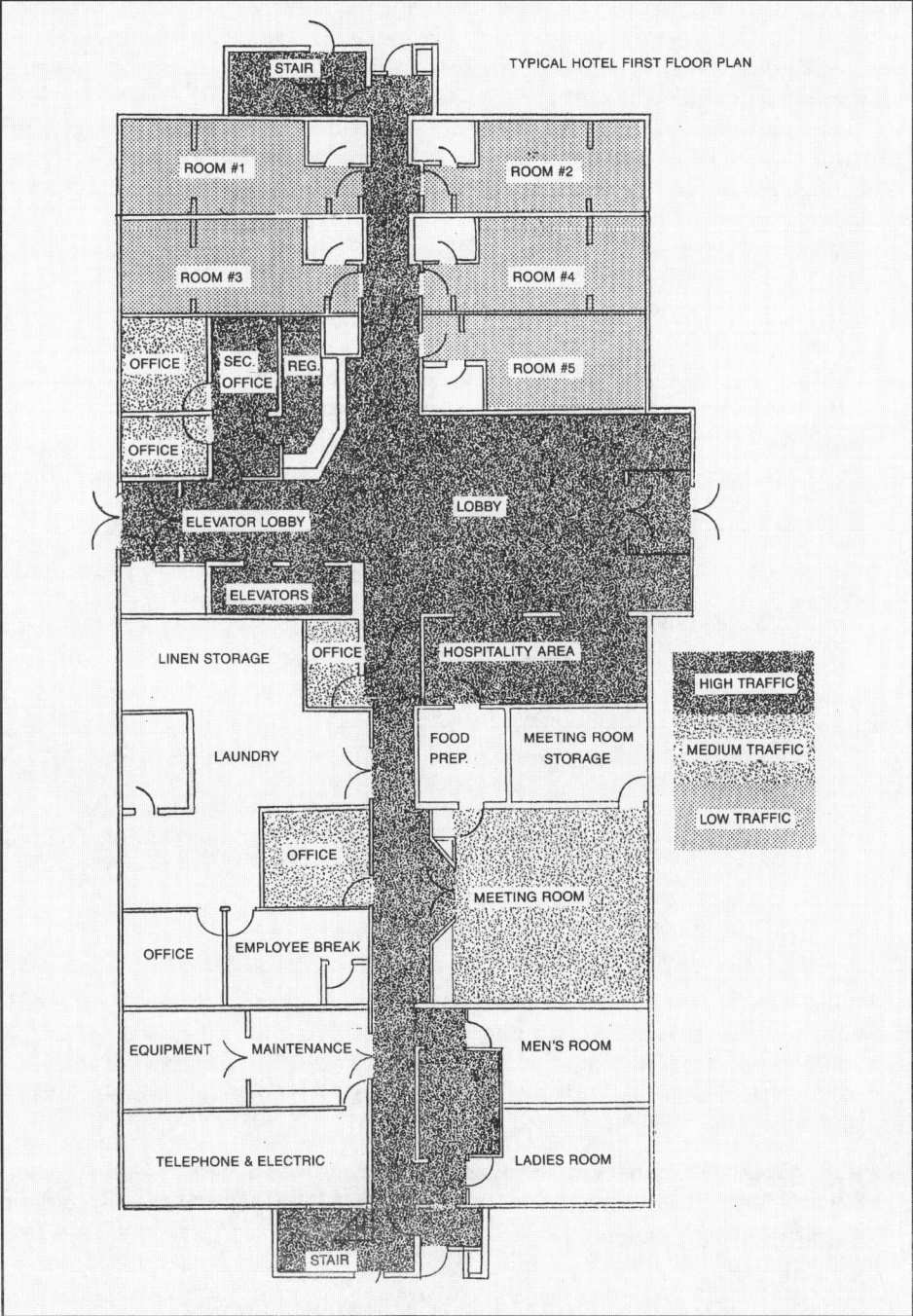

TYPICAL HOTEL FIRST FLOOR PLAN

STAIR

ROOM #1   ROOM #2

ROOM #3   ROOM #4

OFFICE   SEC. OFFICE   REG.   ROOM #5

OFFICE

ELEVATOR LOBBY   LOBBY

ELEVATORS

LINEN STORAGE   OFFICE   HOSPITALITY AREA

HIGH TRAFFIC

LAUNDRY   FOOD PREP.   MEETING ROOM STORAGE   MEDIUM TRAFFIC

LOW TRAFFIC

OFFICE

MEETING ROOM

OFFICE   EMPLOYEE BREAK

EQUIPMENT   MAINTENANCE   MEN'S ROOM

TELEPHONE & ELECTRIC   LADIES ROOM

STAIR

Courtesy of Flagship Cleaning Services, Newtown Square, Pennsylvania

The executive housekeeper should develop a manual for spot and stain removal, stating the proper procedures and cleaning fluids to use on the carpets throughout the property. This task can be simplified by contacting the carpet supplier for care and maintenance instructions and spot-removal techniques especially suited for carpets on-site.

A sample spot and stain removal chart is provided in Exhibit 4. Some general guidelines for spot-cleaning are included here. Because all cleaning techniques are not suitable for all carpets, the following instructions are presented for illustrative purposes only.

1. Identify the source of the spot or stain. This makes removal easier.

2. If you can't identify the source of the stain, carefully remove solid particles from the spot or stain with a hand-held scraper or soup spoon and vacuum thoroughly. Use a wet vacuum whenever possible to pick up large wet spills. After removing as much of the wet spill as possible, blot up remaining moisture by pressing a clean, dry towel against the spot or stain.

3. If vacuuming the spill is impossible, blot carefully from the edges to the center of the spot or stain to remove as much as possible of the staining material before adding a cleaning solution. *Never rub to remove a spot or stain;* you may cause irreversible fiber distortion, such as fuzzing, loss of twist, or loss of fiber.

4. Always try to remove the spot or stain with plain water alone, which will remove many spills. Gently blot excess water as noted above.

5. If unsuccessful with water, spot-test an inconspicuous portion of the spot or stain with another cleaning agent. Blot some of the agent on the affected area with a clean rag. Be sure to apply the solution *to the rag only,* then blot.

6. If a small amount of cleaning solution works, do not add more; it will only increase the amount of rinsing required to remove the residual cleaning agent. If you are using a water-soluble detergent cleaning solution, you will need to rinse the spill to remove the cleaning agent. If you are using a solvent, water rinsing is not necessary. Be sure to rinse out detergent residues with water before trying solvent cleaning agents.

7. If the spot or stain lifts with a water-soluble solution, rinse the area frequently with water to hasten the cleaning without adding more chemical. After the spot or stain is gone, a minimum of three rinses is suggested to prevent any residual detergent from attracting new soil to the clean area. Use the blotting procedure previously described.

8. If the spot or stain lifts with a solvent, continue to work with solvent on a rag. Do not pour solvent directly on the area. Do not introduce water to a spot or stain that is lifted with a solvent. Solvents are volatile and will dry without a residue.

9. In all cases, do not overwork the spot or stain. Work only for short periods and allow the area to dry before returning later to continue spot or stain removal.

Exhibit 4   Sample Stain Removal Chart (Carpets)

Spot removal procedures*

HOW TO USE THIS CHART:

1. Identify the spot and locate it on the chart below.
2. Apply first solution suggested.
3. If it does not remove all of the spot, proceed to second suggested solution. Rinse treated area with water before applying next solution.
4. Proceed with solutions in the sequence outlined until all of the spot is removed.
5. Rinse treated area thoroughly with water to remove any residual spot remover.

	GROUP 1 Oil and Grease	GROUP 2 Liquids	GROUP 3 Food and Body Waste	GROUP 4 Dyes, Inks, Medicines	GROUP 5 Chewing Gum, Rust	GROUP 6 Unknown Spots
TYPE OF SPOT	Asphalt Copy Powder Cosmetics Crayon Duco Cement Grease India Ink Paints Oils Rubber Cements Shoe Polish Tar	Beer Cocktails Coffee Colas Fruit Juices Soft Drinks Tea Tobacco Urine	Animal Glues Blood Catsup Chocolate Cream Eggs Feces Gravy Ice Cream Starch Vomit	Colored Paper Food Furniture Inks Marking Pens Medicines Soft Drinks	Chewing Gum Rust	
SPOT REMOVER	Dry Cleaning Fluid Non-Oily Paint Remover Amyl Acetate Dry Cleaning Fluid Wet/Dry Spotter Detergent Solution 5% Acetic Acid 5% Ammonia Detergent Solution Water	Detergent Solution Wet/Dry Spotter 5% Acetic Acid 5% Ammonia 1% Hydrogen Peroxide Detergent Solution Water	Detergent Solution 5% Ammonia Wet/Dry Spotter Detergent Solution Digestor Detergent Solution Water	Detergent Solution Alcohol 5% Acetic Acid 5% Ammonia Wet/Dry Spotter 1% Hydrogen Peroxide Detergent Solution Water	**Chewing Gum:** Chemical Freezing Compound Dry Cleaning Fluid **Rust:** Rust Remover Detergent Solution Water	Dry Cleaning Fluid Paint Remover Amyl Acetate Wet/Dry Spotter Alcohol Detergent Solution 5% Ammonia 5% Acetic Acid Digestor 1% Hydrogen Peroxide Detergent Solution Water

Note: For specific concentrations and amounts of spot removal agents to be used, consult instructions issued with your spot removal kit or contact a carpet care professional.

*Reproduced in cooperation with the Carpet & Rug Institute

Du Pont Company
Carpet Fibers Division
Fibers Marketing Center
Wilmington, DE 19898

Courtesy of Du Pont Company, Wilmington, Delaware

Exhibit 5   Commonly Used Small Equipment for Carpet and Floor Cleaning

Small waterproof tarp—used for mixing cleaning solutions

Nylon nap brush—used to restore carpet pile after cleaning

Carpet rake—used after cleaning or used to restore very long carpet pile that has become crushed

Hand shampoo brush—used on stairs and edges

Mixing buckets—used to mix and carry cleaning solutions

Stirring paddle—used to mix cleaning solutions

Traffic lane paper—used to protect just-cleaned carpets from guest traffic; can be used to protect floor-length drapes from wet carpet

Clip-on floodlights—used to light dark hallways and stairwells that must be cleaned

Sprayers—may be manually pumped or electric-powered; used to spray solutions and water onto carpet

Measuring cups—used with cleaning solutions

Mop—used for daily cleaning of floors or for rinsing and applying polishes and strippers

Mop bucket—may be a simple stainless steel bucket or a bucket on casters with a wringer attached

Hand wringer—used to squeeze excess moisture out of mop heads and bonnets

Squeegee—a tool with a rubber blade, like a windshield wiper, used for picking up excess moisture from floors

Pickup pan—works like a dustpan; used with a squeegee

Mats, runners—used to protect carpets and floors from traffic or to protect guests and employees from slippery surfaces

Furniture glides—typically small disks or squares that fit under furniture legs to protect carpet pile

Pill shearer—small shaver-like machine that shaves pills from carpets

Carpet and Floor Care Equipment

Deciding when and how carpets and floors should be cleaned is an important task. This task is complicated by the great number of cleaning machines, supplies, and solutions now available. Commonly used small equipment items are described in Exhibit 5. Major equipment items are discussed in the sections which follow.

Because carpet and floor cleaning equipment is a major expense, equipment should be cared for properly. A sample equipment inspection checklist is provided in Exhibit 6.

Outlining the kinds of equipment individual properties should have is beyond the scope of this book. However, a property's arsenal of dirt-fighting equipment often includes wet vacuums or wet extractors and rotary floor machines. Each of these equipment items can be used on carpets and floors. To maintain the warranty on a carpet, be sure to follow the manufacturer's suggested cleaning guide.

Exhibit 6   Sample Equipment Inspection Checklist

Casters and wheels:
- [ ] Clean wheels of litter and dirt, string, hair, etc.
- [ ] Replace broken wheels that could snag or tear carpet.

Hoses:
- [ ] Check for leaks.
- [ ] Check for loose dirt or cleaning solutions caught inside the hose.

Electrical cords:
- [ ] Check cords and plugs. Do not use machines with frayed cords or cords without insulation.

Brushes and mops:
- [ ] Replace brushes that are worn down to less than $1/8$ of an inch.
- [ ] Be sure all machine and hand-held brushes are clean.
- [ ] Make sure the mop handle is in good condition and the mop head is clean and dry.
- [ ] Use a new mop only after soaking for 30 minutes in clear water to remove sizing.
- [ ] Use only clean, dry bonnets.

Tanks, buckets, and bags:
- [ ] Make sure extractor and wet vacuum tanks are clean and empty before using.
- [ ] Make sure extractor tanks are rinsed of solutions before using.
- [ ] Replace torn vacuum sweeper bags or those that are more than half full.

Wet Vacuums

Wet spills will damage many carpets and floors. They also pose safety hazards for guests and employees. Wet vacuums are used to pick up spills or to pick up rinse water used during carpet or floor cleaning. Many wet vacuums can also be used on floors to pick up dry soils. Conventional vacuuming equipment must *never* be used to pick up wet spots because electrical shock can occur and the machines can be destroyed.

Wet vacuums may have only suction or both suction and a water sprayer which can rinse the soiled area immediately. Squeegee attachments on wet vacuums can make floor cleanup, stripping, and scrubbing more efficient.

Wet vacuums come in a variety of shapes and sizes. Some canister models are small enough to be strapped onto the cleaner's back. Canister vacuums have a collection tank which stores the water, the hose, and cleaning head pick up. Typically, the tank rests on casters for easy mobility. Many canister models can be adapted to pick up dry soil on floors. Some manufacturers and most cleaning experts recommend canister models with bypass motors. These units prevent moisture from condensing on the machine's motor, thus reducing mechanical problems over time.

Walk-behind wet vacuums pick up water quickly in very large areas. These models are a little larger than a grocery cart and are often self-propelled for easy handling. Some models come with cords; others are battery-powered.

Wet Extractors

While some wet vacuums have only suction capability, wet extractors have a capacity for both suction and water injection. These units can simultaneously rinse and vacuum the soil from carpets or floors.

Wet extractors spray water and detergent onto the carpet. The extractor then uses suction to remove the water, detergent, and dirt from the carpet. Some machines have a special tool that agitates the carpet before spraying to loosen dirt. Such attachments as specialty drapery and upholstery cleaning tools are also available.

Wet extractors come in a variety of shapes and sizes. Tank and walk-behind models are available. These machines can often be used on floors for dry pickup. Self-contained extractors, which resemble upright vacuums, cannot be adapted to floor use. There are no hoses on a self-contained extractor; this makes the machine more compact to use in small areas such as guestrooms.

Rotary Floor Machines

Rotary floor machines can be used for a wide variety of surface cleaning jobs. Those that accommodate both brushes and pads are the most versatile. On carpets, rotary floor machines can be fitted with pads or brushes to perform dry foam cleaning, mist pad cleaning, rotary spin pad cleaning, bonnet shampoos, or brush shampoos. On hard surface floors, rotary floor machines can be used to buff, burnish, scrub, strip, and refinish. Most manufacturers and floor care businesses offer specially made pads for particular tasks, which makes the floor cleaner's job easier.

Housekeepers should note that manufacturers use a color code for floor pads according to the job they perform. Stripping pads are black and brown, scrubbing pads are blue and green, polishing pads are white, and spray cleaning pads are red. Burnishing pads are not universally color-coded.

Some rotary machines come with a cleaner-dispensing tank and disk holders or "blocks" that are used to hold bonnet pads. By changing from a bonnet block to a brush block, the same machine can be used to shampoo carpets. One machine can even be fitted with a small vacuum and used to vacuum the foam while shampooing. Bonnet pads are made for both hard floor and carpet maintenance, and may be used on the same machine and bonnet block after changing chemicals.

High-speed machines (300 rpms to 1,000 rpms) reduce buffing time and create a more durable buff coat. Burnishers run at extremely high speeds (700 rpms to 1,500 rpms), further reducing buffing time. Rotary floor machines that exceed 175 rpms should *never* be used on carpets; high-speed machines will cause enormous damage.

Rotary floor machines can be tricky to use. Inexperienced employees, for example, may overwet the carpet when shampooing. This can lead to a host of problems, such as seam separation, delamination of the backing, buckling, shrinkage, premature face fiber wear, mildew formation, and other damaging conditions.

Improper use of rotary machines on floors can cause abrasions or inadvertent removal of sealant and/or finishes. Many properties allow only experienced housekeeping personnel to operate rotary machines and other large pieces of equipment.

Carpet Cleaning Methods

There are a number of different kinds of carpet cleaning methods, from simple vacuuming to hot or cold water extraction. Carpet experts frequently disagree over which cleaning methods are most effective and which methods produce less wear on the carpet itself. For example, some say that shampooing can wear down the carpet and leave soap residues that will actually attract dirt.

Complicating the situation is the fact that the many different types of carpets and floors sold today have very different care and maintenance requirements. For example, simple ammonia solutions which work on most synthetic carpets will cause immediate and irreversible damage to natural fibers. Similarly, olefin carpets can be cleaned with bleaching solutions that would damage nylon carpets.

Executive housekeepers should carefully follow the carpet manufacturer's recommendations regarding cleaning methods. Improper carpet cleaning procedures can not only void the manufacturer's warranty, but can also cause accelerated soiling and deterioration of the carpet. Carpet suppliers are able to provide specific data on carpet care requirements that will help executive housekeepers select the correct cleaning methods, products, and services offered by carpet care suppliers and/or carpet cleaning contractors.

The following sections describe basic carpet cleaning methods. The method(s) appropriate for a property will depend on the specific requirements set by the manufacturer.

Vacuuming

Carpet experts do agree on one thing: you can't vacuum too much. Daily vacuuming prevents abrasives (gritty materials such as sand and gravel) from working into the carpet and causing stains and wear over a long period. It also helps restore the carpet's pile. The most effective vacuuming equipment agitates the carpet to loosen the dirt and removes it with suction.

Many different types of vacuums have been developed for different types of carpets. Beater-bar vacuums, for example, use a bar to agitate the carpet to loosen dirt. These vacuums are best for carpets installed over pads. Brush vacuums agitate the carpet with a brush and are best for carpets glued to the floor. Pile-lifter machines have a strong suction capacity and a separate brush motor which helps restore crushed carpet pile. Beater-bar, brush, and pile-lifter vacuums generally come in upright models. An upright sweeper has a large agitator/suction head that pulls dirt up into a bag attached along the handle of the machine. Canister models collect dirt in a tank which moves on casters for easy maneuvering. Some canister models are small and can be strapped to the cleaner's back. Canister models provide little agitation to loosen dirt, but they are handy for cleaning hard-to-reach corners and spaces that upright models cannot reach.

Dry Powder Cleaning

With the dry powder carpet cleaning method, dry powder or crystals are sprinkled onto the carpet and worked into the pile with a hand brush or a special machine that dispenses and brushes in the powder. The powder absorbs oily soils which are then removed by vacuuming. Since no drying time is required, carpets in high-traffic areas need not be closed off. Properties may schedule dry powder cleaning between deeper cleaning activities.

Dry powder cleaning may be a good way to clean carpets which should not be cleaned with water. Dry powder cleaning also does not leave a soap residue on the carpet or excess water that could give rise to mold or mildew. The brushes in the dry powder machine, however, may cause cut pile fibers to flare. Periodic extraction/wet cleaning is suggested to remove residue from carpets. Some experts also caution that dry powder cleaning will not remove some types of dirt very well.

Dry Foam Cleaning

With another dry method, dry foam is sprayed on the carpet and a rotary floor machine brushes the foam into the carpet. The foam is then removed with a wet vacuum. Machines are available that will apply the foam and vacuum away the moisture. Dry foam can also be sprayed and brushed into carpets by hand. Since this method requires little drying time, housekeepers often use dry foam cleaning in high-traffic areas as frequently as once a day.

Dry foam cleaning can cause flaring in some cut pile carpets. If proper cleaning procedures are not followed, overwetting can occur, which could lead to shrinking, mildew, fading, and other conditions. A sample dry foam carpet maintenance program from a carpet care product manufacturer is shown in Exhibit 7. Some manufacturers provide similar materials in languages other than English.

Bonnet Spin Pad Cleaning

Bonnet spin pad cleaning is similar to the dry foam method in that it can be used daily for surface cleaning. With this method, a rotary floor machine with a special holder and pad agitates the tips of the carpet fibers as it moves across the carpet. The bonnet pad lifts and absorbs soil as it rotates. Pads are made of synthetic or natural fibers, and can be laundered and reused as often as necessary. Bonnet spin pad cleaning should be followed by vacuuming after the carpet has dried.

Rotary Shampoos

Rotary or brush shampoo offers more effective cleaning than dry powder or dry foam cleaning. With this method, a rotary bristle brush is used instead of a pad or bonnet. (Before using new bristle shampoo brushes on fine carpets, the brushes should be broken in by running them on concrete floors.) The brush design allows the shampoo to drain down through the bristles directly onto the carpet. The shampoo is agitated by the machine into a foam which can then be left to dry or be vacuumed by a wet vacuum or extractor. When vacuumed, it is necessary to use a defoamer in the wet vacuum or extractor tank. As with all carpet cleaning

Exhibit 7   Sample Dry Foam Carpet Maintenance Program

PREPARED FOR:					
PREPARED BY:					

CARPET MAINTENANCE PROGRAM
using Dry Foam Method

TYPE OF CARPET	Natural	Synthetic	Cut Pile	Wire/Level Loop	Shag
SOIL CONDITION	HEAVY		MEDIUM	LIGHT	

REMOVE FURNITURE AND VACUUM CARPET

Remove all furniture to assist total cleaning of carpet and save time. Thoroughly vacuum carpet.

PRESPOT

Use Carpet Stain Remover for spots and stains.

PRETREAT

For heavily soiled traffic lanes, doorways, etc. use trigger sprayer or Spartasprayer and apply Plus 5 diluted 1:4-1:6.

MIXING

For light to medium soil conditions, add 1 part Plus 5 to 10 parts luke warm water. When carpet is heavily soiled or greasy, dilute Plus 5 1:8. Put water in first to avoid excessive foaming.

SHAMPOOING/NORMAL SOIL CONDITIONS

Release foam and move machine in the direction the pile lays. Overlap slightly and repeat.

SHAMPOO/HEAVY SOIL

Release foam and make forward and backward pass in the **same** path. Overlap slightly and repeat.

WET VAC (OPTIONAL)

Vacuum immediately with wet vac to remove soil-laden foam and speed drying.

SET PILE (OPTIONAL)

Using pile rake – rake the carpet against the way pile normally lies. This sets pile and decreases drying time. Replace furniture, using protective pads under metal casters.

DRY VAC

After carpet is completely dry, vacuum thoroughly with dry vac to remove soil ash. **Vacuum daily to prolong carpet life and extend the time between shampooing.**

PRODUCTS

PLUS 5 CARPET SHAMPOO
CARPET STAIN REMOVER

TOOLS

DRY FOAM MACHINE
WET/DRY VACUUM WITH CARPET TOOL
PILE RAKE
BUCKET
MEASURING CUP
TRIGGER SPRAYER OR SPARTASPRAYER

©SCC-1986

DISTRIBUTED BY

Spartan Chemical Co., Inc.
110 N. Westwood Avenue • Toledo, Ohio 43607
Spartan®

Courtesy of Spartan Chemical Co., Inc., Toledo, Ohio

machinery, caution should be exercised so as not to damage the carpeting by using too little solution, too much solution, or too much agitation.

A sample rotary brush carpet maintenance program from a carpet care product manufacturer is shown in Exhibit 8. Some manufacturers provide similar materials in languages other than English.

Water Extraction

Water extraction is the deepest cleaning method available for most carpets. Hot water extraction is sometimes inappropriately called steam cleaning. Actually, hot water extractors should never be filled with water that is higher than 150°F [66°C] in temperature. Since wool carpets can shrink, they should only be cleaned with warm or cool water.

Hot water extractors spray a detergent and water solution onto the carpet under low pressure (less than 200 pounds per square inch), and, in the same pass, vacuum out the solution and soil. A good extractor can pull 70% to 90% of the water out of the carpet. Some extraction machines have a special tool called a power head that agitates the carpet before the solution is extracted. Other tools, such as upholstery and stair tools, can be attached to the extractor, making the unit useful for a variety of cleaning functions.

With proper extraction, a carpet will dry in an hour or two. However, housekeeping staff using hot water extraction should be careful to control the amount of water that goes into the carpet. Overwetting can be a problem if the equipment is under-powered or if the operator does not take care to thoroughly vacuum the cleaned area.

Hot water extraction requires a great deal of hot water. This can be hard on some carpets. Properties trying to cut their hot water consumption may want to consider models that use cold water in this process. In some cases, cold water extraction works as well as hot water extraction and helps reduce color fading, running, and shrinking.

Excessively soiled carpets are most effectively cleaned with a combination of rotary shampoo and water extraction methods. A manufacturer's sample carpet maintenance program using extraction is shown in Exhibit 9. Some manufacturers provide similar materials in languages other than English.

Special Carpet Treatments

Antimicrobial Treatment

Antimicrobial treatment in carpets kills many different kinds of bacteria, fungi, and the odors they cause. First introduced in hospitals, antimicrobial-treated carpets are appearing more and more often in lodging properties.

Tight building syndrome has increased concern over bacteria and fungi growth. A "tight" building is one that is more or less sealed against the outside and requires ventilation systems to freshen the air supply. In tight buildings, bacteria and fungi can grow in carpets and then circulate through the air supply. As a result, many lodging properties are now considering antimicrobial treatments for carpets.

Exhibit 8   Sample Rotary Brush Carpet Maintenance Program

CARPET MAINTENANCE PROGRAM
using Rotary Method

PREPARED FOR:					
PREPARED BY:					
TYPE OF CARPET	Natural	Synthetic	Cut Pile	Wire/Level Loop	Shag
SOIL CONDITION	HEAVY		MEDIUM	LIGHT	

REMOVE FURNITURE AND VACUUM CARPET

Remove all furniture to assist total cleaning of carpet and save time. Thoroughly vacuum carpet.

PRESPOT/ PRETREAT

Use Carpet Stain Remover for spots and stains. For heavily soiled traffic lanes, doorways, etc., apply Plus 5 diluted 1:4 to 1:6.

MIXING

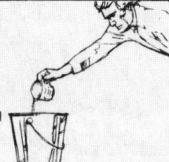

For light to medium soil conditions, add 1 part Plus 5 to 16 parts luke warm water. When carpet is heavily soiled or greasy, dilute Plus 5 1:12. Put water in first to avoid excessive foaming.

BREAK IN NEW BRUSHES

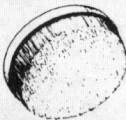

A. New nylon bristle brushes — run 1-3 minutes on concrete to knock off sharp edges.
B. New bassine bristle brushes — soak to soften.

SHAMPOOING/INITIAL PASS

Start machine on smooth surface to generate foam. Begin at upper right hand corner with the shampoo feed open. Move machine from right to left.

RETURN PASS

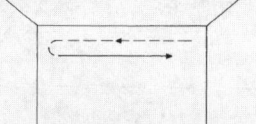

Drop down one machine width being careful to overlap initial pass slightly and move to the right with the shampoo feed open. Repeat—overlapping the two previous passes with shampoo feed closed.

EXTREMELY SOILED AREAS/ REPEAT AT RIGHT ANGLE

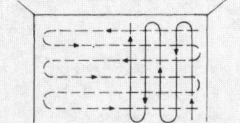

For best results, after small section is shampooed, repeat steps 5 and 6 at a right angle with *shampoo feed closed* to make sure foam is evenly worked into the pile.

WET VAC (OPTIONAL)

Vacuum immediately with wet vac to remove soil-laden foam and speed drying.

SET PILE (OPTIONAL)

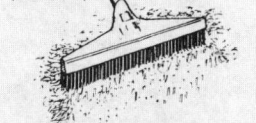

Using pile rake — rake the carpet against the way pile normally lies. This sets pile and decreases drying time. Replace furniture, using protective pads under metal casters.

DRY VAC

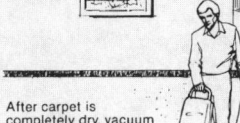

After carpet is completely dry, vacuum thoroughly with dry vac to remove soil ash. **Vacuum daily to prolong carpet life and extend the time between shampooing.**

PRODUCTS
PLUS 5 CARPET SHAMPOO
CARPET STAIN REMOVER

TOOLS
ROTARY MACHINE W/SOLUTION TANK
CHANNEL FEED BRUSH
WET/DRY VACUUM WITH CARPET TOOL
PILE RAKE
BUCKET
MEASURING CUP

DISTRIBUTED BY

SCC-1986

Spartan Chemical Co., Inc.
110 N. Westwood Avenue • Toledo, Ohio 43607

Spartan®

Courtesy of Spartan Chemical Co., Inc., Toledo, Ohio

Exhibit 9   Sample Carpet Extraction Maintenance Program

CARPET MAINTENANCE PROGRAM
using Xtraction Method

PREPARED FOR:					
PREPARED BY:					
TYPE OF CARPET	Natural	Synthetic	Cut Pile	Wire/Level Loop	Shag
SOIL CONDITION	HEAVY		MEDIUM		LIGHT

REMOVE FURNITURE AND VACUUM CARPET

Remove all furniture to assist total cleaning of carpet and save time. Thoroughly vacuum carpet.

PRESPOT

Use Carpet Stain Remover for spots and stains.

PRETREAT

For heavily soiled areas, doorways, etc., use trigger sprayer or Sparta-sprayer and apply Xtraction II diluted 1:16 or 1:32.

MIXING

Add 2 oz. of Xtraction II for every gallon of water in solution tank.

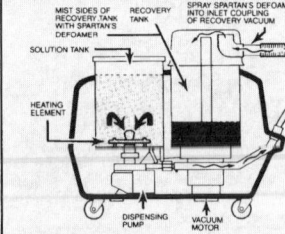

MIST SIDES OF RECOVERY TANK WITH SPARTAN'S DEFOAMER / RECOVERY TANK / SPRAY SPARTAN'S DEFOAMER INTO INLET COUPLING OF RECOVERY VACUUM
SOLUTION TANK
HEATING ELEMENT
DISPENSING PUMP / VACUUM MOTOR

USE OF DEFOAMER

If carpet has been previously shampooed, spray Defoamer into inlet coupling of recovery hose or mist sides of tank. Normally 1-2 oz. of Defoamer per each 5 gallons recovery tank capacity is sufficient. When foam is excessive, spray Defoamer directly onto foam in recovery tank as needed.

CARPET CLEANING PROCEDURE

1. Begin carpet cleaning at corner of room farthest from the door.
2. Reach out with wand at moderate distance, making certain floor tool is always perpendicular to the carpeting.
3. Open cleaner dispensing valve and slowly pull cleaning wand toward you. Close cleaner dispenser valve just prior to completion of first "pass" to eliminate "puddling" of cleaning solution.
4. Lift wand and return to original position. With cleaner dispensing valve closed — vacuum up excess solution. (Repeat for wool carpet.)
5. Always slightly overlap "passes".
6. Always avoid overwetting carpet.

SET PILE

Using pile rake — rake the carpet against the way pile normally lies. This sets pile and decreases drying time. Replace furniture, using protective pads under metal casters.

DRY VAC

After carpet is completely dry, vacuum thoroughly with dry vac. **Vacuum daily to prolong carpet life and extend the time between cleaning.**

PRODUCTS

CARPET STAIN REMOVER
XTRACTION II
SPARTAN DEFOAMER

TOOLS

HOT AND/OR COLD WATER EXTRACTOR
MEASURING CUP
PILE RAKE
TRIGGER SPRAYER
OR SPARTASPRAYER

DISTRIBUTED BY

©SCC-1986

Spartan Chemical Co., Inc.
110 N. Westwood Avenue • Toledo, Ohio 43607

Spartan®

Courtesy of Spartan Chemical Co., Inc., Toledo, Ohio

Many carpets come from the factory treated with a permanent antimicrobial solution. Carpets can also be treated periodically with solutions that offer temporary or permanently bonded antimicrobial protection. Keep in mind that antimicrobials will only be effective when the carpet is properly maintained.

Even though bacteria and fungi are less likely to live in synthetic carpet fibers, they may flourish in the soils that collect in these fibers. Locker rooms and areas around hot tubs and pools are especially good breeding grounds for bacteria and fungi that can cause carpet damage and odors.

Electro-Static Dissipation

Static electricity in carpets irritates guests and employees and can cause real harm to computers. Microchips used in electronic equipment are more sensitive to static electricity than people, especially when computers are opened for repair or other operations. Static electricity can wipe out information stored by the microchips or reduce a computer's memory capacity.

Many carpets receive an **electro-static dissipation** treatment from the manufacturer. Antistatic solutions can be applied by housekeeping staff. Humidity control will also help reduce static.

Types of Floors

The term **hard floor** is sometimes used to describe floor coverings other than carpets. In truth, some so-called hard floors are harder than others. Concrete, for example, is the hardest floor material and is reserved chiefly for areas that receive extremely heavy use. Cork, by contrast, is a springy natural material that is easy to walk and stand on for long periods. This section will discuss various floors and aspects of caring for them.

Floors have some disadvantages when compared to carpets. Floors are generally noisier, harder, and slipperier. They also provide less insulation. But floors also have some advantages over carpets. Floors are more durable and more sanitary, and they do not conduct static electricity.

Many people think that a floor's biggest advantage is that it is easier to clean and maintain than carpeting. This is not entirely true. Floors in lodging properties require frequent care just as carpets do. Housekeeping staff usually mop floors daily, buff them often, and, occasionally, strip and wax them. Floors do repel stains far better than carpets, but they are not completely stain-resistant.

As with carpets, the material a floor is made of dictates the kind of care it should receive. Floors may be made of natural and/or synthetic materials. Because they are less vulnerable to moisture and stains than carpet, floors are most often used in areas where water, soils, and stains frequently collect. These areas include guest bathrooms, public areas (lobbies and restrooms), and some back-of-the-house work spaces (kitchens, garages, and repair shops).

There are three basic types of floors: **resilient** (or springy), **wood**, and **hard** floors. Floors are judged on resilience (how easy the floors are to stand and walk on), cleanability, resistance to stains and abrasions, and safety.

Resilient Floors

Resilient floors are easier to stand and walk on and may reduce noise better than hard floors. Resilient floors are often, however, less durable. Types of resilient floors include:

- Vinyl

- Asphalt

- Rubber

- Linoleum

Vinyl. Vinyl floors may be made of pure vinyl or a blend of vinyl and some other material. Pure vinyl floors cost about twice as much as vinyl blend floors.

Vinyl may be a conventional or no-wax type. Conventional vinyl is used most often in commercial operations like hotels because the wax and finish on the floor protects heavy use areas from undue abrading and soiling.

The vinyls best suited for commercial use are thick and have **homogeneous color**—that is, color that permeates the entire layer of vinyl so that the color does not wear off with use. White vinyl may yellow if it gets too much direct, strong sunlight or if it is covered by a rug or appliance for a long period. Not much can be done about this condition. Yellowing may also occur with soiling; this can sometimes be corrected with a good cleaning.

Vinyl is impervious to most stains and substances. Vinyls with cushioned backing add to the material's natural resilience and noise absorption. Vinyl is also easy to care for and provides good traction.

Asphalt. Asphalt flooring is decay- and mildew-proof, resists ink stains, and is very fire-resistant. It also resists damage from alkaline moisture, which means that it can be installed directly over concrete. Asphalt is relatively inexpensive and durable. It also cracks and chips easily, which calls for careful installation. Harsh cleaners should be avoided.

Some asphalt floors contain asbestos. Asbestos has been outlawed in many states due to health risks. When asphalt floors with asbestos (or vinyl asbestos tiles) are torn up, the remains are considered hazardous waste and must be treated as such. Disposal of this hazardous waste can be expensive.

Rubber. Rubber, whether natural or synthetic (synthetic rubber is far more common today), is the most resilient, sound-reducing flooring available. It is also relatively expensive. Rubber is durable and provides good traction, even when wet. Oil, grease, high heat, and detergent may injure rubber flooring.

Linoleum. Linoleum may be made from a variety of materials. It has been around so long that it has become an almost generic term for any kind of resilient floor. It is made from linseed oil, ground cork or wood, mineral fillers, and resins. It may be bonded to a burlap or felt backing for extra resilience and strength. Linoleum is inexpensive, easily installed, and easy to maintain, but it can be damaged by strong detergents.

Wood Floors

Oak is the wood of choice for most wood floors because it is hard and attractively grained. Maple, walnut, and teak, which tend to be more expensive than oak, may also be used. Softer woods such as pine are sometimes used. Depending upon the finish, soft woods may be more susceptible to dents and scratches. Cork, another soft wood, may be shaved, pressed, and baked into tiles that make a very resilient, but somewhat vulnerable, floor surface.

Because all types of wood are porous and absorbent, they are especially susceptible to water damage. Such alkaline substances as sudsy cleaners or ammonia can cause dark spots that must be removed with vinegar applications. Proper installation, **sealing**, and **finishing** are essential to the durability of a wood floor.

Wood may be laid on the floor in a number of ways. **Parquet floors** are made of wood "tiles," usually oak or maple. **Wood block** floors resemble butcher block surfaces and also come in tiles. Strips of end grain, which can weather heavy blows and use, are laid together to make up the floor's surface. **Plank floors** are long strips of wood fitted together to make a smooth surface. The planks may be of the same or varying widths to make an attractive pattern on the floor.

Wood costs considerably more than most other kinds of floors. Much of its use in commercial operations is probably due to its luxurious appearance.

Hard Floors

Hard floors are made from natural stone or clay. They are sometimes called stone and masonry floors. Hard floors are among the most durable of all floor surfaces, but also among the least resilient. Types of hard floors include:

- Concrete
- Marble and terrazzo
- Ceramic tile
- Other natural stone

Concrete. Concrete floors are most often found in utility areas or areas that will receive a great deal of traffic from heavy equipment. Lodging properties often have concrete floors in parking areas, garages, and trade show areas. Concrete floors may be covered, painted, or sealed.

Marble and Terrazzo. Marble is a type of crystallized limestone and comes in many colors and patterns—white, black (onyx), gray, pink, green (verd antique), brown, orange, red-orange, banded (serpentine), or mottled. Marble is relatively durable, but lighter colors often yellow with age. Oil stains are extremely difficult to remove.

Marble can be finished in a variety of ways. Polished interior marble has a high-gloss finish and is usually used on tabletops, vanities, or other furniture. Since polished marble usually requires a great deal of maintenance, it is generally considered impractical for commercial floor use. Honed marble (satin finished with little or no gloss) is recommended most often for commercial floors. Sand-blasted or abrasive-finished marble has a matte or non-reflective surface and is best suited for exterior use.

Sheet marble is very expensive because it is difficult to mine. Even with careful mining techniques, about half the marble that is quarried crumbles to rubble.

Marble rubble is often salvaged and used in **terrazzo** floors. The rubble is made into small tiles or pieces which are embedded into mortar like a mosaic. Terrazzo can also be made from **granite**, a less expensive natural stone. Granite may be pink, gray, or black. The mortar, not the stone, gives terrazzo its durability. Like most porous surfaces, terrazzo floors must be sealed.

Ceramic Tile. Ceramic tile is made from combinations of clay, marble, slate, glass, and/or flint. It is very durable and easy to maintain. It does not require sealant or wax.

Other Natural Stone. Slate, a gray or blue-gray stone, is another popular natural stone floor. Slate forms when layers of mud and silt build up and solidify over millions of years. These layers allow slate to be "sliced" rather easily into floor pieces. But these floor pieces may split with excessive use or if heavy objects fall on them.

Quarry tile is less durable than slate and has a rougher texture. It is related to slate, however, being made from clay or shale.

There are two types of russet or brown clay floors. **Terra cotta** is a hard-baked tile, while **brick** is clay that is molded into rectangular blocks and hard-baked. Terra cotta tiles and bricks are typically left their natural color.

General Floor Maintenance

Floors require regular cleaning and finishing to retain their appearance and durability. Some housekeeping departments caution that soap should not be used on floors. Hard water often fails to rinse soap residue adequately; this residue can soften finishes or make the finishes unnecessarily slippery.

While floors generally resist stains far better than carpets, housekeeping staff should note floor spills and correct them quickly. Some general rules are:

- Identify the stain and determine how to treat it. (See your property's cleaning manual.)

- Remove solid particles with a hand-held scraper. Be careful not to scratch or gouge the floor.

- Use a wet vacuum on large wet stains.

- Spot-test the stain remover and apply it according to the manufacturer's directions.

In today's world of no-wax floors, floor polish may seem like a thing of the past. However, floor care experts advise that proper waxing not only makes the floor more attractive, but is vital to its survival in commercial operations. Waxing protects all floor surfaces—even no-wax surfaces—from abrasions. It also strengthens porous floor materials such as wood.

Despite their stain-resistance, floors still need protection against scuff marks and abrasives. Mats can help reduce scuffing and scratching—especially in entryways where the greatest number of abrasives are found. As with carpets, routine floor maintenance involves a regular cleaning schedule and prompt attention to problem spots. Exhibit 10 provides a sample form that distributors and/or manufacturers may supply to their customers. These kinds of forms can be used not only to set up a regular maintenance schedule, but also to train employees in the importance and techniques of basic floor care. Some manufacturers provide similar materials in languages other than English.

Floor safety is also a major part of floor maintenance—especially in the face of an increasing number of lawsuits resulting from falls. Housekeepers may want to consider purchasing commercial floor slip testers which can help determine floor safety. In addition, asking about liability coverage of floor care products can help determine which products are safer—and even reduce the insurance premiums for users of those products. Exhibit 11 provides a sample floor safety checklist.

Floor Cleaning Methods

Mopping

Floors in most hospitality operations must be mopped daily, either with a damp mop or, on floors that cannot tolerate much water, with a chemically treated dust mop. Staff should be careful not to over-treat mops because the chemical on the mop head will transfer to the floor. This can create a haze or dulling effect, and may destroy the finish. Mop heads come in a variety of natural or synthetic fibers. Some properties recommend rayon. Soaking new mop heads in water for 30 minutes before use will remove the sizing chemicals.

Floor cleaners should make sure mop heads are rotated so they can be cleaned and dried after each use.

Buffing and Burnishing

Buffing involves spraying the floor with a polishing solution and buffing the floor with a rotary floor machine. Some rotary machines can spread the polishing solution as well as buff the floor. Spray buffing effectively removes scuff marks, heel marks, and restores the gloss to the floor. High-speed rotary machines are available that will make buffing quicker and the buff coat more durable.

Burnishing (polishing) is a relatively new kind of floor cleaning method. Burnishing is something like buffing except that it is a dry method. Another difference between buffing and burnishing is the speed of the rotary floor machine. Burnishing requires faster rotation of the machine head. Some properties recommend burnishing only in low-traffic areas. Burnishing can only be used on hard floors.

Whether to buff or burnish a floor depends on the wax, sealer, or finish on the floor itself. Also, before a floor is buffed or burnished—it must be clean.

Scrubbing

Scrubbing usually requires a stiff scrubbing brush or pad, a suitable cleaning mixture, and a rotary floor machine. Scrubbing is often followed with buffing or

Exhibit 10 Sample Floor Care Maintenance Schedule

Courtesy of Spartan Chemical Co., Inc., Toledo, Ohio

Exhibit 11   Sample Floor Safety Checklist

All employees have been trained to call housekeeping promptly when they discover a spill and to set up "Wet Floor" signs and direct others around the area.
Yes ☐   No ☐

Adequate lighting will prevent guests and employees from slipping and bumping into obstructions.
Yes ☐   No ☐

"Wet floor" signs are used during cleaning.
Yes ☐   No ☐

Floors are stripped on schedule to avoid slippery wax build-up.
Yes ☐   No ☐

Floors are kept dry.
Yes ☐   No ☐

Wet spills are cleaned up immediately.
Yes ☐   No ☐

Floors are swept, mopped, and regularly inspected for nails and loose obstructions that could cause tripping.
Yes ☐   No ☐

Damaged tiles and baseboards that could cause trips are replaced.
Yes ☐   No ☐

Non-slip finishes are used.
Yes ☐   No ☐

Soap that could soften the finish and make it slippery is rinsed during cleaning.
Yes ☐   No ☐

Electrical cords are in good repair, grounded, and do not obstruct the floor space.
Yes ☐   No ☐

Absorbent walk-off mats are used at all entrances.
Yes ☐   No ☐

Source: Adapted from "Safety Checklist for Floor Maintenance," *Programmed Cleaning Guide for the Environmental Sanitarian* (New York: The Soap and Detergent Association, 1984), p. 169.

burnishing, depending upon how much of the old wax comes off the floor during scrubbing.

Stripping and Refinishing

Housekeepers agree that stripping and refinishing are expensive and time-consuming tasks. However, to ensure proper floor care, they should be done on a regularly scheduled basis. Stripping solutions may be water- or ammonia-based. Ammonia is a very powerful chemical and should be used carefully on floors. A rotary floor machine can be used to strip the old finish and spread the new finish on the floor. Exhibit 12 provides a sample form that distributors and/or manufacturers may supply to their customers. These kinds of forms can be used to train employees in floor stripping and refinishing techniques. Some manufacturers provide similar materials in languages other than English.

Finishes come in two types—**wax-based** or **polymer finish**. Wax-based polishes require at least two coats of wax to attain maximum protection for the floor.

Exhibit 12 Sample Stripping and Refinishing Techniques

PREPARED FOR:				
PREPARED BY:				
TYPE OF FLOOR	Vinyl Asbestos	Pure Vinyl	Rubber	Terrazzo
TYPE OF TRAFFIC	HEAVY	MEDIUM	LIGHT	

INITIAL TREATMENT
To Rejuvenate Floors and for Newly Laid Floors

SWEEP FLOOR
Sweep floor using broom, treated dust mop, or industrial vacuum.

APPLY STRIPPER (Mop #1)
Use Spartan _____ diluted _____ part to _____ parts warm or hot water. Apply the solution liberally to the floor.

LET STRIPPER WORK
Allow the solution to stand for three to five minutes. Do not allow solution to dry.

SCRUB FLOOR
Preferably using a machine equipped with an abrasive stripping pad or stiff scrubbing brush.

PICK UP SOLUTION
Pick up solution with mop or wet vac.

CLEAN WATER RINSE (Mop #2)
Keep changing rinse water to insure a clean floor. Minimum of two rinses suggested.

ALLOW TO DRY
Allow to dry thoroughly. For best results allow a minimum of 30 minutes.

DRY BUFF & DUST MOP (OPTIONAL)
Dry buff with machine using soft buffing pad. Dust mop with non-oil base dust mop treatment.

SEAL FLOOR (OPTIONAL)
Apply Spartan _____ undiluted using a clean damp mop or applicator. Two thin coats are desirable. **ALLOW TO DRY**

APPLY FIRST COAT OF FINISH (Mop #3)
Apply thin coat of Spartan _____ undiluted using clean damp mop, or standard applicator. **ALLOW TO DRY** (approx. 30 minutes)

APPLY SECOND COAT OF FINISH
Apply thin coat of Spartan _____ undiluted using clean damp mop, or standard applicator. **ALLOW TO DRY** (approx. 30 minutes)

APPLY THIRD COAT OF FINISH
Apply thin coat of Spartan _____ undiluted using clean damp mop, or standard applicator. **ALLOW TO DRY** (approx. 30 minutes)

SELF POLISHING FINISHES/SPRAYBUFF
While the last coat of Spartan finish will dry to high luster without buffing, floors may be spraybuffed after 24 hours as needed.

WAX ONLY/DRY BUFF
Dry buff second and subsequent coats when dry to desired luster.

GENERAL REQUIREMENTS:

CHEMICALS
STRIPPER: SQUARE ONE   RINSE FREE STRIP
NAD-75   DA-70
FINISHES: SUNNY SIDE   ON AN' ON
PRO-SHINE   SHEEN 17
SHEEN 100
WAXES: HEAVY DUTY WAX 16%
SEALERS: ON BASE   TERRA GLAZE
SPARTAN   SUPER SPRAYBUFF
SPARTAN   SPRAYBUFF
BOUNCE BACK FINISH RESTORER
SPARTAN   DUST MOP/DUST CLOTH TREATMENT

MACHINES
FLOOR POLISHER
W/PAD HOLDER
WET/DRY VAC

OTHER
STRIP PAD
S. BUFF PAD
3-5 MOPS/STICKS
BUCKET/WRINGER
DUST MOP
TRIGGER BOTTLE
W/SPRAYER

DISTRIBUTED BY

© SCC-1985

Spartan Chemical Co., Inc.
110 N. Westwood Avenue • Toledo, Ohio 43607

Spartan®

Courtesy of Spartan Chemical Co., Inc., Toledo, Ohio

Many manufacturers and housekeeping departments recommend three or more coats. Almost all finishes can be spray-buffed. Wax-based finishes are buffable.

Metal interlocking (or **cross-linking**) polymer finishes contain a dissolved metal, usually zinc, that strengthens the floor finish. Some properties use only solutions with at least 18% to 20% solids. This kind of finish is virtually impervious to heel marks, detergents, and abrasions. A polymer finish is also easily touched up with fresh coats to keep the floor glossy and restore the protection of the original finish. Metal interlocking finishes also make stripping easier because ammonia (the active ingredient in many strippers) attracts the metal. This unseals the finish, making removal easier.

Key Terms

acetate
acrylic
antimicrobial treatment
backing
brick
bulk continuous filament (BCF) fibers
ceramic tile
cross-linking polymer finish
delustered
electro-static dissipation
face
face fibers
face weight
finishing
granite
hard floor
homogeneous color
hot or cold water extraction
marble
metal interlocking polymer finish
modacrylic
olefin
parquet floors
pile
pile distortion

plank floors
polyesters
polymer finish
polypropylene
primary backing
quarry tile
rayon
resilient floor
rotary floor machine
sealant
sealing
secondary backing
shading
shampoo (bonnet and brush)
slate
staple fibers
terra cotta
terrazzo
tufted or looped carpets
wax-based finish
wet vacuum
wicking
wood block floor
wood floor

Discussion Questions

1. What are the differences between a carpet and a floor?

2. What are the basic components of a carpet?

3. What kinds of carpets would be most practical in a guestroom? A lobby? A dining room? Why?

4. How can such carpet problems as pile distortion, wicking, and fading be prevented?

5. What are some of the main considerations in planning floor and carpet maintenance schedules and procedures?

6. What are some of the basic pieces of equipment used to clean carpets and floors and how are they used?

7. What are various kinds of carpet cleaning methods?

8. What should be considered when choosing floor surfaces for a lodging operation?

9. What are three basic types of floor coverings?

10. What are various kinds of floor cleaning methods and their applications?

REVIEW QUIZ

When you feel you have covered all of the material in this chapter, answer these questions. Choose the *best* answer.

True (T) or False (F)

T   F   1. In general, the greater the density of the face fibers, the better the grade of carpet.

T   F   2. The backsize of a carpet's primary backing prevents the carpet tufts or loops from shifting or loosening after installation.

T   F   3. Tufted carpet constructed with bulk continuous filament fibers is more likely to shed or pill then carpet constructed with staple fibers.

T   F   4. Most carpets manufactured in the United States are made from wool or other natural fibers.

T   F   5. Wicking is a normal feature of almost all types of carpets.

T   F   6. Pile distortion results from face fibers that were trapped in the carpet during the manufacturing process.

T   F   7. Many wet vacuums can be used on floors to pick up dry soils.

T   F   8. Beater-bar vacuums are best for carpets installed over pads.

T   F   9. No-wax vinyl floors are more common in hotels than conventional vinyl floors.

T   F   10. Alkaline cleaners are especially well-suited for use on wood floors.

Alternate/Multiple Choice

11. Which of the following is the more effective carpet cleaning method?

   a. dry foam cleaning
   b. rotary shampoo

12. Bleach solutions are effective cleaning agents to use on:

   a. nylon carpets.
   b. olefin carpets.

13. A solution-dyed carpet that resists static electricity is made of:

   a. acrylic.
   b. nylon.
   c. olefin.
   d. wool.

14.   Which of the following is the hardest floor material?

a.   concrete
b.   vinyl
c.   asphalt
d.   terrazzo

15.   A type of hard floor surface is:

a.   asphalt.
b.   parquet.
c.   ceramic tile.
d.   vinyl.

15/15

Chapter Outline

The Design Team
Elements of Interior Design
    Design Components
    Style and Color Scheme
    Trend
Designing with a Purpose
    Multipurpose Areas
    Special Services

Learning Objectives

1. Identify and describe various aspects of the interior design process.

Note: This chapter is presented strictly for your information and enjoyment. The test materials for the course include no questions on this chapter.

15

Interior Design

This chapter was written and contributed by Robert Di Leonardo, Ph.D., President, Di Leonardo International, Inc., Hospitality Design, Warwick, Rhode Island.

INTERIOR DESIGN CAN MEAN many different things. For the purpose of this chapter, it means the selection and placement of various elements—furnishings, window coverings, wall and floor coverings, and amenities—to create an atmosphere which serves a particular purpose. The goal is to incorporate the design components into attractive, comfortable, and easy to clean areas.

To explain the design process, the chapter discusses the types of decisions that need to be made and why, and it outlines what housekeeping management and staff should know in order to assist in the selection, purchase, and installation of the design elements.

The Design Team

Lodging facilities are designed with one thing in mind: to attract guests. And usually, a facility has a certain type of guest—or market—in mind. There are many types of lodgers. They range from families on vacation to businesspersons on the road. Each market has different needs and expectations—different preferences and values with regard to room sizes, amenities, and prices. To attract a certain market, a property's design must reflect these preferences and values. The responsibility for design generally rests with a design team.

A design team is typically made up of not only actual designers but also the project architect, appropriate staff members, representatives of the owners, and sometimes the owners themselves. Owners provide guidelines on the target markets, what the company is prepared to offer in terms of services and amenities, and how much the company has available to spend for setup. These guidelines—more than anything else—determine what the design team develops as its unifying ingredient—the theme.

The design team reviews the guidelines provided by the owners and meets to discuss ideas for a theme. A great deal of thought goes into the selection of a theme since it is the framework for the hotel's image. All the design elements must come together via the theme to make a cohesive package; otherwise, the disparate parts will be obvious and unattractive.

Carefully selected design elements work together to evoke a particular mood or style in the guestroom. (Courtesy of Di Leonardo International, Inc., Hospitality Design, Warwick, Rhode Island)

Theme ideas might be based on the region where the hotel is located, on an image the hotel owners wish to convey, or on a novelty in the dining area that is extended throughout the hotel.

Elements of Interior Design

Once the team knows its guidelines and theme, the selection of design elements begins to take place. At this time, especially in the case of a renovation, housekeeping managers and staff are consulted to discuss what items are available and what items will be kept (such as linens, fixtures, and so on). Discussion may also focus on how housekeeping staff can maximize the cleaning program and minimize the problems of damage and theft.

Design Components

Design components include furnishings, wallcoverings, floors and carpets, window coverings, fixtures, and amenities. These items are basic to every area; their layout or placement is critical to the way the room works. Convenience is a basic consideration of interior design. People should not be tripping over furniture to get to the bathroom or knocking over lamps to reach the phone.

There is often no choice as to where bathroom fixtures or electrical and telephone outlets may be placed—especially in the case of renovations. And because of the expense, furnishings and floors are not changed very often. When changes must be made, it is commonly to the "soft" items—linens, draperies, wallcoverings, and amenities, for example. These are items which, if changed, can effectively improve the atmosphere at relatively low cost.

Comfort and ease of maintenance are two major factors to consider when in doubt about selecting any design component. Sometimes it is more cost-effective to pay a higher price for an item that is more comfortable, longer lasting, and easier to clean, than to buy an item of lesser quality that may need frequent replacement. The following sections provide guidelines for selecting various design items.

Furnishings. Furnishings—chairs, tables, desks, beds, and so on—are perhaps the most obvious components of a room. Furnishings also establish a room's purpose. Most guestrooms include one or two beds or a pull-out couch, one or more chairs, a dresser, lamps, and, generally, a table which can double as a desk. A lobby area will have chairs, tables, sofas, lighting, and perhaps a writing surface. The furnishings provided depend on the clientele the property wishes to attract. For the most part, it is the job of the interior design team to make the area as attractive as possible.

Depending upon the shape of the guestroom and its purpose, the designer might group chairs near a window to take advantage of natural daylight and to provide a separate conversation/reading area. Tables and dressers are usually placed opposite the bed or next to a chair out of the way of traffic. There must be enough room to move around easily even if drawers are open and chairs are pulled away from tables and desks.

While choice of furnishings is determined by the area's style, quality selections that avoid large prints and fashion-of-the-day colors will generally have a longer life. If selected with longevity in mind, furnishings can last through one or two redecorating periods before being replaced.

Wallcoverings. Wallcoverings include various kinds of surfaces that are composed of natural or synthetic materials. Three basic types are discussed here: tile, paint, and wallpaper.

Tile, either ceramic or stone, adds a distinctive flavor to any area. Tile comes in many colors and can be arranged to form patterns that evoke certain time periods or moods. Tile is long-wearing and bears up well in areas where there is a great deal of moisture and grime. Tile is fairly easy to maintain. On the other hand, when used for wallcoverings, tile can give rooms a cold look and is expensive. When considering tile, designers must think about how much time will elapse before the next renovation. If maintained well, tile can last for years; it should be incorporated into two or three redecorating cycles to maximize the investment.

Paint is inexpensive to use, extremely versatile, and easy to maintain. However, any scuff or mar on the wall surface shows immediately and distinctly. Painting the walls in hallways and in other areas where furnishings and handprints might leave a mark is often discouraged. Frequent touch-ups take time and create unnecessary inconvenience for both the staff and guests.

Wallpaper is probably the best wallcovering because of its versatility, relatively low cost, and ease of maintenance. Wallpaper not only hides flaws in the wall's surface, but adds to the atmosphere of an area—it makes the area cozier. Wallpaper is available in many colors and prints, can be matched to any number of fabrics and styles, and can change a room's overall appearance.

Wallpaper can usually be cleaned with a quick dusting or, in the case of vinyl papers, with a damp sponge. Patterned wallpaper tends to hide small nicks and is generally replaced every few years—just about the time of many scheduled renovations.

Floors and Carpets. Floors and carpets do more than simply cover surfaces; they help channel traffic, act as soundproofing agents, and add color and texture to the general atmosphere. Like walls, floors can be designed to attract people to various areas. For example, contrasting colors or patterns can be used in hallways to lead a person to a new area in the way that light strips are sometimes used to highlight stairways and glamorize dining and dancing areas. Patterns on the floor can break up the tedium of long hallways or even designate particular areas for meetings, check-in, or waiting for an elevator. Patterned or multitoned carpet also prevents minor spots or stains from being too noticeable.

The selection of the floor material depends on such criteria as guest and employee traffic, acoustics, and the image the property wants to convey. For example, marble is durable, easy to maintain, and has a certain aura. There is something dramatic about hearing footsteps on a marble floor. But, because of its cost and its inability to absorb sound, marble is not appropriate for all lodging facilities or even in all areas of a luxury hotel.

Other types of floors—wood, tile, and linoleum, for example—are used in such heavy traffic areas as hallways, restrooms, and kitchens. Wood floors are rarely seen because they need a great deal of care. Then again, wood floors add a warmth and charm to an area which can't be realized with any other floor material. The same can be said for ceramic or stone. While these types of floors are expensive to install and do nothing to deaden footsteps and chatter, they add significantly to the atmosphere. Selection, again, must concur with the image the owners wish to convey at a price they want to pay.

Economically speaking, this is where linoleum makes its mark. With today's technology, linoleum can replicate ceramic, stone, and wood flooring at a fraction of the cost and provide a greater sound barrier, easier maintenance, and more comfort underfoot. Linoleum does not wear as well as stone or ceramic tiles, but can be updated more easily than the real thing. Linoleum is also available in a multitude of colors and patterns unlike other floor types.

By far, carpeting is the first choice for guestrooms, where people like to kick off their shoes and relax. Carpeting muffles sounds so it is also ideal for hallways and lounge areas. Because noise is a critical factor to guests' comfort, thick carpeting is used more often than not. Economy properties often use a lower grade of carpeting which usually has a shorter pile, thus lessening the sound-deadening barrier.

Carpeting is generally easy to maintain, requiring only a daily vacuuming and a periodic cleaning. Wool carpeting lasts longer and keeps its color well but builds up static electricity, causing unwanted shocks; it is also more difficult to clean than nylon and other synthetic carpets. Synthetics are treated to reject spills and maintain their shape better than wool.

Window Coverings. Window coverings have come a long way from just closing the room off from the outside world. They are now an integral part of the room's look and do more than just open and close to keep out daylight, city lights, and noise. Most guestrooms have a heavy black-out curtain on a traverse rod. But in some cases, window coverings consist of a swag at the top of a window or floor-length draperies with sheer panels. Some areas have potted plants hanging from the window frame. In all instances, the idea is to make the most use of natural light in keeping with the room's image. For example, half-window (cafe) curtains might be used in a bistro setting. Shutters may be used in a room where the atmosphere is more casual, perhaps simulating a country or southwestern motif.

Selection of window coverings is based on the style of the room. What the housekeeping staff needs to know is that while draperies can be kept dust-free with regular vacuuming, they have to be sent out for periodic cleaning. Extras must be kept on hand for use during the cleaning time; housekeeping should know how to install the spares.

Most guestrooms, too, have heating and air conditioning units by the windows. This not only directs the style of window covering, but also affects the way in which the heating and air conditioning unit works, since it may be blocked by the window covering. To make it easy for guests to adjust draperies in their rooms, traverse rod drapes are generally used, with sheer panels or curtains underneath. Traverse drapes can be coordinated with the rooms' other linens and are easily taken down for cleaning and rehung.

Plumbing Fixtures. Plumbing fixtures include toilets, sinks, vanities, and bathtub/shower facilities. When a design team presents its model or drawing of a proposed room design, the fixtures are included based on the location of water and plumbing lines. Once installed, these items are extremely difficult and very uneconomical to change. For the most part, these items are changed only during major renovations.

Generally, there is not a great deal of variation in terms of where fixtures are placed. For example, guests might find two sinks in one long vanity, a standard sink/toilet/shower unit, or a toilet and sink in one area and the tub/shower/Jacuzzi in an adjacent area. Layout depends on the type of property and its service level. But once the fixtures are installed, the layout is extremely difficult to change until major renovations occur.

Amenities. The term amenities refers not only to the shampoo, soap, and complimentary gifts left in guestrooms, but also to such decorative items as fresh flowers. Amenities also include such special services as a chocolate mint placed on the pillow after turn-down service. The number and quality of these items are indicative of a property's service level. To guests, amenities reflect a certain attentiveness to their comfort. Very often, the amenities a property offers affects the guests' decision

Furnishings establish the room's purpose and should be arranged in an attractive and convenient manner. (Courtesy of Di Leonardo International, Inc., Hospitality Design, Warwick, Rhode Island)

to return to a particular lodging facility. While seemingly insignificant in a room's overall design, amenities often reinforce the style of the room and are actually as important as any other element in the design process.

Style and Color Scheme

When the design team develops its theme, a great deal of consideration is given to color scheme and style. Color scheme refers to the use of one or two colors throughout the facility to create a unifying look. Style refers to the "look" of the components which, when assembled, create an atmosphere. For example, a "look" might be labeled "English country" or "southwestern."

Quite often the style will suggest the color scheme, such as the traditional English country look with its dark greens, reds, and blues. In this style, furnishings might include leather sofas, leather-topped cherry wood desks, and large ottomans. On the other hand, the southwestern look usually incorporates rusts, tans, and pale greens, and uses more natural accents like pickled or scrubbed pine wood and pottery lamps. No matter what style is used—and there are dozens of them—components are available in differing dimensions of quality.

Trend

The term trend often appears when discussing interior design. It usually refers to a color or style which has been copied and repeated by every level of lodging facility all across the country. The problem is that overuse of a color or style minimizes impact—and a change of design is often requested after a relatively short time. Choosing harmonious colors and styles that are appropriate to the region and/or the desired clientele will prevent frequent, costly changes.

Designing with a Purpose

With increased competition in the hospitality industry, hotel executives are concerned that they are meeting the needs of the people they wish to attract and that they are doing so in a cost-effective manner. They are therefore asking for accountability in all aspects of lodging management, including the most efficient use of space. More and more, properties are incorporating multipurpose or multifunction areas into their design, and offering special services or facilities to attract a particular market.

Multipurpose Areas

A hallway connects rooms and is a means of getting from one place to another. A hallway can also serve as an art gallery—a place where paintings and other art work are displayed for guest pleasure. Essentially, this is what is meant by multipurpose use of space—looking at the face value of a space and exploring its potential for additional uses.

Another example of multipurpose use is the bar/lounge area in or near a hotel lobby. Lobby bar/lounge areas not only create a pleasant atmosphere where guests can socialize, but also bring in additional revenue for the hotel and its staff.

The efficient use of "dead" space becomes a concern of the interior design team when the team is asked to build options into the initial plan or, in the case of a renovation, to suggest ways to best use already developed space. This means determining whether one area would make a better bar, gallery, or breakfast nook than another. Housekeeping staff should be aware of proposed changes, and should discuss potential problems. Such problems might be related to cleaning and theft and damage control. The housekeeping staff would be likelier than the design team to know whether a particular area attracts people off the street, if it is particularly susceptible to soil from traffic, or whether it is possible to keep the area clean and attractive in a cost-effective manner.

A variation of the multipurpose area is the double-duty room. Since today's business travelers frequently meet outside of the office—either to avoid distraction or to accommodate large groups—lodging facilities have found it necessary to offer rooms where meetings can take place. Often, these rooms are booked through one or more mealtimes, so arrangements for food service are made along with any audio-visual requirements. For the interior design team, this means developing a flexible floor plan of chairs and tables which can be organized quickly and easily. Accommodations for easy conversation, formal discussion, and comfortable viewing of

Enormous potential for creative interior design rests in multipurpose areas such as lounge and bars. (Courtesy of Di Leonardo International, Inc., Hospitality Design, Warwick, Rhode Island)

slides or films must be coordinated with the realization that food may be served throughout any of these activities.

Meeting rooms hold a particular challenge to the interior design team. When selecting furnishings, designers must consider that people will be sitting for long periods, perhaps taking notes, and/or dining. The furniture should also be able to glide over the floor with as little noise as possible. Serving tables, film screens, easels, variable lighting systems, and the like should not call attention to themselves, yet they must be obvious to and convenient for any user.

Acoustics are a critical part of double-duty room design. No one wants the distraction of hallway or kitchen noises, nor is it acceptable for the speaker's words to be swallowed by high ceilings or soundproof panels. Furnishings and wallcoverings, too, must maintain the image and quality found throughout the rest of the hotel.

For the housekeeping staff, the challenge is to ensure that the rooms are clean before and after meetings and that all items in the room are in good working order. This means working closely with the departments that handle banquet or catering, special events, and audiovisual needs, and coordinating their schedules with the regular cleaning routines.

Special Services

A hospitality property's major challenge is to attract guests and keep them returning to the property. More and more, lodging facilities are offering special services as a way to draw in and retain particular guests. Among some of the special services are bars and dance clubs, athletic clubs, and exclusive or specialty restaurants. Sometimes these features are so successful that people will visit the property for the service only—and never set foot inside a guestroom. Office equipment rentals are becoming more prevalent along with such services as computer hook-ups, photocopying, facsimile and telex transmissions, and even clerical assistance. To meet the needs of family travelers, some hotels provide baby-sitters, organize daytime activities such as guided tours to local historic sites, and offer video rentals.

The interior design team becomes involved in these programs by determining the appropriate location for these services. Adjustments in the guest areas may have to be made for the use of additional electronic machinery or for the convenience and safety of the child traveler.

As travel for business and pleasure increases, owners and managers will find it necessary to accommodate people not only for sleeping, dining, and local entertainment, but also for business services, child care, and fitness programs. The housekeeping staff should be kept informed of these changes and instructed in the availability and setup of such new services. By working with other departments in the delivery of these services, housekeeping can help ensure that a property builds and maintains a satisfied clientele.

Glossary

A

ACOUSTICS

Sound absorption quality of certain materials, usually in ceilings, walls, or floors.

ACRYLIC

Synthetic material used in making fabric or molded transparent fixtures or surfaces.

ACUTE HAZARD

Something that could cause immediate harm. For example, a chemical that could cause burns on contact with the skin is an acute hazard.

ALKALIES

Laundry chemicals which help detergents lather better and keep stains suspended in the wash water after they have been loosened and lifted from the fabric. Alkalies also help neutralize acidic stains (most stains are acidic), making the detergent more effective.

ALTERNATIVE SCHEDULING

Scheduling staff to work hours different than the typical 9:00 A.M. to 5:00 P.M. workday. Variations include part-time and flexible hours, compressed work schedules, and job sharing.

AMENITY

A service or item offered to guests or placed in guestrooms for convenience and comfort, and at no extra cost.

ANTICHLORS

Laundry chemicals which are sometimes used at the rinse point in the wash cycle to ensure that all the chlorine in the bleach has been removed.

ANTIMICROBIAL TREATMENT

Carpet treatment in which a solution is applied to the carpet to kill many different kinds of bacteria, fungi, and the odors they cause.

AREA INVENTORY LIST

A list of all items within a particular area which need cleaning by or attention of housekeeping personnel.

B

BACK OF THE HOUSE

The functional areas of the hotel in which employees have little or no guest contact, such as engineering and maintenance.

BALANCE
In a piece of cloth, the ratio of warp to fill yarns.

BAR CODE
A group of printed and variously patterned bars, spaces, and numerals that are designed to be scanned and read into a computer system as label identification for an object.

BASE (TOWEL)
The core of the towel to which the nap or face is attached.

BATH BLANKETS
Extra-large bath towels. Also called bath sheets.

BLEACH
There are two kinds of bleaches: *chlorine* and *oxygen*. Chlorine bleach can be used with any washable, natural, colorfast fiber. Oxygen bleach is milder than chlorine bleach and is generally safe for most washable fabrics. Oxygen bleach should never be used with chlorine bleach as they will neutralize each other.

BOX SPRINGS
Type of bed springs fastened to a wood frame.

BREAK
This is the point in the laundry wash cycle at which a high-alkaline, soil-loosening product is added. The break cycle is usually at medium temperature and low water level.

BUILDERS
Builders or alkalies are laundry chemicals which are often added to synthetic detergents to soften water and remove oils and grease.

BULK CONTINUOUS FILAMENT (BCF) FIBERS
Continuous strands of fiber that are used to construct non-woven or tufted carpet.

C

CALL-BACK LIST
A list of all employees and applicants who possess special skills and interests or who are interested in filling certain positions.

CAPITAL BUDGET
A detailed plan for the acquisition of equipment, land, buildings, and other fixed assets.

CAPITAL EXPENDITURES
Items costing $500 or more that are not used up in the normal course of operations, and that have a lifespan that exceeds a single year.

CARDING (COTTON)

Cotton fiber process that yields poorer grade fabric.

CASE GOODS

Items with tops and sides, such as bureaus and desks.

CHECK-OUT

A room status term indicating that the guest has settled his/her account, returned the room keys, and left the property.

CHRONIC HAZARD

Something that could cause harm over a long period; for example, a chemical that could cause cancer or organ damage with repeated use over a long period.

COACHING

An extension of training which focuses on accomplishments, job duties, and factual observations by providing positive or redirective feedback to an employee.

COMBING (COTTON)

Cotton fiber process that yields better grade fabric.

CORNER DROP

On a tablecloth, the amount of fabric beyond the dimensions of the table itself needed to make a graceful "drop" over the edge.

COUNSELING

A process of one-on-one problem-solving during which a manager helps an employee seek solutions to his/her own problems. Job-related counseling focuses on attitudes or feelings about the job and the work environment.

CRASH CLOTH

Plain weave napery.

CROSS-LINKING

See Metal Interlocking

CROSS-TRAINING

Teaching employees to fill the requirements of more than one position.

D

DAMASK

Type of twill weave used mostly in napery in which a design appears against a satin background.

DEEP CLEANING

Intensive or specialized cleaning undertaken in guestrooms or public areas. Often conducted according to a special schedule or on a special project basis.

DELUSTERED

A process used on nylon carpet to lessen the shine and to give the surface a duller finish that looks more like wool.

DEPARTMENT

A hotel functional area within a division. Typical departments include housekeeping and front office.

DIVISION

A main functional area within a property. Typical divisions include rooms, engineering and maintenance, and human resources.

DOBBY CLOTH

Type of napery fabric.

DOUBLE-LOCKED

An occupied room for which the guest has refused housekeeping service by locking the room from the inside with a dead bolt. Double-locked rooms cannot be accessed by the room attendant using a standard passkey.

DYE LOT

Pieces of fabric dyed in the same vat and therefore of the same shade.

E

EARLY MAKEUP

A room status term indicating that the guest has reserved an early check-in time or has requested his/her room to be cleaned as soon as possible.

ECONOMY/LIMITED-SERVICE

A level of service emphasizing clean, comfortable, inexpensive rooms, and meeting the most basic needs of guests. Economy/limited-service hotels appeal primarily to budget-minded travelers.

ELECTRO-STATIC DISSIPATION

Carpet treatment in which a solution is applied to make the carpet resistant to static electricity.

EMERGENCY KEY

A key which opens all guestroom doors, even when they are double-locked.

EMPLOYEE REFERRAL PROGRAM

A department or property program which influences employees to encourage friends or acquaintances to apply for a position. Such programs usually reward employees who refer successful candidates to the property.

ENAMELED CAST IRON

Common material for bathroom fixtures, especially sinks and tubs.

EXTERNAL RECRUITING

A process in which managers seek outside applicants to fill open positions, perhaps through community activities, internship programs, networking, temporary agencies, or employment agencies.

F

FABRIC BRIGHTENERS

Fabric (or optical) brighteners are laundry chemicals which keep fabrics looking new and colors close to their original shade. These chemicals are often pre-mixed with detergents and soaps.

FACE (CARPET)

The pile of the carpet.

FACE (TOWEL)

Nap.

FACE FIBERS

Yarns which form the pile of the carpet.

FACE WEIGHT

The measure of a carpet's pile. Equal to the weight of the face fibers in one square yard of carpet.

FELTING

Characteristic of some fibers (especially wool) to mat at the surface of the fabric.

FIBERLOCK

Type of unwoven blanket construction.

FILL YARNS

Yarns running the width of the fabric.

FINISH

A liquid applied to floors that dries to a protective coating and enhances the appearance of the floor. Finishes come in wax-based or polymer types.

FIXED STAFF POSITIONS

Positions which must be filled regardless of the volume of business.

FLATWORK IRONER

Flatwork ironers are similar to pressing machines, except that ironers roll over the material while presses flatten it. Some ironers also fold the flatwork automatically.

FLOOR PAR

The quantity of each type of linen that is required to outfit all rooms serviced from a particular floor linen closet.

FLUSHES
Steps in the wash cycle which dissolve and dilute water-soluble soils to reduce the soil load for the upcoming suds step. Items are generally flushed at medium temperatures at high water levels.

FLAME SPREAD INDEX
A number indicting the rate at which flames will spread across a material's exposed, finished surface.

FLAT BED SPRINGS
Bed springs made of metal slats linked with helical hooks.

FOUR-STEP TRAINING METHOD
A training model used to implement an on-the-job training program for newly hired and experienced employees. The four steps include preparation, presentation, trial performance, and follow-through.

FREQUENCY SCHEDULE
A schedule which indicates how often each item on an area inventory list needs to be cleaned or maintained.

FRONT OF THE HOUSE
The functional areas of the hotel in which employees have extensive guest contact, such as food and beverage facilities and the front office.

G

GENERAL COMPANY ORIENTATION
A formal program designed to introduce the company's mission and values to a group of employees, usually conducted by the human resources department.

GUESTROOM KEY
A key which opens a single guestroom door if it is not double-locked.

H

HAND
The feel of a fabric.

HAND CADDY
A portable container for storing, holding, and transporting cleaning supplies. Typically located on the top shelf of the room attendant's cart.

HARD FLOOR
Floors made from natural stone or clay. These floors are among the most durable of all floor surfaces, but also the least resilient. Types of hard floors include concrete, marble and terrazzo, ceramic tile, and other natural stone.

HAZCOMM STANDARD
Hazard Communication Standard; OSHA's regulation requiring employers to inform employees about possible hazards related to chemicals they use on the job.

HELICAL HOOKS
Tightly wound metal springs with hooks on either end.

HIGH DUSTING
Dusting high or hard-to-reach items or areas such as air supply vents, sprinklers, and ceiling corners.

HIRING PERIOD
Begins when the employer makes an offer to a prospective employee, and lasts through the new-hire's initial adjustments to the job. Involves all arrangements necessary to prepare the new-hire and current staff for a successful working relationship.

HOMOGENEOUS COLOR
Color that permeates the entire layer of vinyl flooring so that the color does not wear off with use.

HOPPERS
Openings in washing machines through which detergents can be poured. Also called ports.

HOUSE RULE
Any published company-wide rule for which violation can result in immediate discharge.

HOUSE SETUP
The total number of each type of linen that is needed to outfit all guestrooms one time. Also referred to as one par of linen.

HOUSEKEEPING STATUS REPORT
A report prepared by the housekeeping department which indicates the current housekeeping status of each room, based on a physical check.

I

INCENTIVE PROGRAM
A program offering special recognition and rewards to employees based on their ability to meet certain conditions. These programs vary in structure and design and are a way to award exceptional performance beyond the paycheck.

INCOME STATEMENT
A report on the profitability of operations, including revenues earned and expenses incurred in generating the revenues for the period covered by the statement.

INNERSPRING MATTRESS
Mattress in which springs are sandwiched between layers of padding.

INTERNAL RECRUITING
A process in which managers recruit job candidates from within a department or property. Methods available include cross-training, succession planning, posting job openings, and keeping a call-back list.

INVENTORY
Stocks of merchandise, operating supplies, and other items held for future use in a hospitality operation.

ISSUING
The process of distributing inventory items from the storeroom to authorized individuals by the use of formal requisitions.

J

JACQUARD
Sculpted terry or velvet fabric.

JOB BREAKDOWN
A form that details how the technical duties of a job should be performed.

JOB DESCRIPTION
A detailed list identifying all the key duties of a job as well as reporting relationships, additional responsibilities, working conditions, and any necessary equipment and materials.

JOB LIST
A list identifying all the key duties of a job in the order of their importance.

JOB SAFETY ANALYSIS
A detailed report that lists every job task performed by all housekeeping employees. Each job task is further broken down into a list of steps. These steps are accompanied by tips and instructions on how to perform each step safely.

K

KAPOK
Natural plant fiber used to stuff solid mattresses.

KEY CONTROL
The process of reducing guest and property theft and other security-related incidents by carefully monitoring and tracking the use of keys at a hospitality operation.

L

LATEX MATTRESS
Mattress made of whipped synthetic rubber; foam rubber mattress.

LEASED EMPLOYEES
Employees which a leasing agency hires and leases to businesses. The agency bills the employer on a regular basis for the costs of the workers. Some businesses even enter into contracts with leasing companies whereby the businesses "sell" their employees to the leasing companies. The leasing companies then lease the employees back to the businesses.

LEAD-TIME QUANTITY
The number of purchase units consumed between the time that a supply order is placed and the time that the order is actually received.

LINEN ROOM
Area in a hospitality operation which is often considered the headquarters of the housekeeping department. This is the area where the employee typically reports to work, receives room assignments, room status reports, and keys; assembles and organizes cleaning supplies; and checks out at the end of his/her shift.

LINTERS
Short or waste cotton fibers.

M

MASTER KEY
A key which opens all guestroom doors which are not double-locked.

MAXIMUM QUANTITY
The greatest number of purchase units that should be in stock at any given time.

MERCERIZING
Process used on cotton yarns that makes them more lustrous and durable.

METAL COIL SPRINGS
Bed springs in which metal coils provide support and resiliency.

METAL INTERLOCKING
A type of polymer floor finish that contains dissolved metal, usually zinc.

MID-RANGE SERVICE
A modest but sufficient level of service which appeals to the largest segment of the traveling public. A mid-range property may offer uniformed service, airport limousine service, and food and beverage room service; a specialty restaurant, coffee shop, and lounge; and special rates for certain guests.

MILDEW
An odorous fungus growth that can occur on bathroom surfaces, especially on tile grout, shower curtains, doors, and walls.

MILDEWCIDES
Laundry chemicals added to the wash cycle to prevent the growth of bacteria and fungus on linens for up to 30 days.

MINIMUM QUANTITY
The fewest number of purchase units that should be in stock at any given time.

MITERING
A method for contouring a sheet or blanket to fit the corner of a mattress in a smooth and neat manner. The results are sometimes referred to as "square corners" or "hospital corners."

MODACRYLIC
Acrylic fiber that is less resistant to stains and abrasions.

MOMIE
Type of napery fabric.

MOTIVATION
Stimulating a person's interest in a particular job, project, or subject so that the individual is challenged to be continually attentive, observant, concerned, and committed.

MSDS
Material Safety Data Sheet; a form containing information about a chemical that is supplied by the chemical's manufacturer.

MUSLIN
Cotton fabric made of carded cotton.

N

NAPERY
Table linens.

NETWORKING
Developing personal connections with friends, acquaintances, colleagues, associates, teachers, counselors, and others.

NON-POROUS
Non-moisture-absorbing.

NON-RECYCLED INVENTORIES

Those items in stock that are consumed or used up during the course of routine housekeeping operations. Non-recycled inventories include cleaning supplies, small equipment items, guest supplies, and amenities.

NRC SCALE

Noise Reduction Coefficient; a scale which designates the amount of sound a material absorbs.

O

OCCASIONAL TABLE

Small end table.

OCCUPANCY REPORT

A report prepared each night by a front desk agent which lists rooms occupied that night and indicates those guests expected to check out the following day.

OCCUPIED

A room status term indicating that a guest is currently registered to the room.

OPERATING BUDGET

A detailed plan for generating revenue and incurring expenses for each department within the hospitality operation.

OPERATING EXPENDITURES

Costs incurred in order to generate revenue in the normal course of doing business.

OPTICAL BRIGHTENERS

See Fabric Brighteners

ORGANIZATION CHART

A schematic representation of the relationships between positions within an organization, showing where each position fits into the overall organization and illustrating the divisions of responsibility and lines of authority.

OSHA

The U.S. Occupational Safety and Health Act (OSHA): a broad set of rules that protects workers in all trades and professions from a variety of unsafe working conditions.

OUT-OF-ORDER

A room status term indicating that a room cannot be assigned to a guest. A room may be out-of-order for maintenance, refurbishing, deep cleaning, or other reasons.

P

PAR

The standard number of a particular inventory item that must be on hand to support daily, routine housekeeping operations.

PAR NUMBER

A multiple of the standard number of a particular inventory item that must be on hand to support daily, routine housekeeping operations.

PERFORMANCE APPRAISAL

The process by which an employee is periodically evaluated by his/her manager to assess job performance and to discuss steps the employee can take to improve job skills and performance.

PERFORMANCE STANDARD

A required level of performance that establishes the quality of work that must be done.

PERMANENT ASSEMBLY

In lamps, when the base and light socket are fused together to prevent loosening.

PERPETUAL INVENTORY SYSTEM

An inventory system in which receipts and issues are recorded as they occur; this system provides readily available information on inventory levels and cost of sales.

PH SCALE

A scale that measures the acidity or alkalinity of a substance; according to the scale, a pH of 7 is neutral, acids have values of less than 7 to 0, and alkalies have values of more than 7 to 14.

PILE

The surface of a carpet; consists of fibers or yarns that form raised loops that can be cut or sheared.

PILE DISTORTION

Face fiber conditions such as twisting, pilling, flaring, or matting caused by heavy traffic or improper cleaning methods.

PLAIN WEAVE

Type of weave in which fill yarns are alternately woven under and over warp yarns.

POROUS

Moisture-absorbing.

PORTS

Openings into washing machines through which detergents can be poured. Also called hoppers.

PREVENTIVE MAINTENANCE

A systematic approach to maintenance in which situations are identified and corrected on a regular basis to control costs and keep larger problems from occurring. Preventive maintenance consists of inspection, minor corrections, and work order initiation.

PRIMARY BACKING

The part of the carpet to which face fibers are attached and which holds these fibers in place.

PRODUCTIVITY STANDARD

An acceptable amount of work that must be done within a specific time frame according to an established performance standard.

PRO FORMA INCOME STATEMENT

A report that predicts the results of current or future operations, including revenues earned and expenses incurred in generating the revenues for the period covered by the statement.

PROGRESSIVE DISCIPLINE

A system of discipline that progresses to sterner measures with repeated infractions.

R

RECEIVING

Accepting delivery of merchandise that has been ordered or is expected, and recording such transactions.

RECYCLED INVENTORIES

Those items in stock that have relatively limited useful lives but are used over and over in housekeeping operations. Recycled inventories include linens, uniforms, major machines and equipment, and guest loan items.

REINFORCEMENT

An action that occurs in close proximity to a behavioral response and, when associated with the response, has a tendency to either: (1) strengthen the probability of the response being repeated, or (2) strengthen the intensity of the response.

RESILIENT FLOORS

A type of floor that reduces noise and is considered easier to stand and walk on. Types of resilient floors include vinyl, asphalt, rubber, and linoleum.

REVENUE CENTER

An operating division or department which sells goods or services to guests and thereby generates revenue for the hotel. The front office, food and beverage outlets, room service, and retail stores are typical hotel revenue centers.

ROOM ATTENDANT'S CART

A lightweight, wheeled vehicle used by room attendants for transporting cleaning supplies, linen, and equipment needed to fulfill a block of cleaning assignments.

ROOM INSPECTION

A detailed process in which guestrooms are systematically checked for cleanliness and maintenance needs.

ROOM RACK

An array of metal file pockets designed to hold room rack slips arranged by room number. The room rack summarizes the current status of all rooms in the hotel.

ROOM STATUS DISCREPANCY

A situation in which the housekeeping department's description of a room's status differs from the room status information at the front desk.

ROOM STATUS REPORT

A report which allows the housekeeping department to identify the occupancy or condition of the property's rooms. Generated daily through a two-way communication between housekeeping and the front desk.

ROTARY FLOOR MACHINE

Floor care equipment that accommodates both brushes and pads to perform such carpet cleaning tasks as dry foam cleaning, mist pad cleaning, rotary spin pad cleaning, or bonnet and brush shampoos. On hard floors, these machines can be used to buff, burnish, scrub, strip, and refinish.

ROUTINE MAINTENANCE

Activities related to the general upkeep of the property that occur on a regular (daily or weekly) basis, and require relatively minimal training or skills to perform. These activities occur outside of a formal work order system and include such tasks as sweeping carpets, washing floors, cleaning guestrooms, etc.

S

SAFETY

A condition in which persons are safe from injury, hurt, or loss while present in the workplace.

SAFETY STOCK

The number of purchase units that must always be on hand for smooth operation in the event of emergencies, spoilage, unexpected delays in delivery, or other situations.

SATIN WEAVE

Type of weave in which warp threads interlace with filling threads to produce a smooth-faced fabric.

SCHEDULE
A report which gives supporting detail to a property's financial statements. Examples include departmental income statements such as housekeeping.

SCHEDULED MAINTENANCE
Activities related to the upkeep of the property that are initiated through a formal work order or similar document.

SCHEDULING
A process in which the appropriate number of employees are assigned to fill necessary duties and positions each workday.

SECONDARY BACKING
The part of a carpet that is laminated to the primary backing to provide additional stability and more secure installation.

SECURITY
The prevention of theft, fire, and other emergency situations in the workplace.

SECURITY COMMITTEE
A committee consisting of key management personnel and selected employees that develops and monitors a property's security plans and programs.

SELVAGE
Finished woven edge.

SHADING
A carpet condition that occurs when the pile is brushed in two different directions so that dark and light areas appear.

SILENCE CLOTH
Oilcloth or other padded material placed under the tablecloth to absorb noise.

SIZING
Laundry chemicals added to the wash cycle to stiffen polyester blends.

SOAPS
A kind of detergent. Neutral or pure soaps contain no alkalies; built soaps do. Built soaps are generally used on heavily soiled fabrics; pure soaps are reserved for more lightly soiled items. Soaps are destroyed by sours.

SOLID MATTRESS
Mattress stuffed with hair, cotton, or some other material.

SOURS
Mild acids used to neutralize residual alkalinity in fabrics after washing and rinsing.

SPLINE
Thin wood slat used to attach panels to a wall or ceiling.

STAFFING GUIDE

A system used to establish the number of workers needed.

STAPLE FIBERS

Fibers approximately seven to ten inches long that are twisted together into long strands and used to construct non-woven or tufted carpet.

STAYOVER

A room status term indicating that the guest is not checking out today and will be staying at least one more night.

STEAM CABINET

A box in which articles are hung and steamed to remove wrinkles. Steam cabinets are typically used to remove wrinkles from heavy linens such as blankets, bed-spreads, and curtains.

STEAM TUNNEL

A piece of laundry equipment which moves articles on hangers through a tunnel, steaming them and removing the wrinkles as they move through.

SUITES

Several pieces of furniture of similar design, usually sold together to outfit a complete room.

SUPPORT CENTER

An operating division or department which does not generate direct revenue, but plays a supporting role to the hotel's revenue centers. Support centers include the housekeeping, accounting, engineering and maintenance, and personnel functions.

SUPPORTED EMPLOYMENT PROGRAM

Offers physically and mentally impaired individuals opportunities to obtain jobs in regular work environments. The benefits to host companies include partial government funding, program administration by a non-profit agency, affirmative action assistance, possible turnover reduction, and improved public relations.

SURFACTANTS

Synthetic detergents often contain surfactants, chemicals that aid soil removal and act as antibacterial agents and fabric softeners.

SUSPENSION

A step in the progressive discipline process whereby an employee is sent home from work without pay for a specified period.

SYNTHETIC DETERGENTS

Synthetic detergents are especially effective on oil and grease. Builders or alkalies are often added to synthetic detergents to soften water and remove oils and grease.

T

TABLE SKIRT
Piece of linen that covers the sides of the table.

TERMINATION
The final step in the progressive discipline process which permanently severs the employee from the organization.

THREAD COUNT
Number of warp and fill yarns per square inch.

TICKING
Sturdy fabric used to cover mattresses and springs.

TRAINING PLAN
An outline prepared by a group trainer for his/her own use in planning and conducting a training session or any other informational group meeting.

TUFTED OR LOOPED CARPETS
Carpet constructed by pulling face fibers through the carpet's backing to form a thick pile of tufts or loops.

TUNNEL WASHER
A long, sequential laundry machine that operates continuously, processing each stage of the wash/rinse cycle and extracting in another section of the machine.

TURNDOWN SERVICE
A special service provided by the housekeeping department in which a room attendant enters the guestroom in the early evening to restock supplies, tidy the room, and turn down the guest bed.

TURNOVER
The rate at which a work unit loses workers.

TWILL WEAVE
Type of weave in which diagonal yarn pattern emerges.

TYPE II VINYL
Commercial grade vinyl.

U

UNDERWRITERS LABORATORIES
An independent non-profit organization that tests electrical equipment and devices to ensure that the equipment is free of defects which can cause fire or shock.

V

VACANT
A room status term indicating that the room has been cleaned and inspected, and is ready for the arriving guest.

VARIABLE STAFF POSITIONS
Positions which are filled in relation to changes in hotel occupancy.

VITREOUS CHINA
Common material from which toilets are made.

W

WARP YARNS
Yarns that run the length of the fabric.

WATER EXTRACTION (HOT OR COLD)
A deep cleaning carpet method in which a machine sprays a detergent and water solution onto the carpet under low pressure, and, in the same pass, vacuums out the solution and soil.

WEFT YARNS
See Fill Yarns

WET VACUUM
Floor care equipment used to pick up spills or to pick up rinse water that is used during carpet or floor cleaning.

WICKING
A carpet condition that occurs when the backing of the carpet becomes wet and the face yarns draw the moisture and color of the backing to the carpet's surface.

WOOD FLOOR
A type of floor in which hard or soft woods are cut and laid in planks, blocks, or tiles (parquet) to make attractive patterns.

WORKERS' COMPENSATION
Insurance that reimburses an employer for damages that must be paid to an employee for injury occurring in the course of his/her employment.

WORLD-CLASS SERVICE
A level of service which stresses the personal attention given to guests. Hotels offering world-class service provide upscale restaurants and lounges, exquisite decor, concierge service, opulent rooms, and abundant amenities.

Index

V

W

HOUSEKEEPING MANAGEMENT

REVIEW QUIZ ANSWER KEY

The numbers in parentheses refer to the page(s) where the answer may be found.

Chapter 1	Chapter 2	Chapter 3	Chapter 4
1. F (5)	1. T (29)	1. T (52)	1. T (81)
2. T (5)	2. F (30)	2. F (52–53)	2. F (82)
3. F (9)	3. F (32)	3. T (52)	3. F (82–83)
4. F (9)	4. F (34)	4. T (56)	4. F (84)
5. F (11)	5. F (35)	5. F (56)	5. T (89)
6. F (13)	6. T (35)	6. F (56, 58)	6. F (89)
7. F (13)	7. T (35)	7. F (58)	7. F (89)
8. F (14)	8. T (37)	8. F (61)	8. T (96–97)
9. F (16)	9. T (43–44)	9. F (62)	9. T (97)
10. F (16)	10. F (44, 46)	10. T (74)	10. T (105)
11. a (6)	11. a (32)	11. a (51)	11. b (89)
12. b (9)	12. b (35)	12. a (65)	12. a (105)
13. c (6–7)	13. b (43)	13. b (75–76)	13. a (92)
14. d (8)	14. b (28)	14. c (62–63)	14. c (92)
15. b (15)	15. c (33)	15. b (65, 67)	15. d (94)

Chapter 5	Chapter 6	Chapter 7	Chapter 8
1. F (113)	1. T (145)	1. T (177)	1. T (202)
2. F (113)	2. F (145)	2. F (178–179)	2. F (202)
3. F (114)	3. F (146)	3. T (180)	3. F (203)
4. F (114)	4. T (146)	4. T (179)	4. T (203)
5. T (115)	5. T (146, 148)	5. F (182)	5. T (203)
6. T (115–116)	6. F (149)	6. T (183)	6. F (206)
7. T (118)	7. T (150)	7. T (189)	7. T (211)
8. T (121)	8. T (150)	8. T (190)	8. T (211)
9. F (127)	9. F (154)	9. F (192)	9. T (213)
10. T (133)	10. T (154–155)	10. T (195)	10. F (212)
11. a (113)	11. a (146)	11. b (174)	11. a (206)
12. b (116)	12. b (153)	12. b (180)	12. b (210)
13. c (113)	13. b (157)	13. a (180)	13. b (211)
14. d (132)	14. c (153)	14. d (175)	14. b (212)
15. a (133)	15. c (161)	15. a (189)	15. b (212)

Chapter 9	Chapter 10	Chapter 11	Chapter 12
1. F (231)	1. F (254)	1. F (287)	1. F (311)
2. F (231)	2. F (255)	2. F (289)	2. F (312)
3. T (232)	3. F (256)	3. F (290)	3. T (314)
4. T (232)	4. F (257)	4. F (290)	4. T (317)
5. F (232)	5. T (257)	5. F (291)	5. F (317)
6. T (232)	6. b (254)	6. F (295)	6. T (320)
7. F (232)	7. b (257)	7. F (299)	7. T (327)
8. T (235)	8. a (257)	8. b (288)	8. b (325)
9. T (235)	9. b (257)	9. a (289)	9. b (326)
10. F (241)	10. d (270)	10. b (297)	10. b (314)
11. a (231)			
12. a (232)			
13. b (233)			
14. c (233)			
15. d (241)			

Chapter 13	Chapter 14
1. T (342)	1. T (364)
2. F (344)	2. T (364–365)
3. T (347)	3. F (365)
4. T (347)	4. F (367)
5. F (351)	5. F (369)
6. F (350–351)	6. F (368)
7. T (351)	7. T (375)
8. T (352)	8. T (377)
9. F (352–353)	9. F (384)
10. T (353)	10. F (385)
11. b (341–342)	11. b (378)
12. b (352)	12. b (377)
13. a (347)	13. c (367)
14. b (352)	14. a (383)
15. c (350)	15. c (385)

The
Educational Institute Board of Trustees

The Educational Institute of the American Hotel & Motel Association is fortunate to have both industry and academic leaders, as well as allied members, on its Board of Trustees. Individually and collectively, the following persons play leading roles in supporting the Institute and determining the directions of its programs.

Steven J. Belmonte, CHA
President & COO
Ramada Franchise
  Systems, Inc.
Parsippany, New Jersey

John Q. Hammons
Chairman & CEO
John Q. Hammons
  Hotels, Inc.
Springfield, Missouri

David J. Christianson, Ph.D.
Dean
William F. Harrah College of
  Hotel Administration
University of Nevada,
  Las Vegas
Las Vegas, Nevada

Arnold J. Hewes, CAE
Executive Vice President
Minnesota Hotel & Lodging
  Association
St. Paul, Minnesota

Caroline A. Cooper, CHA
Dean
The Hospitality College
Johnson & Wales University
Providence, Rhode Island

S. Kirk Kinsell
President—Franchise
ITT Sheraton World
  Headquarters
Atlanta, Georgia

Edouard P.O. Dandrieux, CHA
Director
H.I.M., Hotel Institute,
  Montreux
Montreux, Switzerland

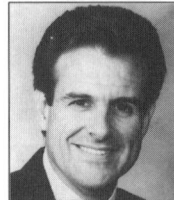

Donald J. Landry, CHA
President
Choice Hotels International
Silver Spring, Maryland

Valerie C. Ferguson
General Manager
Ritz-Carlton Atlanta
Atlanta, Georgia

Georges LeMener
President & CEO
Motel 6, L.P.
Dallas, Texas

Douglas G. Geoga
President
Hyatt Hotels Corporation
Chicago, Illinois

Jerry R. Manion, CHA
President
Manion Investments
Paradise Valley, Arizona

Joseph A. McInerney, CHA
President & CEO
Forte Hotels, Inc.
El Cajon, California

William R. Tiefel
President
Marriott Lodging
Washington, D.C.

John L. Sharpe, CHA
President & COO
Four Seasons-Regent Hotels
    and Resorts
Toronto, Ontario, Canada

Jonathan M. Tisch
President & CEO
Loews Hotels
New York, New York

Paul J. Sistare, CHA
President & CEO
Richfield Hospitality Services
Englewood, Colorado

Paul E. Wise, CHA
Professor & Director
Hotel, Restaurant &
    Institutional Management
University of Delaware
Newark, Delaware

Thomas W. Staed, CHA
President
Oceans Eleven Resorts, Inc.
Daytona Beach Shores, Florida

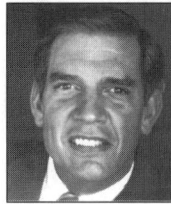

Ted Wright, CHA
Vice President/Managing
    Director
The Cloister Hotel
Sea Island, Georgia

Thomas G. Stauffer, CHA
President & CFO
Americas Region
Renaissance Hotels
    International, Inc.
Cleveland, Ohio